Membrane Technology in Separation Science

Membrane Technology in Separation Science

Mihir Kumar Purkait
Randeep Singh

CRC Press is an imprint of the
Taylor & Francis Group, an **informa** business

CRC Press
Taylor & Francis Group
6000 Broken Sound Parkway NW, Suite 300
Boca Raton, FL 33487-2742

First issued in paperback 2020

ISBN-13: 978-0-367-57210-5 (pbk)
ISBN-13: 978-1-138-62626-3 (hbk)

Library of Congress Cataloging-in-Publication Data

Names: Purkait, Mihir K., author. | Singh, Randeep, author.
Title: Membrane technology in separation science / Mihir K. Purkait, Randeep Singh.
Description: Boca Raton : Taylor & Francis, [2018] | Includes bibliographical references and index.
Identifiers: LCCN 2017045458| ISBN 9781138626263 (hardback : acid-free paper)
| ISBN 9781315229263 (pbk.)
Subjects: LCSH: Membrane separation.
Classification: LCC TP248.25.M46 P87 2018 | DDC 660/.28424--dc23
LC record available at https://lccn.loc.gov/2017045458

Visit the Taylor & Francis Web site at
http://www.taylorandfrancis.com

and the CRC Press Web site at
http://www.crcpress.com

Contents

Preface

Long gone is the era when membranes were used as an analytical tool. Now, because of technical advancements and commercialization, membranes have found their industrial importance. The use of membranes in industry has been attributed to the various advantages associated with membrane processes, such as good product quality, better energy efficiency than conventional separation methods, simple operation, no chemical alteration of constituents, easy scale-up, and better environment suitability. Although the rise in membrane usage has been observed only recently, the membrane phenomenon has been studied for a long time. This fact in itself explains that the field of membrane science is steady and stable like a marathon runner. Membrane science has scope for further growth, which can be seen in the form of dedicated research to further decrease the production cost of the membranes and adding stimuli-responsive character to the membranes for various applications. Intensive research is going on to make low-cost and stimuli-responsive membranes, which further diversifies the area of membrane science. The field of advance membrane science is evolving with these efforts on the said technology. These membranes are responsive to one or more stimuli of various natures, namely pH, temperature, chemical, light, or magnetic field in addition to a better antifouling profile. Not only are they advanced in their nature but they also solved the old-time problem of membrane fouling. That is why these membranes are very promising and have a bright future. These advancements and various future perspectives show that the field of membrane science is not saturated and there is very much to do in this field.

Industries related to food, pharma, textile, tannery, chemicals, and petroleum are using membranes in one way or another. These low-cost ceramic membranes and stimuli-responsive polymeric membranes further dominate the use of membranes in these industries. During the last decade, tremendous progress has been seen in the development and advancement of these two prospects of membrane science. These advanced membranes will find their use in the fermentation industry of the food sector for the production of the end product with less time and energy input; in the pharma industry for the production and separation of enzymes or active pharmaceutical ingredients (APIs); efficient dye and toxin removal from the wastewater of textile and chemical industries; and separation of oil–water emulsions in the case of the petroleum industry. The most promising part for these membranes is in the health sector for accurate delivery of drugs and in the production of artificial organs. These membranes are loaded with all the properties desired for these two applications. All these impacts and applications of these membranes are discussed in individual chapters of the book with every minute detail.

This book can be easily distinguished from other available books on membrane science by the content and writing style. The book contains vast content on the topics and a writing methodology that will inculcate and make the topics very easy to understand. In this book, a very detailed first chapter is dedicated to the introduction of the membrane science field, covering each and every part of the membrane science field from its history to recent developments, methods related to the membrane preparation and characterization, and various fouling mechanisms and modules. This chapter gives detailed insight about the field of membrane science, especially to beginners in the field. The second and third chapters of the book cover the preparation and characterization methods of the membranes, with special emphasis given to the two widely used methods for the preparation of membranes: the interfacial polymerization method for the preparation of ceramic membranes and phase inversion method for polymeric membranes. Chapters 4 and 5 cover the ceramic and polymeric membranes. Emphasis is mainly given to the low-cost ceramic and stimuli-responsive membranes with their various applications. Chapters 6 and 7 cover the detailed analysis of current hot research topics in the field of membrane science and future perspectives of the growth and development of the field. Also each chapter of the book contains engaging exercise problems developed for the better understanding of the topics for the readers. These exercises make the readers think critically about the membrane field and prepare them well so that they can come up with their own ideas and take the field of membrane science to new heights in the near future.

This book is developed in a way that it is applicable to a diverse group of readers related to the field of membrane science. Scientists, engineers, technologists, educators, and students all are the spokes (readers) of this umbrella (book). This also attracts the attention of designers for the development of various modules and membrane setups for efficient use of developed advanced membranes. Also, this book is unique in its content and writing that is not available in any other book, which makes it more prominent for the readers.

With a positive hope in our hearts we believe that this book will help many enthusiasts of the membrane science field, which may lead to further development of the field. We as the authors put our greatest efforts into making this book fruitful in every way possible, but as all of you know no one person belonging to the *Homo sapiens* species is perfect; therefore we expect positive comments as well as suggestions for further improvement of the book.

Mihir Kumar Purkait
Randeep Singh

Authors

Mihir Kumar Purkait is a professor in the Department of Chemical Engineering at the Indian Institute of Technology Guwahati (IITG). Prior to joining the faculty at IITG (2004), he received his PhD and MTech in chemical engineering from the Indian Institute of Technology, Kharagpur (IITKGP) after completing his BTech and BSc (Honors) in chemistry from the University of Calcutta. He has received several awards including the Dr. A.V. Rama Rao Foundation's Best PhD Thesis and Research Award in Chemical Engineering from IIChE (2007); BOYSCAST Fellow award (2009–2010) from the DST; Young Engineers Award in the field of Chemical Engineering from the Institute of Engineers (India, 2009); and the Young Scientist Medal award from the Indian National Science Academy (INSA, 2009). His current research activities are focused in four distinct areas: (1) membrane synthesis and separation technology (both ceramic and polymeric), (2) water pollution control, (3) smart material, and (4) adsorption. In each of the areas, his goal is to synthesize responsive materials and to develop a more fundamental understanding of the factors governing the performance of these technologies. He has more than 15 years of experience in academics and research, and has published more than 125 papers in different reputed journals of importance. He has submitted 7 patents and been involved with more than 12 sponsored projects from various funding agencies. Purkait has guided 12 PhD candidates. He has coedited one book (*Membrane Technologies and Applications*, Taylor & Francis, 2012) and written one textbook (Springer, in press).

Randeep Singh received his undergraduate (BTech) degree from Kurukshetra University in biotechnology in 2011 and masters (MTech) degree in chemical engineering from National Institute of Technology Trichy, Tamilnadu in 2013. Presently he is pursuing doctoral research in the Department of Chemical Engineering at the Indian Institute of Technology Guwahati, Assam. His research work is dedicated to the preparation of various membranes and mathematical analyses of the transport phenomena in membrane separation processes including wastewater treatment, fruit juice clarification, protein separation, and value-added product separation from biogenic source. He is also working on fabrication of different membrane modules for environmental separation. He has already submitted six papers to international journals and presented five papers in national and international conferences. He has received several awards in his field including the Young Scientist Award at the International Science Congress (ISC-2015) held in Nepal. He is an associate member of the Indian Institute of Chemical Engineers.

1 Introduction to Membranes

1.1 AN OVERVIEW OF SEPARATION PROCESSES

Life on Earth depends on various membrane processes and membranes play an important role in the lives of all of the living beings present on this planet. Examples are the process of respiration through the lungs, which continuously allows the diffusion of O_2 and CO_2 to and from our bodies to the outside; the purification of groundwater because of the different layers of earth that act as membranes of different porosities; and the skin of living organisms also works as a semipermeable membrane. The skin does not allow microorganisms to enter into our bodies, but allows diffusion of water in the form of sweat and other toxic substances from the body to the outside. Similarly, there are an infinite number of other examples that confirm the presence of different kinds of membranes around all of us.

The word *membrane* is derived from the Latin word *membrana*, which means "skin." Membrane is a sort of barrier that separates things and allows materials to be passed selectively [1]. There are many definitions available for a membrane. A general definition could be: A membrane is a thin barrier, placed between two phases, or mediums, which allow one or more constituents to selectively pass from one medium to the other in the presence of an appropriate driving force while retaining the rest [2]. This definition is based on a macroscopic level, but it should be taken care that the separation is at the microscopic level. Accordingly, it can be said that a membrane process is a combination of both mass and momentum transfer. A membrane can be homogeneous or heterogeneous, symmetric or asymmetric, solid or liquid; it can carry a positive or negative charge or be neutral or bipolar. Transport through a membrane can be affected by convection or by diffusion of individual molecules, induced by an electric field or concentration, pressure, or temperature gradient. The membrane thickness may vary from as small as 100 microns to several millimeters.

Membrane science is promising and has the upper hand when its advantages are compared to other separation processes like adsorption, distillation, extraction, and crystallization. The important advantages of membranes are their low capital cost, low energy requirement, high separation efficiency, compact design, easy organization with other separation processes, and no requirement of secondary separation processes. The properties of a membrane mainly depend on factors like porosity, pore diameter, pore size distribution, particle size distribution of the solutes, and the affinity between the feed and the membrane material for solubility/diffusivity. There is vastness in the methods available for the preparation of different membranes by using different types of materials. Membrane science, even with these many advantages and options available for preparation and applications, has enough scope for further improvements.

Membranes and membrane processes were first introduced as an analytical tool in chemical and biomedical laboratories. However, their industrial importance increased quite rapidly with significant technical and commercial impact [3]. Membranes now find wide applications in the food/beverage industry; in desalination processes; in the medical industry for the purification of body fluids, antibiotics, and so on; and in separation of industrial effluents, among other processes. This increase in the use of membranes in industry has been attributed to the various advantages associated with membrane processes, such as good product quality, better energy efficiency than conventional separation methods, simple operation, no chemical alteration of constituents, easy scale up, and better environment suitability. Although the rise in membrane usage has been observed only recently, the membrane phenomenon has been long studied.

1.1.1 Principle of Membrane Separation

There exists different mechanisms of separation for different membranes; membranes can also be classified based on the mechanisms. Porous membranes separate the feed from permeate on the basis of size and nonporous membranes on the basis of affinities between the feed components and membrane materials. The physical and chemical interactions between the feed components and membrane material also control the rate of mass transfer, that is, the flux in common terms. In general, the membrane flux can be given as

$$Flux = \frac{Membrane\ permeability}{Membrane\ thickness} \times Driving\ force \quad (1.1)$$

Expression 1.1 shows that the membrane flux is proportional to the membrane permeability and driving force. The driving force differs for different membrane processes. It can be pressure, concentration, chemical, electrical, or thermal potential. The membrane flux is inversely proportional to the membrane thickness, which means the lesser the membrane thickness, the higher will be the membrane flux, and vice versa. Further, the membrane transport can be of three types, namely, passive, active, and facilitated or carrier mediated. In a passive system, the transport will be totally under the influence of the driving force and is not energy intensive. The transport usually takes place from a higher potential region to a lower potential region. The transport takes place until equilibrium is reached between the feed and permeate side for the driving force to be neutral. Active transport is totally opposite to passive transport as it is an energy-intensive process. Active transport is energy intensive because here the transport is opposite to the conventional transport that requires energy. The transport will be from a lower potential to a higher potential, and to carry out transport in this way energy is required. The process like osmosis is an example of passive transport and, on the other hand, reverse osmosis is an example of active transport. Facilitated or carrier-mediated transport accomplishes transport with the help of a carrier that interacts with the feed components and carries them across the membrane. The carrier works as a boat for passengers in a stream of water. Similar to a boat, it carries the feed components across the membrane and comes back again after dissociating the previously attached components to the feed side. The carrier undergoes no changes and can be retrieved after the completion of the membrane process. Facilitated or carrier-mediated transport is mainly used in liquid membrane processes.

In a membrane separation process, the membrane allows selective passage of feed components to the permeate side. The feed components passed through the membrane are known as permeates or filtrates, and the feed components that remain on the membrane surface are known as retentates or filtrates. Figure 1.1 shows the membrane separation process with process variables and entities clearly in two common configurations modes: (a) dead-end filtration and (b) cross-flow filtration.

Some of the common terminologies used in membrane separation processes are mentioned next.

Flux—The permeate flux of a membrane is a very important parameter. It can be defined as the permeate volume (or mass) through the membrane per unit of membrane area. The permeate flux or simply the flux, F, of a membrane is given by the general relation [3]

$$F = \frac{1}{A}\frac{dQ}{dt} \quad (1.2)$$

where

F = flux (L/m^{2}h)
Q = permeate volume (L)
t = time (h)
A = membrane area (m^2)

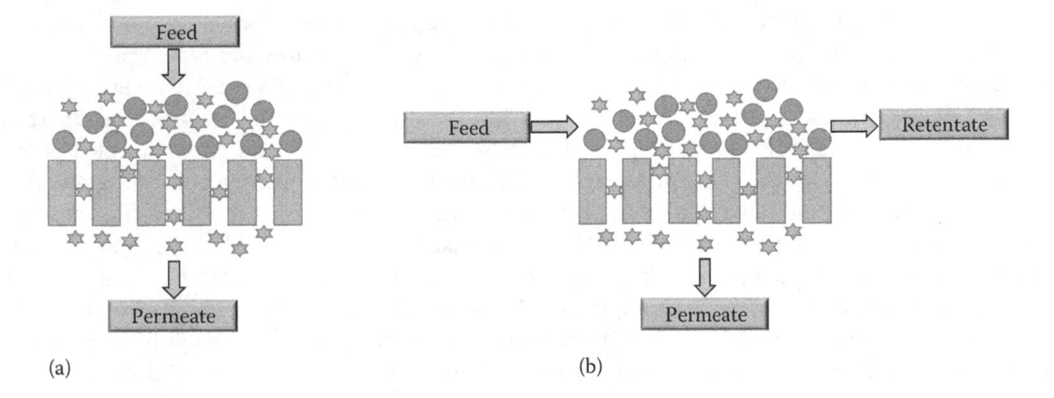

FIGURE 1.1 Schematics of a membrane process in two common configurations: (a) Dead end and (b) cross flow.

TMP—The pressure difference between the feed side and the permeate side across the membrane is known as the transmembrane pressure (TMP). The relationship between J and TMP can be given by a modified form of Darcy's law [4]:

$$F = \frac{\Delta P}{\mu R_t} \tag{1.3}$$

where

ΔP = pressure difference (kPa)
μ = viscosity (kPa.s)
R_t = total resistance to permeate flow (m^{-1})

Rejection—An important feature of a membrane is its rejection, which is a measure of membrane selectivity [3]. Selectivity can be expressed in terms of rejection (or retention), R, which expresses the extent to which a solute is retained by the membrane and is defined as

$$R = 1 - \frac{C_p}{C_f} \tag{1.4}$$

where

R = retention
C_p = concentration in the permeate (mg/L)
C_f = concentration in the feed (mg/L)

In the case of total retention of the feed components by the membrane, the R will be unity.

1.1.2 Classification of Membranes

In general, membranes are barriers that selectively separate two phases. There are different types of membranes available: They may be solid or liquid, charged or neutral, homogeneous (uniform in composition and structure) or heterogeneous (nonuniform mainly structurewise). Further, membranes can also be classified on the basis of their nature, structure, or mechanism of separation. Naturewise, the membranes may be biological or synthetic, which is further subdivided as living and nonliving in the case of biological membranes, and organic (polymeric or liquid) and inorganic (ceramic or metal) in case of synthetic membranes. On the basis of structure or morphology, the membranes are categorized under two categories: symmetric (isotropic) and asymmetric (anisotropic). The symmetric membranes are divided into porous, nonporous (dense), and charged membranes. The symmetric membranes are firm with voided structures and interconnected pores. The size exclusion mechanism is apt to explain the separation mechanism of porous membranes. That means the feed components are separated on the basis of their size; microfiltration and ultrafiltration are the most common examples. On the other hand, nonporous membranes separate the feed components on the basis of diffusion driven by the driving force of pressure, concentration, thermal, or electrical potential gradient, for example, nanofiltration and gas separation. The separation efficiency of two or more feed components depends upon their transport rate through the membrane, which further depends upon their diffusivity and solubility in the membrane material, for instance, pervaporation. The charged membranes carry a fixed positive or negative charge, and may be porous or nonporous. The separation is based on the diffusivity and selectivity of the ions across the membrane, such as electrodialysis.

In the case of asymmetric membranes there are two layers of membranes with different thicknesses, pore sizes, and porosity. A dense top layer, also known as skin layer, with thickness of 0.1 to 0.5 μm is present on a porous sublayer of thickness around 50 to 150 μm. The asymmetric membranes with their top and sublayers can be prepared in a single step, like in the phase separation method, or separately, like in the coating method. The membrane layers when prepared with different polymers are known as composite membranes. Here, both the layers can be optimized individually. The function of the dense top layer is to provide selectivity to the membrane and the porous sublayer to provide mechanical strength. This type of membrane is the most famous and well established commercially.

Liquid membranes are also of two types: supported liquid membranes and emulsion liquid membranes. In the case of supported liquid membranes, a porous structure is used as a support for the membrane phase (liquid). The membrane phase is supported by the porous structure in its pores. The porous structure provides the needed mechanical strength and the membrane phase the needed selectivity for the process. The separation process takes place under the influence of the concentration or chemical potential gradient from the feed phase to the permeate phase via the membrane phase, where the membrane phase acts like a connecting channel for the feed components. In the case of emulsion liquid membranes, the membrane phase is stabilized with the help of an emulsifier between the feed and permeate phase. Figure 1.2 presents the classification of membranes based on morphological aspects.

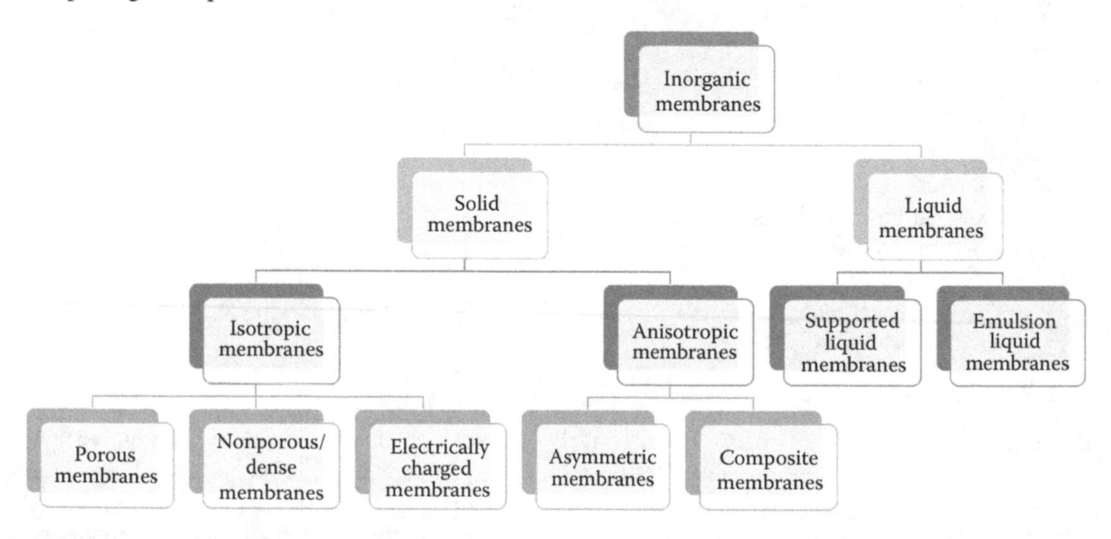

FIGURE 1.2 Classification of membranes based on morphological aspects of membranes.

1.1.3 Classification of Membrane Separation Processes

Membrane separation processes differ from one another in various ways, including type and configuration, mechanism of separation or transport, and nature of the driving forces. Many of the membrane processes are well established on an industrial scale for the separation, purification, and concentration of various feed components. Industries like chemical, food, pharmaceutical, biotechnology, oil, and wastewater treatment plants employ membranes on a large scale. Still some membrane processes are on the experimental stage, which might very soon show up on the industrial scale.

Membrane separation processes can be classified on the basis of various criteria. Membrane processes are filtration techniques in which the membrane acts as a selective barrier between two phases. Due to this, the specific feed components under the influence of a driving force transported to the permeate phase and the other feed components remain on the membrane surface as retentate. The membrane processes are available for different applications depending upon their separation characteristics and driving forces present. Further, the membrane processes can be classified on the basis of different aspects of membrane processes and membranes in particular. Table 1.1 gives details on different processes like membrane type used, mechanism of separation, and range of applications available for the particular membrane process.

Pressure-driven membrane processes are mainly categorized on the basis of their membrane pore size. Therefore, in these processes separation occurs based on the size exclusion mechanism of separation. The different membrane processes under this category are microfiltration, ultrafiltration, nanofiltration, and reverse osmosis. The membrane processes are named in the order of decreasing membrane pore size with microfiltration having the maximum pore size range and reverse osmosis with the least, as shown in Table 1.2. Then comes the membrane processes based on concentration or electrical potential or vapor pressure gradient like dialysis, electrodialysis, membrane distillation, gas separation, and pervaporation. These membrane processes are well established and used on an industrial scale for various applications like hemodialysis, desalination of brackish water, or separation of azeotropic mixtures. In these processes, the feed components are transported across the membrane to the permeate side based on the different driving forces depending upon the process type. The membranes used also differ for the different processes, like porous in the case of dialysis and electrodialysis, nonporous in the case of pervaporation, and porous or nonporous in the case of gas separations.

The membrane processes can also be classified according to the mode of operation like batch or fed batch, or continuous like in the case of reactors. Dead-end filtration is used for batch recycling of permeate in the process and can be said to be fed batch when the permeate still contains a good amount of the feed component, and the continuous filtration process is one in which the feed is given continuously to the membrane. Other than this, based on the design symmetry of the membrane process, they can also be categorized into dead-end filtration and cross-flow filtration. For dead-end filtration, the feed is permeated through the membrane and the retentate keeps on accumulating on the membrane surface. This accumulation of the retentate on the membrane surface results in a flux decline, since the retentate accumulated over time on the membrane surface fouls the membrane. On the other hand, with the cross-flow method the feed flows parallel to the membrane surface and the stream is separated into a permeate stream and a retentate stream. The advantage with cross-flow is that membrane fouling is not a concern as the retentate is not accumulated over the membrane surface and continuously dissipated due to the operation of the process. Dead-end filtration is better for a batch process, where the membrane can be cleaned after a run and cross-flow filtration is best for continuous membrane separation processes, since membrane fouling is not of major concern.

1.1.3.1 Pressure-Driven Processes

Pressure-driven membrane processes are widely used in different industries like chemical, pharmaceutical, biotechnology, desalination, food, and dairy. In this membrane process, pressure is the source of the driving force for the separation of feed into permeate and retentate. These membranes

TABLE 1.1
Important Membrane Processes

Driving Force	Membrane Process	Membrane Type	Separation Mechanism	Applications	Pore Size (nm)
Pressure difference	Microfiltration	Symmetric and asymmetric microporous membrane	Sieving	Sterile filtration clarification	100–10000
	Ultrafiltration	Asymmetric microporous membrane	Sieving	Separation of macromolecular solutions	10–100
	Nanofiltration	Asymmetric "skin type" membrane	Solution diffusion	Separation of divalent ions from solutions	0.5–5
	Reverse osmosis	Asymmetric "skin type" membrane	Solution diffusion	Separation of salts and microsolutes from solutions	<1
Concentration difference	Dialysis	Symmetric microporous	Diffusion	Separation of salts and microsolutes from macromolecular solutions	<1
	Supported liquid membrane	Microporous membranes supporting adsorbed organic liquid	Solution diffusion	Separation and concentration of metal ions and biological species	<1
	Pervaporation	Asymmetric membrane	Solution diffusion	Separation of organics	<1
Electric potential difference	Electrodialysis	Cation and anion exchange membrane	Electric potential gradient	Desalting of ionic solutions	<1
Temperature difference	Membrane distillation	Microporous membrane	Temperature gradient	Ultrapure water concentration of solutions	1–10

TABLE 1.2
Pressure-Driven Membrane Processes

Membrane Process	Transmembrane Pressure (kPa)	Pore Size (nm)	Removable Components
Microfiltration	100–200	100–10000	Suspended solids, bacteria
Ultrafiltration	200–1000	1–100	Macromolecules, viruses, proteins
Nanofiltration	1000–3000	0.5–5	Micropollutants, bivalent ions
Reverse osmosis	3500–10000	<1	Monovalent ions, hardness

are generally polymeric, ceramic, metallic, or organometallic. The processes can be classified on the basis of pore size, charge, or the pressure exerted on the membranes. These characteristics define what type of retentate and permeate one will get. The separation efficiency of a membrane depends on the rejection of a compound and can be calculated using Equation 1.4. The rejection ranges from 0 (for complete permeation) to 1 (complete rejection). In industries it ranges from 50% to 90%, and the average is 80%. Total membrane resistance plays a vital role in all membrane processes. It affects the membrane process and is a determining factor in membrane fouling. The total membrane resistance can be calculated by using the following relation:

$$R_t = R_r + R_{ir} \tag{1.5}$$

where, R_t, R_r, and R_{ir} represent the total membrane resistance, and reversible and irreversible fouling resistances, respectively. Reversible fouling is the type of membrane fouling that can be removed with simple hydraulic washings from the membrane, and irreversible fouling is membrane fouling that cannot be removed by simple hydraulic washings. These two types of foulings give rise to total membrane fouling, which in turn affects the complete membrane process. Table 1.1 shows the different membrane separation processes with their properties of operating principle, driving force, applications, and pore size. Figure 1.3 shows different pressure-driven membrane processes in dead-end filtration mode.

1.1.3.1.1 Microfiltration (MF)

Microfiltration (MF) is the pressure-driven membrane process that is capable of retaining particles of size 1000 nm and molecular weight higher than 100 kDa. The separation or retention capabilities depend on the membrane pore size. From Table 1.2 it can be seen that the MF membrane pore size ranges from 100 nm to 10000 nm. Since, the MF pore size is quite high, the pressure needed for the separation is low and ranges from 10 kPa to 300 kPa. MF is commonly used to separate suspended solids, sediments, algae, protozoa, and bacteria. Compounds that are more microscopic such as monovalent species like sodium (Na^+) or chloride (Cl^-) ions, dissolved organic matter, colloids, and viruses cannot be separated via MF. The feed passes through the membrane at a velocity of 1 to 3 m/s either in a dead-end (flat-sheet membrane) or a cross-flow (tubular membrane) assembly at low (10 kPa) pressures to moderate (300 kPa) pressures. A simple MF assembly is shown in Figure 1.4, where it can be seen that a pump is used as the source of the driving force for the separation to take place via the MF membrane. A pressure gauge is also installed so as to measure the operating or applied pressure. MF is mostly used in combination with ultrafiltration or reverse osmosis, as can be seen in desalination or wastewater treatment plants. MF also plays a vital role in the dairy, biotechnology, pharmaceutical and beverage industries.

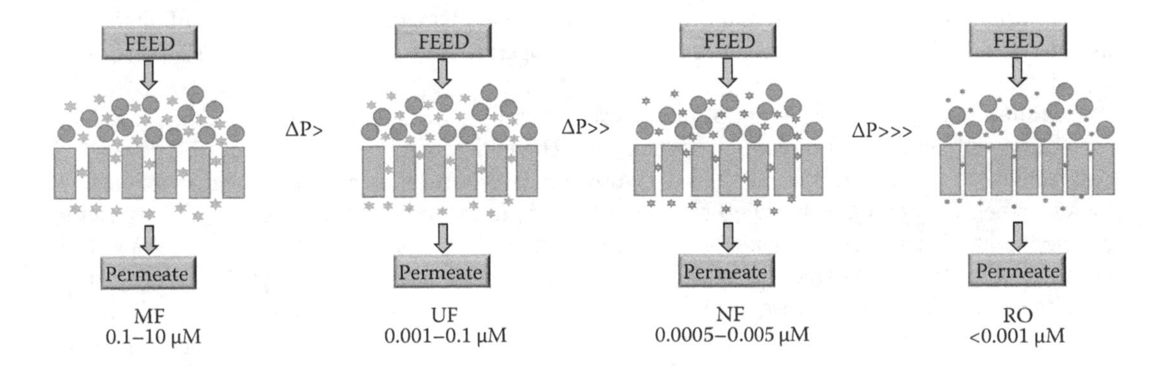

FIGURE 1.3 Pressure-driven membrane processes in dead-end configuration.

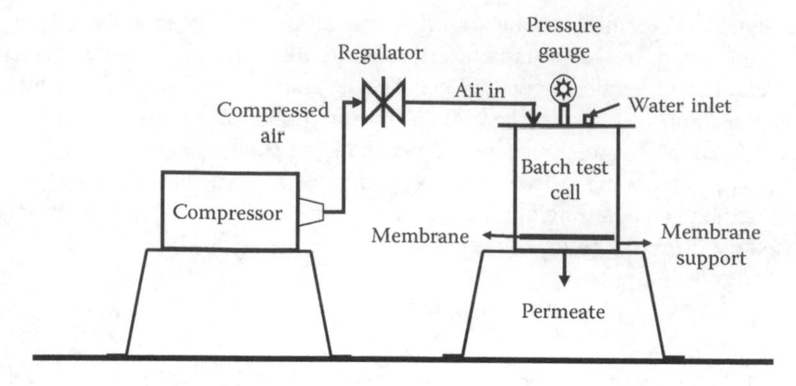

FIGURE 1.4 Schematic representation of a microfiltration membrane setup.

1.1.3.1.1.1 Advantages and Disadvantages of MF

Advantages

- Low operating pressure
- Low energy consumption among other membrane processes
- Fouling is not severe due to two reasons: larger pore size and use of low pressures
- Easy to maintain
- Cheap

Disadvantages

- Suspended matter and bacteria can only be removed
- Sensitive to oxidizing agents
- Hard and sharp particles can damage the membrane
- Damage to the membrane can also occur if cleaning pressures are in excess of 100 kPa

1.1.3.1.2 Ultrafiltration (UF)

Ultrafiltration (UF) is the pressure-driven process with pore size in the range of 1 to 100 nm. The separation is based on the size-exclusion-based sieving type of mechanism in which feed components are separated based on their size. The larger and wider components are not able to pass through the UF membrane as compared to the smaller and narrower components. The operating pressure required for UF is lower than other pressure-driven membrane processes like nanofiltration and reverse osmosis. This is because of the porous nature of the UF membranes with pores of larger size as compared to these other processes and thus can be operated in the range of 200 to 1000 kPa with high fluxes.

Ultrafiltration in its initial years of origin was generally used as a laboratory technique for the separation and purification of various components. The initial membranes were not well developed on chemical and mechanical fronts so as to be used on the industrial level. It was the development of the asymmetric membranes by Loeb and Sourirajan [5] in the 1960s that has given a thrust to the development of better UF membranes to be used on the commercial level. These membranes were both selective and permeable at the same time and gave the much needed push to the field of membrane science to be successful on the commercial front. The first cellulose acetate UF membranes lacked extended chemical resistance but were great for permeability and thus are used in industrial applications. The first UF membrane-based industrial plant was installed by the end of the 1960s. This is the boom and need of membranes at the industrial level.

The conditions of use for both MF and UF are similar in many ways with only one difference: pore size. The UF membrane is good for the removal of micron- and submicron-sized components from the feed due to its small pore size. Species like colloids, suspended solids, and organic molecules of higher molecular weight can be separated from a feed by using a UF membrane. The lower osmotic pressure of these species also plays a role in the low pressure operation of UF membranes. The separation efficiency of UF membranes also depends on the molecular structure, size, shape, and flexibility of the species to be separated.

1.1.3.1.2.1 Advantages and Disadvantages of UF

Advantages

- Perfect pore size range, thus can be applied for the separation of most of the feed components
- Low energy consumption due to the fact that separation takes place without phase change
- Operation at low pressures for the separation and concentration of feed components
- Best available membrane separation process for temperature-sensitive materials from the biological, food, and pharmaceutical industries
- Easy to operate due to its simple and compact nature

Disadvantages

- Low molecular weight species or dissolved salts cannot be separated by UF, and thus cannot be used for desalination of seawater.
- UF is not efficient in the separation of macromolecular mixtures; it can only be efficient if the species has 10 times or more difference in their molecular weight.

1.1.3.1.3 Nanofiltration (NF)

Nanofiltration (NF) is the newest pressure-driven membrane process. NF is a better option than reverse osmosis (RO) in many applications due its lower energy consumption and higher flux [3]. Considering the properties, NF lies in between size exclusion and in some cases charge-based UF and solution–diffusion mechanism-based RO. The pore size of NF lies in the range of 0.5 nm to 5 nm. Due to a very small pore size, NF is capable of rejecting even small, uncharged particles and multivalent ions, but due to the surface electrostatic properties, monovalent ions can pass through easily. These properties make NF eligible for fractionation and the selective removal of solutes from complex process feeds. Since NF has properties of both UF and RO, both charge and particle size play a role in NF process. NF is stated as a charged UF system in terms of UF and as a low pressure RO system in terms of RO. Therefore, NF has a lower operating pressure than RO and a higher organic matter rejection than UF. Thus, for colloids and larger particles, sieving will be the dominant separation mechanism, and for ions and smaller particles solution diffusion and charge effect will be the separation mechanism. NF is a very tenable process in the paper and pulp, textile, dairy, and pharmaceutical industries; metal recovery; and virus removal.

1.1.3.1.3.1 Advantages and Disadvantages of NF

Advantages

- Lower operating pressure than RO
- Lower cost of operation, as it performs at room temperature and low pressure
- Ability to continuously process large volumes
- No need to add extra ions for water softening as is in the case of ion exchangers

Disadvantages

- UF and RO are preferred over NF.
- Concentration polarization.
- Cost and maintenance are high.
- Short life span.

1.1.3.1.4 Reverse Osmosis (RO)

In general, reverse osmosis (RO) is the reverse of the osmosis process. In osmosis, a solvent flows from lower to higher solute concentrations when a semipermeable barrier is placed between two solutions. On the contrary, if the solvent is made to flow from a higher to lower solute concentration by virtue of an external force, then it is known as reverse osmosis. In the normal osmosis process, the driving force is the reduction in the free energy of the system, which decreases because the system tries to attain equilibrium. When equilibrium is reached by the system, then the process of osmosis stops. In the case of RO, the driving force for the process is the external force applied greater than the osmotic pressure of the system. RO is similar to other pressure-driven membrane processes; however, other processes employ size exclusion or straining as the mode of separation and RO employs diffusion.

In the year 1748, Jean-Antoine Nollet [6] discovered osmosis. Osmosis was only a phenomenon until the year 1950, when researchers from the University of California at Los Angeles first investigated desalination of seawater by using semipermeable membranes. The process was not viable on a commercial level due to low flux. Loeb and Sourirajan [5] made an asymmetric membrane, which made the RO process commercially viable. Further, John Cadotte [7] of the FilmTec Corporation filed a patent for the process of interfacial polymerization of m-phenylene diamene and trimesoyl chloride for the preparation of asymmetric membranes with high flux and low salt permeations. Almost all commercial RO membranes follow this procedure for their preparation.

RO membranes are usually dense membranes having pore size less than 1 nm. They are generally a skin layer in the polymer matrix. The membrane material (polymer) forms a layer and a weblike structure. The water follows a tortuous path to get permeated through the membrane. RO membranes are capable of rejecting the smallest entities from the feed. These include monovalent ions, dissolved organic content, and viruses, almost everything that other membrane processes are not capable of. RO membranes can also be used in both dead-end and cross-flow configurations, but usually cross-flow is preferred for its low energy consumption and low fouling properties. Mostly they are available as spiral wound modules, where the membrane is wound around the inner tube. RO has several applications, of which desalination is the most important and widely used. RO is also used in wastewater treatment, and dairy and food products.

1.1.3.1.4.1 Advantages and Disadvantages of RO

Advantages

- Desalination of sea and brackish water can be performed compared to other membrane processes.
- Separation without phase change as with other membrane processes.
- Compact and thus less space consuming as compared to other desalting systems.
- Easy to maintain.
- Easy to scale up.

Disadvantages

- High pressure requirements.
- Pretreatment of the feed is required.

- Lower flux.
- Energy-intensive process.
- Fouling.

1.1.3.2 Concentration-Driven Processes

There is not much difference between pressure- and concentration-driven membrane processes. The operation under both processes is based upon concentration or activity difference in the permeant across the membrane. Therefore, many of the processes that are discussed under concentration-driven membrane processes can also be discussed under pressure-driven membrane processes. In other words, increasing the total pressure increases both the applied pressure as well as the concentration of the permeant. An interesting fact to note is that the concentration and applied pressure are almost proportional to each other for membrane-based gas separations. The difference between pressure- and concentration-driven membrane processes lies in the fact that in pressure-driven membrane processes the applied pressure increases the chemical activity of the permeant on the feed side of the membrane. On the other hand, in concentration-driven membrane processes the chemical activity remains the same on the feed side but reduces significantly on the permeate side.

1.1.3.2.1 Gas Separation through Dense Membranes

Gas separation is a prominent application of membrane technology. Generally, polymeric membranes are used for gas separation applications. In general, a gas has the tendency of solubility in a polymeric membrane, which certainly depends upon the type of gas and polymer. A gas at a higher partial pressure on the feed side of the membrane will diffuse through the membrane to the permeate side due to the concentration as well as the partial pressure gradient. There are three steps in a membrane gas separation process. These are (1) sorption, which is the dissolution of a gas into the polymer (the membrane material); (2) diffusion, which is the permeation of the gas through the membrane from the feed side to the permeate side; and (3) desorption, which is the occurrence of the gas on the permeate side of the membrane. This mechanism of membrane gas separation is well known as the solution–diffusion mechanism. The driving force for the process as discussed earlier is the difference in the concentration and partial pressure of the gas across the membrane. This technique is further discussed in detail in the subsequent sections.

1.1.3.2.2 Pervaporation (PV)

The term *pervaporation* was coined by Kober in the year 1917 [8]. It is the combination of two processes: permeation and evaporation. The permeant (liquid phase) permeates through the membrane and evaporates (gas phase) to the other side of the membrane. In other words, a liquid feed comes in contact with the feed side of the membrane and the permeant is collected on the other side (permeate side) of the membrane in vapor form. The volatility difference of the components in the feed plays no role. The process mechanism is shown in Figure 1.5. The driving force for the separation is the partial pressures of the feed and permeant components. Several ways are available to maintain the required vapor pressure difference. For example, on the laboratory scale a vacuum pressure is used for this purpose, and on the industrial level the permeate vapor is cooled so as to create a partial pressure instead of a vacuum, which is economical.

At American Oil, Binning and coworkers systematically studied pervaporation for the first time in the 1950s. Their efforts remained constrained to laboratory levels only due to the nonexistence of high-performance membranes and modules. Monsanto in the 1970s also actively worked on this process and got many patents but without any commercial use of this process. It was the academic research that was carried out at the University of Toulouse that enabled this process to be economically viable, and developments in membrane science led to its commercial viability [6].

At present, pervaporation is used for various commercial applications, including in the chemical, pharmaceutical, and food industries. The process is used for the concentration of heat-sensitive products, removal of aromatic compounds or volatile organic contaminants, or for the detection of

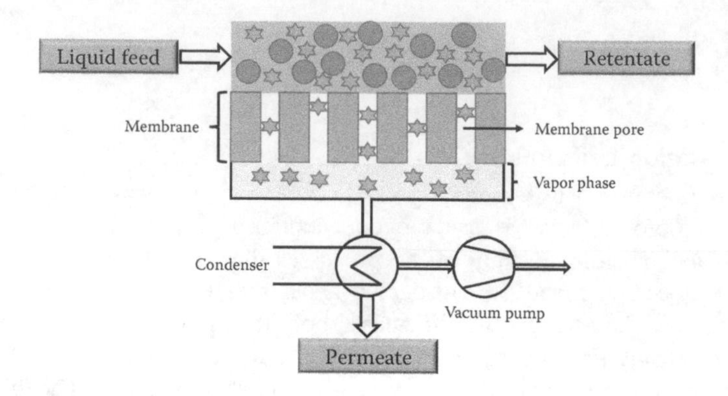

FIGURE 1.5 Separation mechanism of pervaporation membrane process in cross-flow configuration.

analytes in analytical applications. On broader terms, the applications can be classified as dehydration, volatile organic compounds removal, and separation of azeotropic mixtures or isomers.

1.1.3.2.3 Dialysis

Dialysis is a process where diffusion of solutes takes place based on their concentration gradients. The solutes are separated based on their diffusion rates, which depend on the difference in their solubility and molecular size. Figure 1.6 represents the dialysis process. Thomas Graham [9] is acknowledged as the discoverer of dialysis; he worked on the separation of small molecules from colloids. The development of the Cerini dialyzer [10] made dialysis the first membrane process to be used on an industrial scale. It was used in the separation of sodium hydroxide from hemicellulose in the process of rayon production from cellulose. It scored 90% sodium hydroxide recovery of the original solution, and due to such good economics was widely accepted and used. In the 1950s–1960s, dialysis was also used at the laboratory level for the purification of biological samples or the fractionation of macromolecules. At present, the major use of dialysis is as an artificial kidney. Other than this application, dialysis is not used much because of its dependency on diffusion-based separation. Since diffusion is unselective and a very slow process, most of the potential dialysis separations are taken care by ultrafiltration or electrodialysis. These processes are opted because of their selectivity (due to the availability of better selective membranes) and faster separations. In dialysis, it is better to have a thin membrane, as it results in higher flux.

In 1994, Kloff and Berk [11] developed the first operational artificial kidney for the removal of urea, uric acid, creatinine, and other protein metabolism products from blood. Their dialyzer consisted of a drum wrapped around by a cellophane sausage, which is a rotating drum dialyzer. The twin coil dialyzer was introduced by Kolff and coworkers in the year 1956. It was Kiil [12] who introduced the first plate-and-frame hemodialyzer in 1960 based on same principles as of early industrial dialyzers. In 1967, a breakthrough occurred regarding the design of a hemodialyzer: a hollow fiber artificial kidney was developed. Its working principle was like a shell and tube heat exchanger; the blood flows in the fibers and the dialysate outside of the tubes [6]. The process of hemodialysis including its main components is shown in Figure 1.7. A plasma-water-resembling

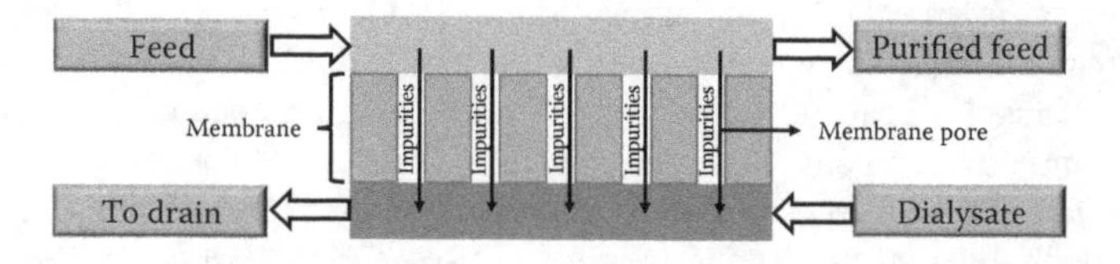

FIGURE 1.6 Mechanism of dialysis membrane process in countercurrent flow configuration.

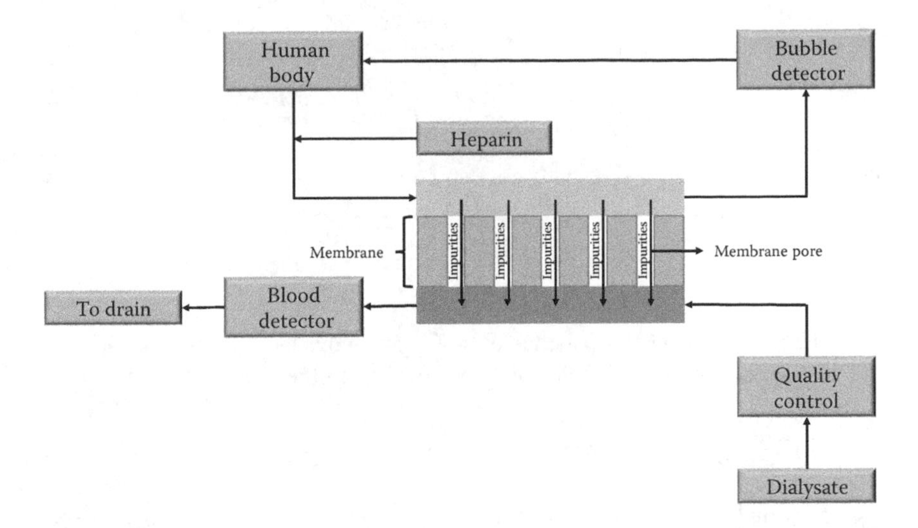

FIGURE 1.7 Schematic representation of hemodialysis.

solution and the dialysate were delivered to the dialyzer by the delivery system at a rate of 200 to 500 ml/min. The dialysate is heated to 38°C prior to its delivery to the dialyzer for the removal of dissolved gases if any and is also checked for its composition. Heparin, an anticoagulant, is added to the blood stream. The blood flow rate through the dialyzer was kept at 300 ml/min. Generally, the hemodialysis procedure would last up to 4 to 6 h. For chronic uremia patients this procedure has to be repeated twice per week. The most common material for the hemodialysis membrane is cellulose. A number of noncellulosic membrane materials have also been studied and now polycarbonate materials are very famous. These membrane materials are prominent because of their capability to allow the passage of solute molecular weights in the range of 1000 to 2000 Da as compared to cellulosic materials.

The middle molecule hypothesis states that the artificial kidney is not able to clear the higher molecular weight molecular species as compared to human kidneys. This is considered as the potential reason for the incomplete well-being of artificial kidney users. This led to the search for new competitive membrane materials like polycarbonate, as previously discussed. Also, it gave rise to a new technique known as hemodiafiltration. In this the blood is not dialyzed but ultrafiltered and the toxins are removed. This helps in the removal of small as well as medium molecular sized species from the blood along with the ultrafiltrate. The blood is diluted before or after this ultrafiltration. Polysulfone and polyacrylonitrile membrane materials have shown some favorable results.

The dialysis in which electrolytes are separated by using neutral or charged membranes, then Donnan effects come into effect is known as Donnan dialysis [13]. It will be explained in the electrodialysis section (Section 1.1.3.4.1). The diffusion process of dialysis can be explained by Fick's law [14]. Figure 1.8 schematically shows the concentrations, and inlet and outlet flow rates in a dialyzer, where *F* represents the volumetric flow rates and *C* represents the solute concentrations. The subscripts *in* and *out* represent the inlet and outlet conditions, respectively. Applying mass

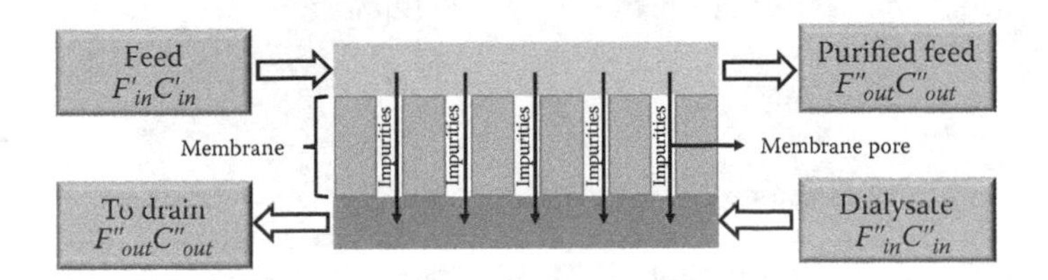

FIGURE 1.8 Schematic representation of a dialyzer inlet and outlet flow rates with concentrations.

balance to the system, we have

$$M = F'_{in}C'_{in} - F'_{out}C'_{out} = F''_{out}C''_{out} - F''_{in}C''_{in} \tag{1.6}$$

Dialysance (D), the efficiency of a dialyzer, can be defined as

$$D = \frac{M}{C'_{in} - C''_{in}} \tag{1.7}$$

Also, the extraction ratio (E) is a better option than dialysance, as it represents the actually achieved fraction of maximum attainable solute removal from the feed. It can be written as

$$E = \frac{D}{F'} \tag{1.8}$$

Clearance (ϕ), which is the volumetric flow rate at which complete clearance of the feed from solute is achieved, can be written as

$$\phi = \frac{M}{C'_{in}} \tag{1.9}$$

At, $\phi = 0$, clearance is equal to dialysance. This is common with single pass dialysate systems.

Also, the transport resistance in dialysis is the combination of membrane resistance and boundary layer resistances. Therefore, the overall mass transfer coefficient (K_o) can be obtained as

$$\frac{1}{K_o} = \frac{1}{R_m} + \frac{1}{K'} + \frac{1}{K''} \tag{1.10}$$

K_o is assumed to be constant, but since K' and K'' change with position under the laminar flow conditions, a length-averaged overall mass transfer coefficient, K, is used in further equations.

Now, assuming volumetric flow rate across the membrane negligible, compared to feed and dialysate flow rate, F is constant in the two phases.

Therefore, the rate of mass transfer, across a differential element of area dA, of a solute is given as

$$dM = K\left(C' - C''\right)\, dA \tag{1.11}$$

The differential mass balance on both sides of the process yields

$$dM = -F'dC' \tag{1.12}$$

$$dM = -F''dC'' \tag{1.13}$$

Taking the concentration difference between the two liquids:

$$\Delta C = C' - C'' \tag{1.14}$$

The differential form of Equation 1.14 gives

$$d(\Delta C) = dC' - dC'' \tag{1.15}$$

Substituting the concentration terms from Equations 1.12 and 1.13:

$$d(\Delta C) = -\left(\frac{1}{F'} - \frac{1}{F''}\right) dM \tag{1.16}$$

Integrating Equation 1.16 results in

$$\Delta C_{in} - \Delta C_{out} = \left(\frac{1}{F'} - \frac{1}{F''}\right) M \tag{1.17}$$

Adding mass flow from Equation 1.11 to Equation 1.16 gives

$$\frac{d(\Delta C)}{\Delta C} = -\left(\frac{1}{F'} - \frac{1}{F''}\right) KdA \tag{1.18}$$

Integrating Equation 1.16 across the entire system gives

$$\ln\left(\frac{\Delta C_{out}}{\Delta C_{in}}\right) = -K\left(\frac{1}{F'} - \frac{1}{F''}\right) A \tag{1.19}$$

Combining Equations 1.17 and 1.19 gives the relation for the mass flow through the total membrane area:

$$M = KA(\Delta C)_{lm} \tag{1.20}$$

where the log-mean concentration is defined as

$$(\Delta C)_{lm} = \frac{\Delta C_{in} - \Delta C_{out}}{\ln\left(\frac{\Delta C_{in}}{\Delta C_{out}}\right)} \tag{1.21}$$

ΔC_{in} and ΔC_{out} represents the feed and dialysate concentrations at the inlets and outlets of a dialyzer, respectively.

Equation 1.10 can be utilized for other flow geometries, by giving correct $(\Delta C)_{lm}$ values, like for cocurrent and cross-flow geometries

$$\Delta C_{in} = C'_{in} - C''_{in} \tag{1.22}$$

countercurrent and well-mixed dialysate flows

$$\Delta C_{in} = C'_{in} - C''_{out} \tag{1.23}$$

cocurrent, cross-flow, and well-mixed dialysate

$$\Delta C_{out} = C'_{out} - C''_{out} \tag{1.24}$$

and countercurrent flow

$$\Delta C_{out} = C'_{out} - C''_{in} \tag{1.25}$$

The overall performance equations, in terms of extraction ratio to mass transfer parameters, can be given by combining Equations 1.6, 1.7, 1.20, and 1.21 as shown next:

Cocurrent

$$E = \frac{1 - exp[-N_T(1+Z)]}{(1+Z)} \tag{1.26}$$

Countercurrent

$$E = \frac{1 - exp[-N_T(1-Z)]}{1 - Z\ exp[-N_T(1-Z)]} \tag{1.27}$$

Well-mixed dialysate

$$E = \frac{1 - exp[-N_T]}{1 + Z(1 - exp[-N_T])} \tag{1.28}$$

Cross-flow

$$E = \frac{1}{N_T Z} \sum (S_n[N_T] S_n[N_T Z]) \tag{1.29}$$

The extraction ratio can be calculated from the two dimensionless parameters given by

$$N_T = \frac{KA}{F'} \tag{1.30}$$

and

$$Z = \frac{F'}{F''} \tag{1.31}$$

where N_T represents the transfer units that ultimately tell about the size of a dialyzer [5].

1.1.3.2.4 Supported Liquid Membrane (SLM)

Liquid-membrane-based separations are the combination of solvent extraction and stripping processes. Liquid membranes are great in terms of low energy consumption, low capital and operating costs, and requirement of small amounts of membrane phase. These characteristics of liquid membranes make them special. A supported liquid membrane (SLM) is a membrane system in which the capillary forces in the microporous polymeric or ceramic membrane pores hold the membrane phase (liquid). Generally, in SLMs two liquids are separated by a polymeric membrane containing immobilized membrane phase.

SLMs were first reported in 1960 by Scholander [15]. He used hemoglobin solution as the membrane phase for the transport of oxygen through cellulose acetate polymeric membrane. Wittenberg also reported a similar system explaining the molecular oxygen transport mechanism [6]. It was the 1980s when SLMs started getting attention because of their easy addition to a continuous flow system. Applications of SLMs were found in wastewater treatment, greenhouse gas capture, analytical chemistry, environmental science, biotechnology, and the pharmaceutical industry.

The mechanism of SLM and ELM (emulsion liquid membrane) is shown in Figure 1.9. In the SLM separation process the membrane phase is immobilized in a porous polymeric membrane or a ceramic support by the virtue of capillary forces separating two liquids: the feed (donor) and the strip (acceptor). Based on the concentration difference of the solutes in the two liquids (donor and acceptor), the solutes are transferred from the feed side to the strip side via membrane phase. In some

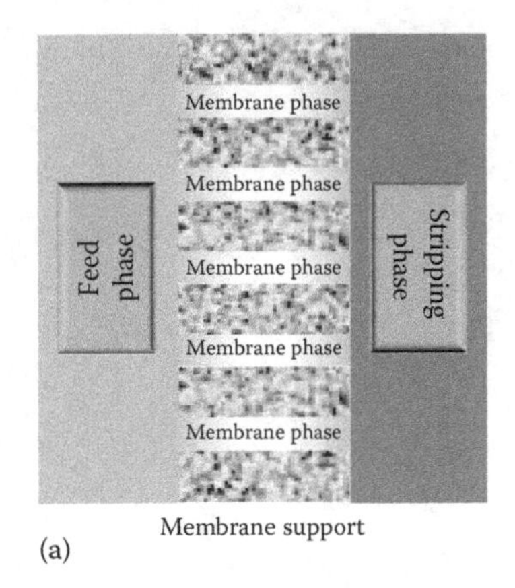

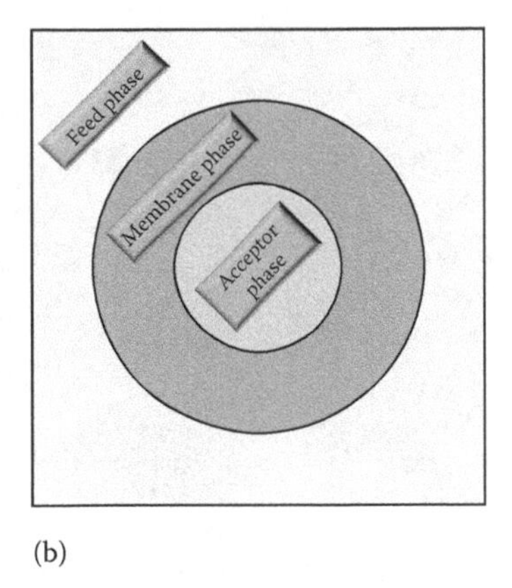

FIGURE 1.9 Schematic representation of (a) supported and (b) emulsion liquid membrane.

cases a carrier can also be used to enhance the selectivity of the separation process, where the solute selectively binds to the carrier and transports from the feed side to the acceptor side via the membrane phase. This process was also called pertraction by Schlosser, based on the mechanism of operation of the process [16]. Pertraction is derived from the Latin word *pertrahere*, which means "extraction," which is derived from *extrahere*. The success of SLMs is based on their advantages like requirement of membrane phase and carrier in small amounts, one step mass transfer, potential for higher separation, and low cost. However, SLMs are not without limitations, like stability issues due to leak or loss of membrane phase. Still, by choosing the support wisely, selecting organic solvents as the membrane phase, the stability issues can be taken care of.

1.1.3.2.5 Emulsion Liquid Membrane (ELM)

Liquid membranes are techniques with great potential for diverse separation applications. Emulsion liquid membrane (ELM) is a very attractive liquid membrane separation process. It is a simple to use technique with high efficiency and one-stage extraction and stripping. It also holds strong as a continuous process with a large interfacial area. It also does great where other conventional separation processes fail, like in the removal of metal ions and hydrocarbons from wastewater.

ELMs were first reported for the separation of hydrocarbons by Li [17]. The emulsions used in ELMs are of two types. The first is water-in-oil emulsion dispersed in an aqueous phase, usually a water-in-oil-in-water (W/O/W) type, where the oil separates the aqueous phases and acts as the membrane phase for the separation process. The second type is the oil-in-water emulsion dispersed in an organic phase, usually an oil-in-water-in-oil (O/W/O), where the water separates the two organic phases and acts as the membrane phase for the separation process. The solute diffusion is the principle rate-determining step in liquid membranes. Nevertheless, the use of carriers, additives, reagents, or external stimuli can be used to enhance the rate of separation.

There are two broad separation mechanisms in ELMs: simple and facilitated. In the simple permeation mechanism, the membrane phase (aqueous or organic) is embedded between the feed and acceptor phase (Figure 1.9b). The solute having affinity for the membrane phase diffuses from the feed phase to the acceptor phase via the membrane phase based on the concentration difference of the solute in the feed and acceptor phase. The simple permeation due to its dependency on the rate of solute diffusion, generally, is used for the study of emulsion stability. On the other hand, the facilitated transport presents separation with higher rates by increasing the flux and capacity of the solute diffusion through the membrane phase. In facilitated transport, a carrier or reactive component is used for the separation of the solutes at higher rates (Figure 1.10). The solutes react or form a complex with the reactive component or carrier, respectively, and thus diffuses through the

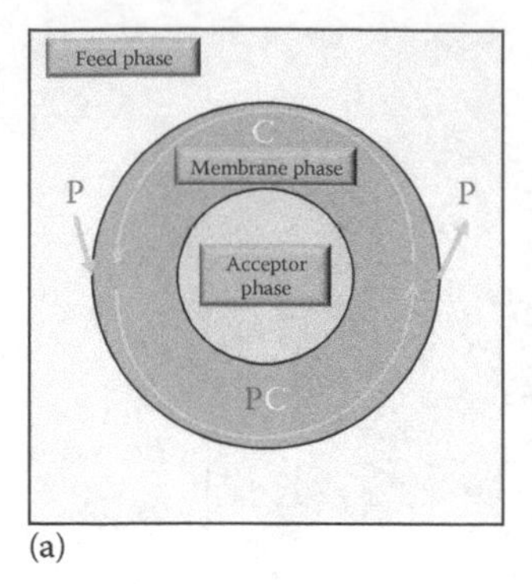

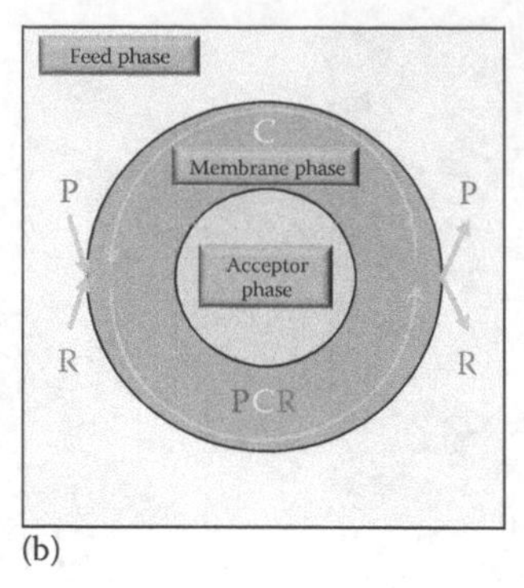

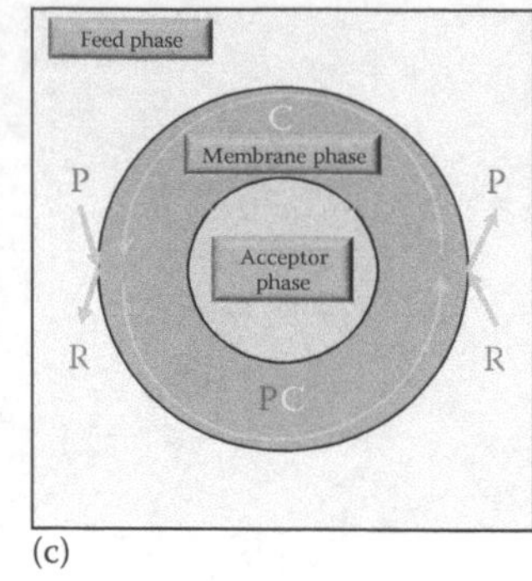

FIGURE 1.10 Schematic representation of different carrier-mediated transport mechanisms: (a) simple, (b) cotransport, and (c) countertransport in liquid membranes.

membrane phase from the feed to the acceptor phase. To increase the separation rate, the concentration gradient can be increased by inducing carrier transport with a chemical reaction at the acceptor phase or by using coupled mass transport. In coupled mass transport, two different solutes are permeated at the same time either in the same direction, known as codiffusion, or in different directions, known as counterdiffusion. The concentration gradient can also be increased by using feed and acceptor phases of different pH values. It will change the ion-exchange process to the opposite directions on the two liquid membrane surfaces. Some examples of carriers used for the separation of metal ions are –COOH, $–SOH_3$, or chelating groups, which are acidic in nature. Basic carriers like amines or quaternary ammonium are also used.

1.1.3.3 Temperature-Driven Processes

The membrane processes where temperature plays an important role comes under this category. In these membrane processes the temperature gradient is the source of the driving force for the separation. These processes are important where size, concentration, and compositional differences of two different species are not enough for effective separation, and the membranes are generally ineffective for the separation process. The difference in the vapor pressures of the two species is utilized by the virtue of temperature gradient for effective and efficient separation in these membrane processes. Therefore, the separation efficiency of these membrane processes is determined on the basis of the distribution coefficient of the species in two phases (liquid–liquid or liquid–gas) and the membrane just acts as an interface between the two phases.

1.1.3.3.1 Membrane Distillation

Membrane distillation is a process where two liquids at two different temperatures are separated by a membrane, as shown in Figure 1.11. A critical condition for this process is that the two liquids should not come in contact with the membrane, and thus a hydrophobic porous membrane is used in this process. The vapor pressure difference of the liquids allows the vapors to cross the membrane. This is due to the temperature difference of the liquids, which in turn results in different vapor pressures. The more the temperature difference, the more flux obtained. The presence of salts in the feed also play a vital role, since the presence of salts decreases the vapor pressure, which is the driving force for the separation in this process. Salts present in minimum concentration do not affect the process much; it becomes point of concern when the salt concentration is very high.

In membrane distillation, the membrane acts only as a barrier and plays a small role in the separation process. In this technique, the separation process or its selectivity vitally depends upon the vapor–liquid equilibrium, that is, the entity having the highest vapor pressure will have the highest flux. Like in case of water and a liquid with higher vapor pressure compared to water, the flux of the liquid will be higher as compared to water. Similarly, for water and a liquid with lower vapor

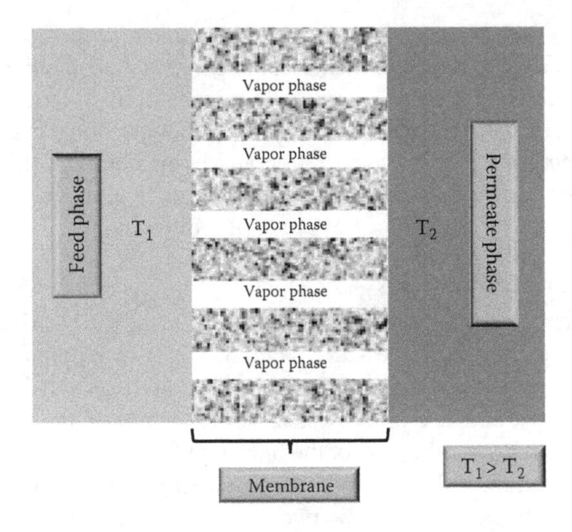

FIGURE 1.11 Schematic representation of membrane distillation mechanism.

pressure as compared to water, water will have higher flux. The main application of this process is the separation of salt from wastewater. This process was developed by Enka, an Akzo division, on the commercial scale in the 1980s [6]. Membrane distillation contains many advantages over reverse osmosis, an alternative for membrane distillation. An advantage is the use of low-grade heat to initiate the temperature gradient, the driving force for the process, as compared to the expensive high-pressure pumps. A high membrane area is not required and the membrane fluxes are also comparable. The process is even effective with high concentrations of the feed, which is a limitation of reverse osmosis. The process is beneficial for large applications like seawater desalination considering the energy savings, but the modules required are very expensive. Therefore, in spite of the technical upper hand, significant commercial success has not been attained. Currently, membrane distillation is not used on a commercial level but it is still a process of interest with academicians.

Phenomenological equations can be used to describe the transport of the volatile compounds in the process. These phenomenological equations state that the flux is proportional to the temperature difference across the membrane. The vapor pressure difference with temperature may be calculated using the Antoine equation [18].

The flux in terms of a phenomenological equation can be given as

$$F_x = \beta \; \Delta P_x \tag{1.32}$$

where β and Δp are the membrane and system-based parameters, respectively. The flux is related to these two parameters. The proportionality factor, β, is dependent on the membrane parameters like membrane material (hydrophobic/hydrophilic), pore structure, porosity, and membrane thickness. The higher the porosity, the better is the process as membrane porosity and thickness are the two main structural parameters of the membrane. The membrane pore size should also be higher and uniform, since the large pores are the first to be wetted. The system parameter, Δp, solely depends upon the temperature gradient, ΔT, of the system. Other important factors for the process are the hydrodynamic conditions and module design. These parameters tell about the temperature polarization effect, which in turn affects the driving force of the process.

1.1.3.3.2 Thermo-Osmosis

Thermo-osmosis is a process where a nonporous (dense) membrane separates two phases of different temperatures. The driving force for the process is the thermal gradient as it is in membrane distillation. Thermo-osmosis is quite similar to the membrane distillation process when it comes to the required driving force. In both processes, the temperature gradient is the driving force. However, the two processes differ on the basis of type of membrane used, its role, and phase transitions. In thermo-osmosis a nonporous membrane is used, whereas in membrane distillation a porous membrane is used. Phase transitions take place in membrane distillation but not in the case of thermo-osmosis. In thermo-osmosis, the membrane performs separation and in membrane distillation a membrane acts only as a barrier between the phases, which are at different temperatures. Therefore, the separation features and mechanisms of the two processes are totally different.

1.1.3.4 Electrically Driven Processes

For membrane processes that come under the category of electrically driven processes, electrical potential is the driving force for the successful separation of the feed components. These membrane processes utilize the current-conducting potential of charged ions and molecules. Electrically charged membranes are used in these processes, which allow a specific charge-carrying species to permeate through it and retain the other. Also, the uncharged species can be separated from charged species because the driving force (electrical potential gradient) for these membrane processes has no effect on the uncharged species. There can be two types of charged membranes, namely, (1) positively charged anion-exchange membranes (or anionic membranes) for the permeation of negatively charged species or anions and (2) negatively charged cation-exchange membranes (or cationic membranes) for the permeation of positively charged species or cations. The combination of the driving force (i.e., electrical potential gradient) and electrically charged membranes is used in different combinations for a better membrane separation process.

1.1.3.4.1 Electrodialysis

Electrodialysis is a process that uses ion-exchange membranes. These membranes are membranes with charged groups; these charged groups exclude the same charge ions either partially or completely. In simple words, the membrane charged with positive groups excludes cations and is permeable to anions or negatively charged species; this membrane type is known as an anionic membrane. Similarly, a membrane charged with negative groups excludes anions and is permeable to cations or positively charged species; this membrane type is known as a cationic membrane. The electrodialysis process is shown in Figure 1.12.

The ion-exchange membranes are arranged in a plate-and-frame model. The anionic and cationic membranes are arranged in between the anode and cathode in an alternating fashion. The pair of an anionic and cationic membrane is known as a cell. There used to be more than 100 cells in an electrodialysis stack. While maintaining the electrical potential across the electrodes, the salt solution passes through these individual cells. Due to this, the anions will migrate toward the anode and the cations to the cathode. The anions simply cross the anionic membrane but are blocked by the cationic membranes; similarly, the cations are able to cross the cationic membrane but not the anionic membrane. In turn, the ions are separated from the feed. Therefore, electrodialysis is widely used for the removal of ions from water.

Ostwald in the 1890s explained ion-exchange membranes experimentally for the first time [19]. The phenomenon of Donnan exclusion was given by Donnan [13] with the concept of membrane potential a few years later. In 1939, Manegold and Kalauch [20] proposed ion-exchange membranes for the separation of ions from water for the first time. Subsequently, Meyer and Strauss explained the concept of multiple cells within a single pair of electrodes, and to date this arrangement is used for electrodialysis process. The developments in polymer chemistry yielded better ion-exchange membranes. Kressman, Murphy et al. and Juda and McRae [21–23] developed some of the better ion-exchange membranes of their time at Ionics Inc. These developments made Ionics the first developer and installer of an electrodialysis plant in the year 1952, and by 1956, Ionics had eight plants installed [6]. These installed plants were unidirectional in operation and thus faced the problem of fouling. Also, due to the scaling, the solutes and salts start to deposit on the membrane surface and reduce the efficiency of the process. This increases the burden on the whole procedure,

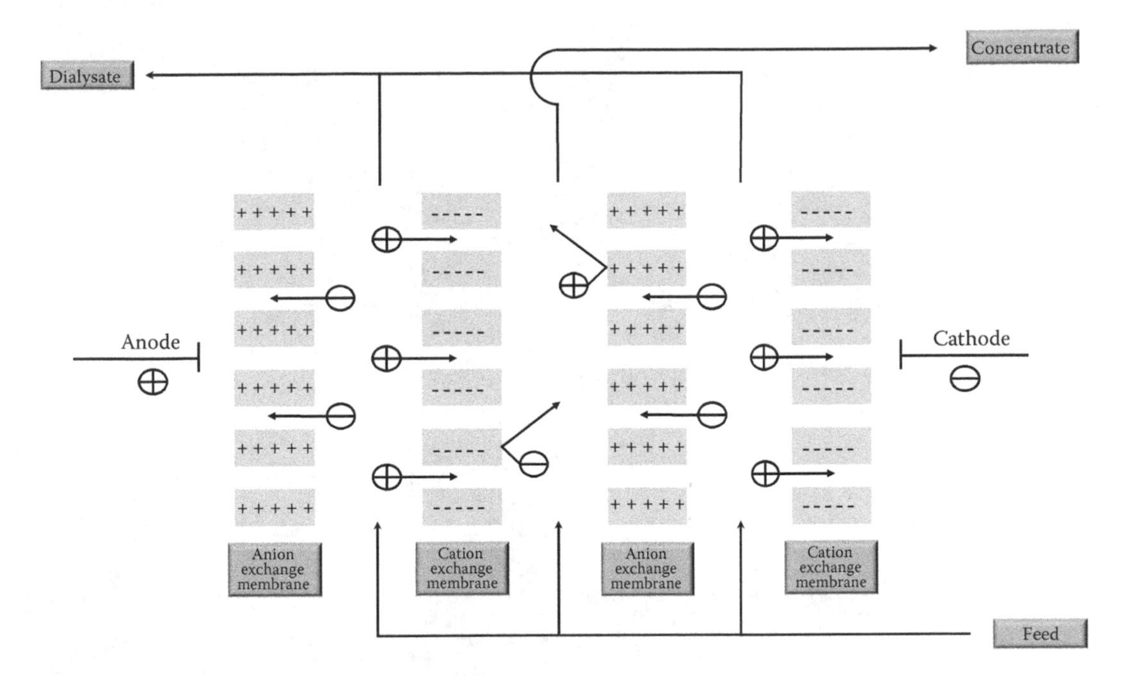

FIGURE 1.12 Schematic representation of electrodialysis mechanism.

and additional preventive measures have to be used like pH adjustment of the feed, antiscaling chemicals in the process, and cleaning of the membranes with detergents and descaling chemicals. Still, the problem remains severe for the process. It was in 1970s when Ionics developed a concept to modify the system design. This idea was to regularly change the polarity of the system for a short time period. This means that the usual flow starts to be reversed and thus the freshly settled scale on the membrane surface will flush out before its solidification. Thus, this change came as a boon for the electrodialysis industry and has been used since.

Electrodialysis is now an established process on an industrial scale. Ionics is the major shareholder for the process throughout the world. Electrodialysis is generally used in the production of boiler feed water, so as to decrease the scaling in the boilers' internal surface, desalination of brackish water, and production of industrial process water. There came a decline in the use of the electrodialysis process after the development of better reverse osmosis membranes. However, it is still used for the desalination and deionization of water in ultrapure water plants. This process is known as electrodeionization, which is derived and is a combination of electrodialysis and ion exchange. Other than these, some applications where electrodialysis is employed are water softening, desalination of food products like in case of whey protein, in the removal of ionic impurities from industrial wastewaters, in advanced fuel cells and batteries, and in the production of acids and alkalis from salts by using bipolar membranes. The bipolar membranes are the combination of anionic and cationic membranes bonded together. In simple words, it is a layered membrane that consists of two layers, in which one layer is anionic and the other is cationic in nature.

1.1.3.4.2 Electrofiltration

Electrofiltration is symbiosis between membrane filtration and electrophoresis. An electric field is applied to the dead-end membrane filtration so as to avoid membrane fouling. This technique is mainly used with colloidal substances (biomaterials), especially biopolymers for dewatering and fractionation. These biopolymers, for example, proteins, possess high filtration resistances, which hinder the membrane filtration process. The electric field can be used to decrease this resistance, since the biopolymers commonly contain charge on their surfaces. Therefore, the electric field can easily influence their movement. The electric field is applied in a manner that it is in the opposite

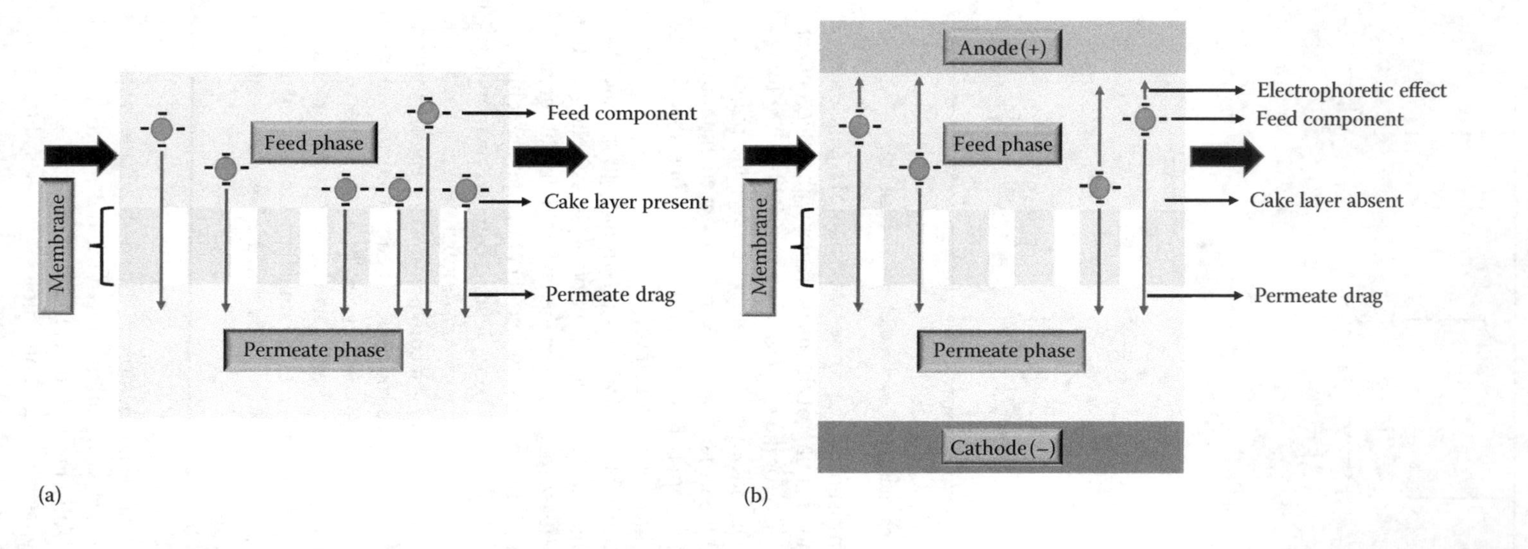

FIGURE 1.13 Schematic representation of (a) simple membrane filtration and (b) membrane electrofiltration mechanism along with antifouling advantage of electrofiltration membrane process.

direction to the flux for the molecule to be separated. Thus, it directs the molecule in the opposite directions, since the electrophoretic force is opposite to the resistance force of the filtrate flow. This significantly decreases the process time and fouling of the membrane by the molecule. The process delivers a high level of selectivity and separation. This also reduces further steps down the downstream process line. The other added advantages with this process are increased flux and reduced shear force stress on the molecules. Therefore, it can be considered as a mild process for the separation of biomaterials, as high shear forces used to have negative impact on the biomaterials and in the process might be degraded. The process is shown in Figure 1.13.

1.1.3.4.3 Electrochemical Ion Exchange (EIX)

Electrochemical ion exchange (EIX) is a novel separation ion exchange process. In this process, an exchange material is introduced into the EIX electrode using a binder, and an externally applied potential controls the absorption of ions into an electrode. The EIX process doesn't need any eluents for elution of the material of interest. Simple polarity change is sufficient for the elution of the material, therefore, this makes the EIX process to use ion exchange in various ways, such as anionic or cationic based on the material to be eluted. Anions as well as cations of IA, IIA, transition, and posttransition metals are absorbed. EIX is explored mainly for the treatment of nuclear waste streams. For example, cesium and cobalt can be selectively removed from sodium- and lithium-bearing feeds, respectively, by using an inorganic ion exchanger. The properties of the ion exchangers matter in the removal of anions including nitrate, borate, chloride, and sulfate. Other potential applications of EIX are removal of heavy metals from industrial wastewaters and conventional water softening.

1.2 DEVELOPMENTAL HISTORY OF MEMBRANES

The development of membrane utilization and membrane processes technical utilization took place just five decades ago. The study conducted by Nollet in the middle of the 18th century is the first available record that explains the membrane phenomenon and discovery of osmosis. In the study it was discovered that a pig's bladder passes ethanol when it was brought in contact with a water–ethanol mixture on one side and pure water on the other side. Thomas Graham studied mass transport in semipermeable membranes systematically. He studied the diffusion of gases through different media and discovered that rubber exhibits different permeabilities to different gases.

In the early phase of membrane studies, mostly natural materials like animal bladders or gum elastics were used for better understanding of membrane permeation. The first person to use an artificially prepared semipermeable membrane was Traube [6,24]. He took a thin layer of porous porcelain in which he precipitated cupric ferrocyanide to convert the thin layer of porous porcelain into a semipermeable membrane. Pfeffer used membranes of this type for his fundamental studies on osmosis. Later, Fick and van't Hoff explained the membrane phenomenon by interpreting the diffusion in liquids as a function of concentration gradients, and provided thermodynamic explanations and quantitative measurements for the osmotic pressure of dilute solutions [6]. The flux equation for electrolytes under the driving force of a concentration or electrical potential gradient was introduced by Nernst and Planck. Donnan published the theory of membrane equilibria and membrane potentials in the presence of electrolytes, which satisfactorily described and theoretically interpreted the basic phenomena of membrane permeation.

The first method for synthesizing synthetic membranes was developed by Bechhold [25]. He impregnated a filter paper with a solution of nitrocellulose in glacial acetic acid. These membranes could be prepared and accurately reproduced with different permeabilities by varying the ratio of acetic acid to nitrocellulose. Zsigmondy also used nitrocellulose membranes so as to separate macromolecules and fine particles from aqueous solutions. Kolff [11] successfully developed the first functioning hemodialyzer, which was the key for applications of membranes on a large scale in the biomedical field.

Prior to the 1950s, membrane science and technology had very few practical applications, but after 1950 the membrane-based industry rapidly developed. The development in polymer chemistry resulted in polymers with better mechanical and chemical properties, which results into the preparation of membranes with unique and novel properties. Staverman, Kedem, and Schlogl [26–28] illustrated the membrane transport properties on the basis of a comprehensive theory, which was based on the thermodynamic properties of irreversible processes. Merten [29] postulated certain membrane transport models such as the model of a solution–diffusion membrane.

The development of a reverse osmosis membrane was a great achievement in the field of membrane technology. This developed membrane was based on cellulose acetate with properties such as high salt rejection and high fluxes at moderate hydrostatic pressures. This development was a boon for the desalination process for producing potable water out of seawater.

The next big step came when Loeb and Sourirajan [5] developed a membrane with an asymmetric structure with a dense skin at the surface that determined the membrane selectivity and flux and highly porous substructure that provided the mechanical strength. The phase inversion procedure was used for the preparation of asymmetric cellulose acetate membranes, which converts a homogeneous polymer solution into a two-phase system, that is, a solid polymer rich phase and a polymer lean phase, which form the solid polymer structure and liquid-filled membrane pores, respectively. Similarly, after some time other polymers, such as polysulfone, polyamides, polyacrylonitrile, and polyethylene, were also used as basic material for the preparation of synthetic membranes. The results of these membranes have shown that these membranes are having better mechanical and chemical stability than the cellulose acetate membrane. Cellulose acetate membranes were not used after Cadotte [30] and Riley et al. [31] developed the interfacial-polymerized composite membranes. Membranes produced by this method are even better in terms of chemical and mechanical stability and have significantly higher fluxes and higher rejections. Usually the first membranes were prepared as flat sheets and then installed in spiral-wound modules for reverse osmosis and other applications. Self-supporting membranes were a different approach to membrane geometry with wall thickness of 6 to 7 microns. Du Pont Corporation developed the first asymmetric hollow fiber membranes mainly for the desalination process [6].

Modules were the next development in the field of membranes after the development of efficient membranes, so as to further increase membrane efficiencies. These modules give the user a higher membrane surface area, as the membrane is compactly packed in an assembly that is known as a module. Other advantages are ease of membrane or module replacement, control of concentration polarization, and low cost. Membranes are produced in three configurations: flat sheets, hollow fibers or capillaries, and tubes. In modern desalination plants, spiral-wound modules are used, whereas hollow fiber membrane modules are utilized in gas separation and pervaporation. Capillary membrane modules are used especially in medical applications such as blood oxygenator and artificial kidney. In ultrafiltration and microfiltration, tubular membranes are used.

Electrodialysis was an industrial-level application prior to the use of reverse osmosis for desalination of sea and brackish water. Meyer and Strauss [32] developed the first multicell stack for electrodialysis. With the development of the first ion-exchange membranes, electrodialysis became practically feasible. Ion-exchange membranes with good electrolyte conductivity and ion permselectivity made it possible. Ionics was the first company to commercially exploit electrodialysis for the desalination of brackish water [6]. Ionics succeeded in the field of electrodialysis because of good membranes, compact stacking of these membranes, and a special mode of operation known as electrodialysis reversal, which allows the membrane to self-clean and thus increase the life of the membrane/membrane stack. Due to this operation, the process can be run in a continuous mode at a high concentration of scaling materials without the need for mechanical cleaning. The early 1980s opened a new era for the field of electrodialysis by the development of bipolar membranes for the separation of acids and bases from a corresponding salt.

Gas and vapor separation on a large scale is also an industrial area relevant to membranes applications. Monsanto Inc. pioneered this field [6]. Its original idea was to recover hydrogen from gases and to produce oxygen- or nitrogen-enriched air. Nowadays, there are large number of other applications such as removal of CO_2 from natural gas or recovery of organic vapors from gases.

A closely related technique to gas separation is pervaporation. It has been extensively studied and a number of interesting potential applications were pointed out. But still there is a lot of scope to make pervaporation applicable to big commercial plants for large-scale production.

Recent applications of membranes that have attained the status of applicability on large technical and commercial scales are the controlled release of drugs in therapeutic devices, and the storage and conversion of energy in fuel cells and batteries. However, to date the most important and commercially successful application of membranes is reverse osmosis in desalination and in hemodialysis and hemofiltration.

1.3 ADVANTAGES AND DISADVANTAGES OF MEMBRANES

1.3.1 Energy Efficient

Membrane technology, unlike other conventional processes, does not use heat, which means it is energy effectual and saves a lot of energy. For example, the common dewatering techniques for liquid products are freezing and evaporation. These processes require very high amounts of energies to carry out the dewatering process. In contrast, membrane processes need no phase change to carry out the dewatering process and thus saves significant amount of energy.

1.3.2 Simple Design

In membrane processes, complicated designs are not required. Membrane processes can work in ambient pressure and temperature conditions. These traits of membrane processes make them competent and convenient.

1.3.3 Product Specific

In this era, technology is developing daily the number of options available and adding to present-day processes. There are processes available where products of high value are produced but the concentration is very low. Therefore, conventional separation processes are not effective for their separation. On the contrary, membrane processes have the potential to separate such products effectively and efficiently. Thus, membranes processes are very important for the present as well as the future.

1.3.4 Open Domain

There is continuous growth and development of present technologies. Over time, new technologies came into existence and the old technologies were refined to a great extent for better product development. Membrane technology is capable of incorporating improvements and modifications effectively with developing technology and processes. Therefore, membrane technology has a sound present and future, where it can be applied to potential new applications. The membrane technology sector is responding superbly and swiftly to the present-day needs of industry. Continuous exploitation of present-day membrane processes is going on for the development of better membrane technologies for future demands and applications. Thus, it can be said that membrane processes are an open domain that welcomes change and incorporates it for its growth.

1.3.5 Flexible

Membrane processes can be easily combined with other processes, both upstream and downstream, effectively. The combination of other processes with membrane processes ultimately makes the overall process more effective and efficient. The production quantity as well as quality increases and improves to a great extent. The combination of processes are also easy, cost effective, and efficient.

1.3.6 Clean Technology

Membrane separation processes require simple and unharmful chemicals, no adsorbents and therefore no regeneration and disposal problem. Thus, membrane processes are potentially good for the environment and provide a clean technology.

1.3.7 Concentration Polarization

When feed components permeate through a membrane, some of the components pass through the membrane and some are retained. Over time, this results in a high concentration of retained feed components on and near the membrane surface as compared to their concentration in bulk. This forms a gelatinous layer over the membrane surface. This phenomenon is known as concentration polarization. The formed layer further acts as a barrier and significantly reduces the selectivity and efficiency of a membrane process. This phenomenon is quiet common with batch processes.

1.3.8 Fouling

Membrane fouling is the biggest problem of membrane separation processes. The high tendency of concentration fouling, which is the formation of a gel layer of feed components with time over the membrane surface, increases the rate of membrane fouling and in result gradually decreases the selectivity and efficiency of a membrane process. The rate and extent of membrane fouling mainly depends upon the feed type. Since it is the feed components that play the role in the membrane fouling, their size, chemical composition, and their affinity toward the membrane material all are crucial for the fouling to take place. Membrane fouling, though present in all types of membrane processes, is persistent in hollow fiber membranes and dead filtration.

1.3.9 Feed Composition

Membrane processes, like reverse osmosis, have limitations for feed containing high solid content or, in simple words, concentrated feeds. Since feed concentrations are very high, they make the membranes prone to fouling very easily and in no time. In the case of reverse osmosis, the osmotic pressure of the feed components plays a role in limiting the overall process efficiency. In case of ultrafiltration and microfiltration, the feed concentration is not of much concern, but in their case the viscosity of the feed plays a role in limiting the overall membrane flux. Therefore, reverse osmosis is led by ultrafiltration or microfiltration to relieve the feed concentration to some extent and make the feed concentration easy on reverse osmosis. The size, chemical composition, and affinity of feed components toward membrane material also plays an important role in membrane processes.

1.3.10 Selectivity

Membrane processes are said to be very selective, but in reality they seldom produce a pure product out of two or more feed components. The product used to be contaminated by the presence of other feed components in minute amounts. Though this is the case, membranes still outperform the selectivities of other available processes. The membrane selectivity can be 100% if prepared without errors of composition and production.

1.3.11 Life

Generally, the life of membranes is low because of membrane fouling. Fouling drastically reduces the lifetime of a membrane in a process. Generally, the life span of a polymeric membrane ranges from 12 to 18 months, and for ceramic membranes it is approximately 10 years, which when compared to an industrial process is very low, especially in the case of polymeric membranes. The reason behind this is the low resistance of polymeric membranes against harsh chemical, mechanical, and other environmental conditions. On the other hand, ceramic membranes are capable of sustaining these harsh process conditions and thus manage to live a little longer. Other things that define the lifetime of a membrane are its capacities to withhold harsh conditions of temperature, pressure, and chemicals. Recent developments in membrane science have brought out some very good membranes with very good lifetimes that can successfully withstand harsh conditions of membrane operations.

Overall, membranes can be expensive because of their fabrication methods, the need of regular replacement, concentration polarization and fouling, and availability of restricted cleaning options. Therefore, there is a wide scope for research and development in this field. Researchers can opt for any of the present limitations of membranes and work on their development by omitting these restrictions from membranes.

1.4 APPLICATIONS OF MEMBRANES IN VARIOUS FIELDS

Recent developments in the membrane science field have given wide recognition to membranes and brought them out from the laboratory domain to the industrial level. Membrane processes are critically acclaimed for their impressive industrial presence and for their resolution-providing powers for industries such as problems associated with liquid mixture, concentration and purification of chemical and biological products, and wastewater treatment. The vital features of various membrane processes are based on several factors including

- Size of the feed species
- Nature of the feed species to be permeated across the membrane (acidic, basic, volatile, or electrolyte)
- Mechanism of separation
- Selectivity of the membrane for two feed components

Developments in the field of membrane technology have brought change to the conventional strategy of use of membrane processes. Now, new membrane processes are developed that are a combination of more than one membrane process. This is to eliminate the drawbacks of conventional membrane processes. The best example is that of reverse osmosis, where it is combined with microfiltration or ultrafiltration in the upstream to relieve the burden of fouling and flux declination thereon.

Processes like reverse osmosis and nanofiltration are good for the removal of organic or inorganic feed components like dye molecules, heavy metals, or salts, but these processes demand very high pressure and the final flux of these processes is also not promising. Therefore, these processes are combined with other membrane processes like micro- or ultrafiltration for better use of energy and results. Nowadays, methods like surfactant-enhanced metal removal with microfiltration, detoxification and removal of a dye by advanced oxidation process, and micro- or ultrafiltration are widely used. Polymer-enhanced methods in combination with ultrafiltration can also be used for heavy metal removal and for protein fractionations. Therefore, it is better to invest in a combination of two or more process as compared to a single process for a particular application. It will give sound results with less time and energy.

1.4.1 Drinking Water and Desalination

In the present world potable water is scarce. Worldwide, 783 million people do not have access to clean and safe water [33]. One in 9 people do not have access to safe and clean potable water. The health, education, poverty, and economy of a country are also linked to water and the water quality available to its people. In developing countries as much as 80% of illnesses are linked to poor water and sanitation conditions. An estimated 443 million school days are lost each year due to water-related diseases. Half of the world's hospital beds are filled with people suffering from a waterborne disease. Over half of the developing world's primary schools do not have access to water and sanitation facilities. Girls drop out of schools at puberty due to these reasons. Women are the most affected by poor water availability, as they have to fetch water for household purposes from distant places. Globally the percentage of such households is 64%, where a women has to go and fetch water for household purposes. This affects both their health and education. Industry and agriculture are the major sectors of water consumption throughout the world and only 10% of the total is used for domestic purposes. The World Health Organization states that for every $1 invested in water and sanitation, there is an economic return of $3 to $34. At the rate at which the use of water consumption and depletion of water resources is taking place, we are not far from the day when drinking water will be a currency of trade. Therefore, it is essential to increase the availability of clean and safe water to the world using technologies and techniques that allow exploitation of alternative water resources like marine and brackish water. At the same time it is also important to maintain the purity of these sources by ensuring that the wastewaters generated are properly purified before entering the ecosystem.

Water treatment with membrane processes is the universal solution; it is adaptable under all situations. The best trait of membrane processes is their capability to operate on site. Therefore, they are the most suitable solution for the problem of water scarcity. Membrane processes in the field of water were popular from their inception days as a promising technology. This is also due to their natural occurrence and working in the near surroundings and that is for the separation purposes only, like the separation of impurities from the blood, gas exchange in lungs, and the removal of impurities from the body in the form of sweat. The membrane processes that are on the upper side of the pore size scale are mainly used for the removal of suspended particles like microfiltration and ultrafiltration. On the other hand, membrane processes that are on the lower side of the pore size scale are mainly used for the removal of dissolved organic contents, salts, and ions. Membrane processes resolve multiple issues related to drinking water treatment like the removal of organic content, precursors and by-products of various disinfectants used, toxic chemicals, salts, and ions. Membrane fouling is the major concern of membrane processes in drinking water treatment. Fouling affects the membrane selectivity, efficiency, and life. This results in the increase of the cost of operations. Because of the feed constitute of different types of materials, it is important to examine a water type for the constituents to be removed prior to employing membrane processes for the final treatment of the drinking water. This will decrease the membrane fouling and indirectly the cost of drinking water production.

Water is also finite, as is fossil fuel. Every day 200,0000 people are added to the world population, and it is very difficult to provide clean and safe water to everyone. About 85% of the world population lives in water-scarce areas. One good solution to address this problem is to turn seawater into freshwater, which is a process well known as desalination. Commonly, desalination refers to the removal of salts and minerals from a feed. Reverse osmosis is the membrane process of choice for desalination globally due to its pore size and easy to scale up. Globally, 120 countries are using desalination plants, including Australia, Cape Verde, China, Cyprus, Gibraltar, Greece, India, Italy, Japan, Malta, Oman, Portugal, Spain, United Arab Emirates, and the United States. Worldwide, the output of desalination plants is 3.5 billion gallons of potable water per day. Approximately 50% of the installed desalination plants use seawater as feed and the rest use brackish water.

In a desalination plant, a number of membrane elements are connected to an automated and efficient water treatment system. Normally, an efficient desalination plant produces 1 gallon of fresh water from 2 gallons of seawater. This efficiency, energy consumption, cost of production, and maintenance of the membrane elements formulize the cost of the desalinated water. The technological and production improvements in the field of membrane science in the last two decades to a large extent made desalinated water affordable to the masses. Pressure 60 to 70 times higher than the atmospheric pressure is used to separate salts from the seawater to produce fresh water. After this process, much of the energy remains with the concentrated seawater and now due to technological innovations it can be recovered and reused. Membrane systems are prone to fouling and naturally deteriorate over time due to wear and tear. The combination of these two limits the life span of the membranes. Current improvements in the membrane material and production processes made membranes more durable and consistent. Presently, their useful life is extended to over 5 years. Also, use of micro- and ultrafiltration membrane processes before RO desalination further improves their life span and reduces the cost of replacement and indirectly the overall cost of desalinated water. These technological innovations in the membrane field and energy recovery equipment coupled with the enhanced efficiency of the pumps reduced the power consumption to desalinate seawater from 114 KWh/1000 gallons in 1979 to 14 KWh/1000 gallons of freshwater today. Considering that the cost of power is normally 20% to 30% of the desalinated water, these technological innovations made a great cut in the total cost of seawater desalination. Novel energy recovery systems and further developments and improvements in the membrane systems may reduce desalination costs further by 10% to 15%.

The future of desalination is bright with efficient processes like forward osmosis. This membrane system is capable of treating water with salt concentrations up to five times more than seawater. This system can achieve up to 85% water recovery. Forward osmosis is similar to osmosis in the way that water flows from lower concentrations to higher concentrations to attain equilibrium. Forward osmosis differs from reverse osmosis as osmosis differs from reverse osmosis in that in reverse osmosis water flows from higher concentration to lower concentrations leaving behind the solutes (salts) by virtue of an external pressure or force. Hence, forward osmosis is energy efficient as compared to reverse osmosis. Another important difference between the processes is in the resultant permeates from the two processes. The permeate from RO is mostly freshwater that is ready to use, but in the case of FO results in a permeate that may contains solutes of the draw solution (draw solution is the source of driving force for FO) and thus the concentration of solutes in the draw solution and the final use of the permeate decide the further process steps. In recent times, Gibraltar and Oman have installed forward osmosis desalination plants. In the year 2010, *National Geographic* magazine counted forward osmosis as one of the three promising technologies for reduction in the energy consumption for desalination. Present-day RO membrane technology is consistent and commoditized in terms of efficiency, productivity, durability, and life span. Nowadays, a number of manufacturers are producing excellent quality RO membranes with proven track records and performance. Leading manufacturers are extensively exploiting further options for the improvement of RO membranes. They are supporting both the desalination industry as well as advancement in membrane technology and playing a vital role in the field of membrane science. Today desalination plants are fully automated with full protection and safety systems. These advancements cut the labor costs and further cut the cost of the desalinated water.

The tremendous change in the desalination industry is seen in the form of increases in both number and size of desalination plants because of cost reductions to a great extent offered by the advantage of size and centralization. Normally, the bigger the desalination plant, the further reduction in desalinated water by 5% to 10%. Advancements in intake and discharge facilities further played a role in desalinated water cost reduction. Today, by virtue of advanced technology, which made available large size, off-the-shelf high-pressure pumps, and large energy recovery systems, it is possible to construct large desalination plants. Developments in desalination technology combined with construction of large capacity desalination plants, coallocation of power generation plants, and improved competition by the build–own–operate–transfer (BOOT) method of project funding and delivery have resulted in further decreases in the cost of desalinated water. Worldwide, governments are using the BOOT method for the installation and operation of desalination plants.

1.4.2 Industrial Wastewater Treatment

Water is a major substrate or raw material for almost all the industrial processes. It is contaminated after going through an industrial process and needs to be cleaned before being discharged into open water bodies. Membrane processes are used successfully on a very large scale for industrial wastewater treatment. Wastewater from different industries is treated with different membrane processes to check the water pollution problem. Membrane processes are vital since they can be used without a phase change or further increasing the complexities of a treatment process. Precious chemicals and products can also be retrieved during the treatment of industrial wastewaters by using membrane processes. Industrial wastewater treatment is a growing application of membrane processes, but high cost is a limiting factor. Recent developments lower the cost of operations by inculcating new designs, membranes, and modules for membrane processes. The most common industries where membrane processes are used for the treatment of wastewater are explained in the following sections.

1.4.2.1 Municipal Wastewater Treatment

Membrane processes are used for the clarification of municipal waste on a large scale throughout the world. Municipal waste is a complex feed for membrane processes and other treatment procedures. Therefore, a combination of techniques, including membrane processes, are used for the treatment of municipal wastewaters. Microfiltration and ultrafiltration are the two widely used membrane processes for municipal wastewater treatment. In the primary and secondary phases of treatment, membrane processes are employed. The membrane processes make the feed fit for the tertiary processes of treatment.

1.4.2.2 Food Industry

Membrane processes are successfully used in the treatment of wastewaters from food industries like cheese, fruit juice, beer, and wine. These industries produce various kinds of wastewaters, like whey in the case of cheese productions. The disposal of wastewaters is a burden on these industries, since it is a source of water pollution. Membrane processes like microfiltration, ultrafiltration, and reverse osmosis are employed in the production process for better production and to relieve the downstream processing of the products. The whey of cheese production, fruit pulp of juice production, and microorganisms (for example, yeast) of the beer and wine industry are separated from the product by means of filtration. The permeate is thus a clear product free from the impurities and the retentate contains the feed components like whey protein, fruit pulp, or microorganisms, which can be used for other purpose or reused. Thus, the problem of disposal of the wastewater containing these impurities is solved.

1.4.2.3 Tannery Industry

Tanneries are one of the highest consumers of water. Gallons of water are used in this industry for various processes. Tannery wastewater is contaminated with organic matter as well as different chemicals used in the different processes of the industry. Therefore, membrane processes are great for the clarification of the tannery industry wastewater. Membrane processes like microfiltration and ultrafiltration are widely used for this purpose. The contaminated water is given as feed to the membrane processes and the permeate after treatment with some other processes is disposed. The membrane processes relieve the load of the contaminated water to a great extent and thus help in decreasing the levels of water pollution from the tannery industry.

1.4.2.4 Biotechnology

Membrane processes are widely used in the biotechnology field, especially in the fermentation industry for the removal and concentration of products from the permeate so as to make the discharge safe for disposal. Processes like enzyme production, cell harvesting, and virus production are a few of the processes where membrane processes are widely used. The biotechnology field still requires a lot of development for membrane processes to be used for a wide range of applications. Therefore, membrane researchers should search for better membranes with chemical, mechanical, temperature, and pressure stabilities and membrane materials of different types to develop better membranes for a range of applications.

1.4.2.5 Pharmaceuticals

The pharmaceutical industry uses membrane processes for a range of applications. The stringent guidelines given by the regulatory agencies for standard operating as well as preparation procedures demand a process of great trust. Membrane processes fulfill all these requirements and are capable of giving a safe, sterile, and consistent product. Membranes processes are important, as the products of pharmaceutical industries are heat labile, and therefore not many processes are available for their effective and efficient production. The membrane processes like ultrafiltration and microfiltration are used to prepare the product and discharge it safely. Pharmaceuticals use bacteria, viruses, and other viable disease-causing microorganisms for the production and testing of their products; therefore, discharge from the pharmaceutical industry may affect the flora and fauna drastically if not tested or treated effectively before discharging it to the environment. Membrane processes make it safe to discharge the waste into the environment because of their capability to retain particles of micro and nano levels. Thus, membrane processes plays a vital role in the successful running of pharmaceutical industries.

1.4.2.6 Automobile Industry

Electrodeposition of paint on a large scale was used by the automobile industry for the first time in the 1960s. The paint solution used for this process is an emulsion of charged paint particles, which is coated on a metal body by making the metal body an electrode of charge opposite the paint particles. The applied voltage influences the deposition of charged paint particles on the metal surface. The painted metal piece is rinsed with water so as to remove the excess paint from the metal body. This rinsed water is one of the sources of wastewater in the automobile industry. The ultrafiltration membrane process is widely used to address this specific problem. Feed containing 15% to 20% solids is given to the ultrafiltration setup, and a clean permeate with ionic impurities is collected on the permeate side. The retained paint is sent back to the paint tank, and the clean permeate is sent to the counter rinsing operation. The fouling of the ultrafiltration membrane is also a major problem due to the high concentration (15%–20%) of solid particles in the feed. These solid particles form a gel layer on the membrane surface and induce fouling quiet swiftly, which results in lower membrane flux. Irrespective of this fouling problem, the ultrafiltration membrane process is widely used because the amount of paint recovered is quiet high and there is no need of other extra steps for the rinsed water clarification. The ultrafiltration process single handedly treats the contaminated water effectively and efficiently.

1.4.2.7 Oil–Water Emulsions

Metal machining processes widely use oil–water emulsions for the purpose of cooling and lubrication. Instead of following the recycling procedure, spent wastewater streams are produced. Therefore, the use of membrane processes for the recovery of oil and safe disposal of the wastewater is carried out. The wider range of application also makes membrane processes more vital. Mostly, the ultrafiltration membrane process is used for this purpose on a very large scale in the automobile, metal, and metallurgical industries. The ultrafiltration membrane process contains many stages for the better treatment and recovery of oil–water emulsion feeds. The membranes are mainly employed in tubular or capillary configurations because of their low fouling potential and for handling a wide range of feeds. The feed is first filtered with a fine filter so as to remove fine metal particles and other solid matter present in the feeds. Then the pretreated feed is given to the ultrafiltration membrane setup for the recovery and treatment of oil–water emulsion wastewater.

1.4.2.8 Nuclear Industry Wastewater

The nuclear industry produces wastes of a broad spectrum. There are low and intermediate levels of liquid radioactive wastes produced. The radioactive wastes are also treated with the methods used for other industrial wastes like adsorption, chemical treatment, filtration, ion exchange, evaporation, and sedimentation. A single conventional method is not able to completely treat the wastes; therefore, a combination of methods are used for better and complete treatment of the wastes. Membrane processes are economical, effective, and efficient in waste treatments, especially the liquid wastes either alone or in combination with other conventional methods. Thus, membrane processes are a stupendous technique that can be used for the treatment of nuclear waste. The membrane processes are combined with other effective conventional methods for complete treatment of the nuclear wastes. Especially, liquid membranes are explored for their better use in nuclear waste treatment. Membrane processes are in their infancy stages of nuclear waste treatment, but the results shown by membrane processes are satisfactory and promising. Membrane processes proved their mettle in other industrial processes and they will prove themselves in the nuclear industry too. This is also a promising field for membrane enthusiasts to explore membrane technology to apply in wider horizons.

1.4.2.9 Process Water

There are industries that use water for cooling, heating, producing various products, and for various other purposes. Membrane processes can be used for the recovery of these waters economically on a large scale. The water recovery from municipal wastewater relieves the burden on the environment. The capability of membrane processes to deal with hot water from different processes saves on the cost of heating, which is a boon for many industrial processes. Since the feed is not required to be cooled for the treatment, the reheating cost is saved. The water treatment costs are also saved because of membrane processes, as they work for the production units and the treatment units at the same time. This technique is used in the food industries where they clarify the product from the contaminants and at the same time treat the discharge for safe disposal. The recycling and product concentration are two other important facets of membrane processes that are widely used to make the industrial process economical and efficient. Therefore, membrane processes are a boon for present industrial processes and waste treatments.

1.4.3 Dairy Industry

The dairy industry is a huge industry on which much of the world depends. The products of the dairy industry need to be purified before their consumption. This needs to be done with a cheap, less energy prone, no phase change, and good recovery of the final products. Membranes come out to be the best option for this purpose. They fulfill all the criteria required by this industry. India and the United States are the biggest users of membrane science in the food and dairy industry along with other European countries. There are various processes used for milk and food-based products that employ membranes. They include

- Lactose, protein, enzyme, gelatin, whole and skim concentration
- Protein and lactose separation
- Fractionation and concentration of animal and fish oils and proteins
- Concentration of various fruit extracts
- Treatment of wastewater from this industries

1.4.4 Biotechnological Applications

Membrane science is an apt field for the biotechnology sector. Membrane processes for both upstream and downstream technology are used for applications related to the field of biotechnology like preparation of emulsions and particles. Membrane science is advantageous for the field of biotechnology because of its high selectivity, high surface area per unit volume, and control of contact/mixing of two phases. The use of low temperature and pressure conditions makes it perfect for the processing of biological molecules, which are otherwise susceptible to high temperature and pressure. This helps in getting these molecules in their intact form without much denaturation, deactivation, or degradation.

Microfiltration and ultrafiltration processes were used for the separation of macromolecules and to retain suspended colloid particles from the feed. Both of these processes are integrated into upstream and downstream processes. The major use of these processes is in protein concentration, buffer systems exchange, cell harvesting, and in sterilizing liquids so as to make them bacteria and virus free. Membrane bioreactors are widely used for the catalytic or noncatalytic bioconversion processes. Enzymes, cells, microorganisms, or antibodies are suspended or immobilized on a substrate (membrane) surface and are compartmentalized by membranes in a reaction vessel. In a membrane bioreactor, raw materials are transformed into valuable products, like the production of amino acids, enzymes, proteins, antibiotics, and antibodies. Membrane chromatography is also an alternative for the purification of products based on affinity, charge, and hydrophilicity or hydrophobicity of the products. Membrane contactors are used for the preparation of emulsions and different particles. The membrane contactors by the virtue of pressure force the dispersed phase to permeate through the membranes into the continuous phase for the preparation of particles or emulsions. Water-in-oil or oil-in-water emulsions and various polymeric particles are prepared by this membrane process.

1.4.5 Pervaporation

Pervaporation, as previously described, is a process that includes both mass and heat transfer. The role of the membrane in pervaporation is that of a barrier between the aqueous and vapor phase, and controls the phase transition from the feed side to the permeate side. The presence of two phases, liquid and vapor, and a membrane as the third component make the process a sort of extractive distillation. If we consider the principles of both processes, they differ because for distillation the separation is due to vapor–liquid equilibrium, whereas for pervaporation it depends on the solubility and diffusivity of the liquid and vapor.

In pervaporation the transport can be explained on the basis of a solution–diffusion model [6]. The chemical potential of the liquid (feed) is equilibrated with that of the membrane at the liquid–membrane interface at same pressure. The flux of a single component x in terms of partial vapor pressure on the liquid as well as the membrane side p_{xo} and p_{xl}, respectively, can be given by the following equation:

$$F_x = \frac{P_x^G}{l}\left(p_{x_o} - p_{x_l}\right) \tag{1.33}$$

where F_x represents the flux, P_x^G the gas separation permeability coefficient, and l the membrane thickness. The equation will be the same for individual or single components like for y. The separation in pervaporation process will be proportional to the fluxes of components x and y.

Equation 1.33 describes the membrane performance perfectly since it separates the two contributions: the membrane contribution (P_x^G/l) and the driving force contribution ($p_{xo} - p_{xl}$). The membrane permeabilities can be calculated by using Equation 1.33; the values of the partial pressures of the components should be known. The partial pressures of the components on the permeate side can be obtained from the total permeate pressure and the permeate composition, but the partial pressures of components on the feed side are not easily available. They have to be obtained from published tables or calculated by using a suitable equation of state. At present, computers with simulation programs of higher efficiency can be used. Partial pressures of complex mixtures can also be obtained with ease and reliability with computers.

In the literature, the performance of a membrane in a pervaporation process is reported mainly in terms of total flux and separation factor ($\delta_{\text{pervaporation}}$). The separation factor can be defined as the ratio of the components on the permeate side divided by the ratio of the feed side components. The separation factor can be written as (in the case of two components):

$$\delta_{pervaporation} = \frac{c_{x_l}/c_{y_l}}{c_{x_o}/c_{y_o}} = \frac{n_{x_l}/n_{y_l}}{n_{x_o}/n_{y_o}} = \frac{p_{x_l}/p_{y_l}}{p_{x_o}/p_{y_o}} \tag{1.34}$$

where c_x and c_y represents the concentrations, n_x and n_y the mole fractions, and p_x and p_y the vapor pressures of the two components x and y.

Factors including membrane intrinsic permeation properties, composition and temperature of the feed, and the permeate pressure influence the separation factor, $\delta_{pervaporation}$. Further, the pervaporation can be divided into three steps. The first is the evaporation to form the saturated vapor phase in equilibrium with the feed liquid. Separation in this step is due to the presence of different volatilities of the feed components. This separation step, $\delta_{evaporation}$, can be defined as the ratio of the component concentrations in the feed vapor to their concentration in the feed liquid:

$$\delta_{evaporation} = \frac{p_{x_o}/p_{y_o}}{n_{x_o}/n_{y_o}} \tag{1.35}$$

The second step is processing the permeation of the feed components across the membrane; this step is the same as in conventional gas separation. The difference in the vapor pressures of the feed and permeate plays the role of the driving force for the process to take place. The separation step, $\delta_{membrane}$, can be defined as the ratio of the components in the permeate vapor to the ratio of the components in the feed vapor:

$$\delta_{membrane} = \frac{p_{x_l}/p_{y_l}}{n_{x_l}/n_{y_l}} \tag{1.36}$$

In Equation 1.36 it is shown that separation achieved in the process is equivalent to the product of the separations achieved in the first (evaporation) and second (membrane) steps:

$$\delta_{pervaporation} = \delta_{evaporation} \times \delta_{membrane} \tag{1.37}$$

The $\delta_{pervaporation}$ term can be derived in terms of $\delta_{evaporation}$ by using Equations 1.34 and 1.37 and a standard solution–diffusion model. Therefore, the membrane flux can be written as

$$F_x = \frac{P_x^G\left(p_{x_o} - p_{x_l}\right)}{l} \tag{1.38}$$

and

$$F_y = \frac{P_y^G\left(p_{y_o} - p_{y_l}\right)}{l} \tag{1.39}$$

where F represents the permeation flux, P_x^G and P_y^G the permeation coefficients for the vapors x and y, respectively, and l the membrane thickness. Dividing Equations 1.38 and 1.39 gives

$$\frac{F_x}{F_y} = \frac{P_x^G\left(p_{x_o} - p_{x_l}\right)}{P_y^G\left(p_{y_o} - p_{y_l}\right)} \tag{1.40}$$

where F_x and F_y represent the weight fluxes (g/cm^2 s), and P_x^G and P_y^G the weight-based permeabilities (g cm/cm^2 s cmHg). Equation 1.40 can be written in molar terms as

$$\frac{f_x}{f_y} = \frac{\acute{P}x\left(p_{x_o} - p_{x_l}\right)}{\acute{P}y\left(p_{y_o} - p_{y_l}\right)} \tag{1.41}$$

Here f_x and f_y represent the molar fluxes (cm^3 (STP)/cm^2 s) and $\acute{P}_x$ and $\acute{P}_y$ the molar permeabilities (cm^3 (STP) cm/cm^2 s cmHg). This ratio of the molar permeabilities is known as gas membrane selectivity, $\alpha_{membrane}$. The ratio of the molar fluxes is comparable to the ratio of the permeate partial pressures:

$$\frac{f_x}{f_y} = \frac{p_{x_l}}{p_{y_l}} \tag{1.42}$$

Combining Equations 1.36, 1.37, 1.41, and 1.42 gives

$$\delta_{pervaporation} = \frac{\delta_{evaporation}\alpha_{membrane}\left(p_{x_o} - p_{x_l}\right)}{\left(p_{y_o} - p_{y_l}\right)\left(p_{x_o}/p_{y_o}\right)} \tag{1.43}$$

Equation 1.43 clearly identifies the main factors on which the performance of the pervaporation process depends. These factors are, first, $\delta_{evaporation}$, the vapor liquid equilibrium. It mainly depends upon the feed composition and temperature. Second is the intrinsic property of the membrane material, $\alpha_{membrane}$, and the third is the vapor pressures of the feed and permeate, which speaks about the influence of the operating parameters on the membrane performance. This equation also informs about the similarity between gas separation and pervaporation. In both of the processes the separation depends on the selectivity and the membrane pressure ratio. There are two limiting cases in pervaporation similar to the ones present in gas separation, in which the separation is based on one out of the said factors that dominate the process. The limiting cases are explained next.

First, in this case the membrane selectivity is much higher than the vapor ratio between the feed liquid and the permeate vapor:

$$\alpha_{membrane} \gg \frac{p_o}{p_l} \tag{1.44}$$

This says that the membrane with very high selectivity for component x will have equal vapor pressure on both sides of the membrane for component x, that is,

$$p_{x_l} = p_{y_l} \tag{1.45}$$

Combining Equations 1.36 and 1.45 gives

$$\delta_{membrane} = \frac{p_{y_o}}{p_{y_l}} \tag{1.46}$$

Again combining Equations 1.37 and 1.46 gives

$$\delta_{pervaporation} = \delta_{evaporation} \times \frac{p_{y_o}}{p_{y_l}} \quad when\ \alpha_{membrane} \gg \frac{p_o}{p_l} \tag{1.47}$$

Similarly, for component x Equation 1.47 gives

$$\delta_{pervaporation} = \delta_{evaporation} \times \frac{p_{x_o}}{p_{x_l}} \tag{1.48}$$

Considering a special case, where component x is a minor component as compared to y in the feed, then p_{yo} approaches p_o and p_{yl} approaches p_l. Then Equation 1.47 gives

$$\delta_{pervaporation} = \delta_{evaporation} \times \frac{p_o}{p_l} \tag{1.49}$$

where p_o/p_l represents the total vapor pressure of the process.

The second limiting case is the one in which the vapor pressure ratio is too large as compared to the selectivity of the membrane. Here, the partial pressure of the permeate will be smaller than the feed partial vapor pressure and p_{xl} and p_{yl} tend to be 0. Thus, Equation 1.43 reduces to

$$\delta_{pervaporation} = \delta_{\text{evaporation}} \times \alpha_{\text{membrane}} \text{ when } \alpha_{\text{membrane}} \ll \frac{p_o}{p_l} \tag{1.50}$$

1.4.6 Electrodialysis

Electrodialysis is a charge-based membrane separation process, as shown in Figure 1.14. The driving force for the separation of components in this process is the concentration and electric potential gradients. The transport in electrodialysis is, generally, discussed in the form of charge transported as compared to material transported in other separation processes. The process can be understood by taking a simple univalent electrolyte, NaCl as an example. NaCl when diluted deionizes completely. The Na^+ (cation) and Cl^- (anion) concentration are given by Na^+_c and Cl^-_c, respectively. The velocities of the two ions is given by u (cm/s) in the case of a cation and $-v$ (cm/s) in the case of the anion under the external applied field of energy, E. The protonic and electronic charge present on the cations and anions is $+e$ and $-e$, respectively. Therefore, the total amount of charge transported through an area of 1 cm^2 is

$$\frac{I}{J} = Na_c^+(u)(+e) + Cl_c^-(-v)(-e) = ce(u+v) \tag{1.51}$$

where I represents the current and J the Faraday's constant. It converts the electric charge transported to the electric current in amps. Therefore, this equation links charge transport to electric current.

It is also evident that the fractions of the current carried by both anions and cations need not be equal. This fraction of current carried by the anion or cation is known as the transport number of the particular ion. Therefore, it can be denoted as t^+ and t^- in case of a cation and anion, respectively.

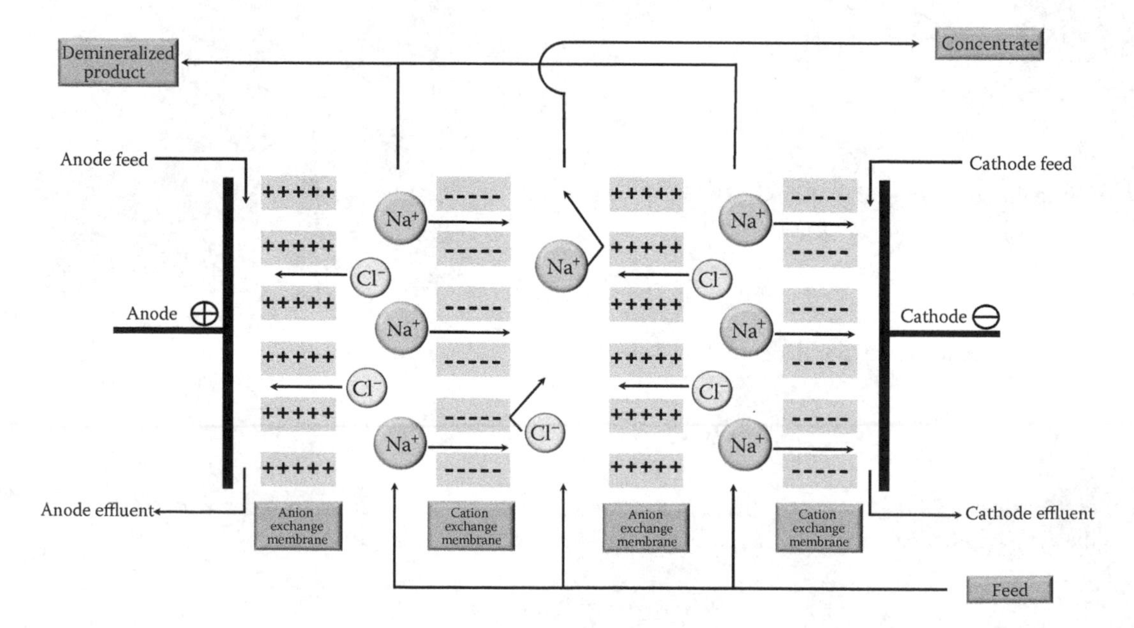

FIGURE 1.14 Schematic representation of electrodialysis membrane process principle shown with NaCl solution.

Also,

$$t^{+} + t^{-} = 1 \tag{1.52}$$

Combining Equations 1.51 and 1.52 for the cation transport number gives

$$t^{+} = \frac{Na_c^{+}ue}{ce(u+v)} = \frac{u}{u+v} \tag{1.53}$$

Similarly, for the anion transport number:

$$t^{-} = \frac{v}{u+v} \tag{1.54}$$

The size of an ion is an important factor in the difference of transport number of different ions. In electrodialysis the ions with the same charge as the fixed ions in the membrane are not allowed to be permeated and therefore the total current flow is because of the counter ions, which carries an opposite charge compared to the fixed ions. These counter ions give a transport number in the range of 0.95 to 1.0. The fixed ions contribute trivially; their transport number comes out to be in the range of 0 to 0.05. This difference between the transport numbers of the fixed and counter ions is the basis of the relative permeability and thus permits the separations in ion exchange membranes. This ability to differentiate the fixed and counter ions in an ion exchange membrane was very aptly expressed mathematically by Donnan in the year 1911 [13]. Figure 1.15 shows an ion exchange membrane with fixed charge Q^{-} placed in front of NaCl solution and shows the ion distribution in the process.

The membrane (*m*) and NaCl solution (*s*) ions equilibrium can be given as

$$c^{+}_{(m)} \times c^{-}_{(m)} = kc^{+}_{(s)} \times c^{-}_{(s)} \tag{1.55}$$

where k represents the equilibrium constant, and c^{+} and c^{-} the cation and anion concentrations, respectively. Imposing charge balance on both sides gives

$$c^{+}_{(m)} = c^{-}_{(m)} \times c_{Q^{-}(m)} \tag{1.56}$$

For NaCl, which dissolves completely, its total molar concentration $c_{(s)}$ is equivalent to the concentration of each of the ions, therefore

$$c_{(s)} = c^{+}_{(S)} = c^{-}_{(S)} \tag{1.57}$$

Combining Equations 1.55, 1.56, and 1.57 and rearranging gives

$$\frac{c^{+}_{(m)}}{c^{-}_{(m)}} = \frac{\left[c^{-}_{(m)} + c_{Q^{-}(m)}\right]^{2}}{k\left[c_{(s)}\right]^{2}} \tag{1.58}$$

Since the membrane is a cationic membrane (fixed negative charges), the negative counter ions concentration will be trivial as compared to the fixed charges, that is,

$$c_{Q^{-}(m)} \gg c^{-}_{(m)} \tag{1.59}$$

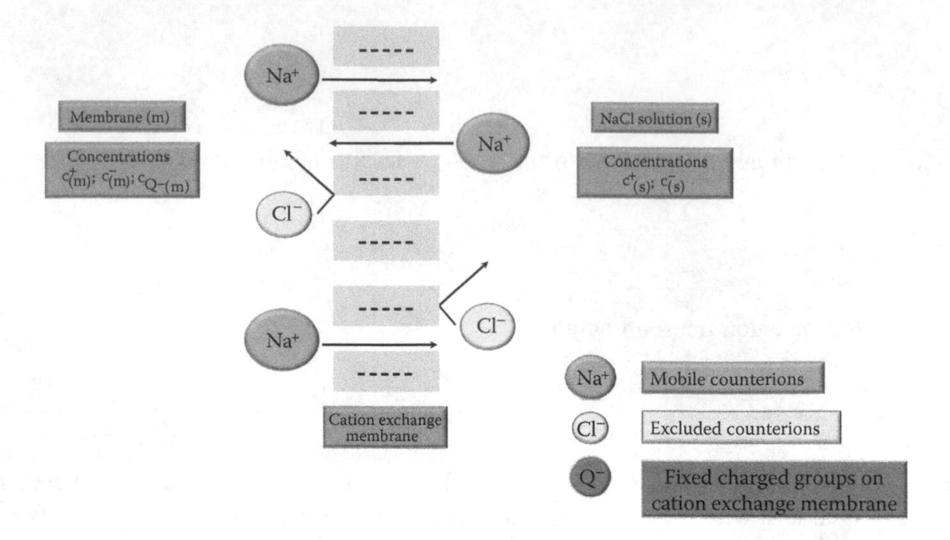

FIGURE 1.15 Schematic representation of an cation exchange membrane with fixed charge Q^- placed in front of NaCl solution.

Therefore, it can be assumed that

$$c^-_{(m)} + c_{Q^-_{(m)}} \approx c_{Q^-_{(m)}} \tag{1.60}$$

Thus, Equation1.58 can be rewritten as

$$\frac{c^+_{(m)}}{c^-_{(m)}} = \frac{1}{k}\left(\frac{c_{Q^-_{(m)}}}{c_{(s)}}\right)^2 \tag{1.61}$$

The expression of Equation 1.61 depicts that the sodium-to-chloride ions ratio is proportional to the square of the ratio of the fixed charge groups in the membrane to the salt concentration in the surrounding solution.

Present-day ion-exchange membranes hold a high concentration of fixed ionic charges, normally 3 to 4 meq/g or more. The membranes have the tendency to absorb water due to the presence of these charges. Upon water uptake, the membrane swells excessively due to the charge repulsion of the ionic groups. Therefore, it is good to impart cross-linking to the membranes so that the swelling can be restricted. The cross-linking should be optimum, as otherwise the membrane will become brittle if the extent of cross-linking is high. On the contrary, membranes will show excessive swelling if the cross-linking is less or absent. The ion-exchange membrane can be further classified as homogeneous or heterogeneous. Homogeneous membranes are the membranes in which the charged groups are uniformly distributed throughout the membrane matrix. Due to the uniform distribution of charge, the swelling in these membranes is also uniform and controlled by imparting cross-linking. As explained earlier, the cross-linking density matters for the control of swelling and thus taken care during the preparation of the ion-exchange membranes. On the other hand, heterogeneous membranes are the membranes in which the charge groups are distributed throughout the inert support matrix in the form of small domains. This uneven distribution leads to the uneven swelling of the membrane and thus leads to mechanical failure. Due to the mechanical failure the membrane leaks at its weak points, such as at the boundary of two domains. Therefore, homogeneous membranes of thickness 50 to 200 μm are commonly prepared and used in the electrodialysis process.

1.4.7 Gas Permeation

Thomas Graham (December 20, 1805–September 16, 1869) is the pioneer of gas separation using both porous as well as nonporous membranes. He is also known as the father of colloid chemistry. Graham worked systematically for almost 20 years on gas separation and measured permeation rates for almost all known gases of his time. Graham's law [9] of diffusion was based on his studies with gases. An interesting thing about Graham's studies is that he used to synthesize the gases for his experiments himself. However, Graham's studies on gas separations did not have any industrial or commercial impact until the end of 20th century. It was not until 35 years ago that gas separation through membranes gained industrial and commercial importance. This may be due to the great industrial and population explosions that resulted in degraded air quality and air pollution. Before this the Graham studies were just used as a theoretical tool for the development of physical and chemical theories, like Maxwell's kinetic theory of gases. Recently, many industrial and commercial processes have been developed where gases such as hydrogen, nitrogen, carbon dioxide, and helium were utilized. The studies of Graham helped the development of present-day gas separation techniques.

The Manhattan Project in the years 1943 to 1945 engaged Graham's law of diffusion first time for the separation of $U_{235}F_6$ and $U_{238}F_6$. The project resulted in the world's first largest membrane gas separation plant, but the secret nature of the project yielded no further advancement in the gas separation technology. It was the development in the work of gas separation in the 1940s that was the footing of modern theories of gas separation. The only limitation in further development of the gas separation discipline was the undeveloped membrane technology. The development of anisotropic membranes for reverse osmosis applications in the late 1960s paved the way for the development of modern membrane gas separation technologies. The first company to make a breakthrough was Monsanto by developing its Prism® membrane in 1980 for hydrogen separation. In the 1990s, Dow launched its Generon®, the first membrane for nitrogen separation from air. The membrane gas separation field is expanding daily with the addition of new systems and technologies and it will be so for the coming 10 to 15 years. There is still need for improvements in the materials, processes, and membranes in particular.

Membranes of both types—porous and nonporous—can be utilized for gas separation. Figure 1.16 represents the gas separation or permeation mechanism. There are three types of gas flow or gas permeation that occur based on the pore size of the membranes, as shown in Figure 1.17. The first one is the convective flow; it occurs in membranes with larger pore size (0.1–10 μm). The second type is the Knudsen diffusion; in this case the pore size of a membrane is smaller than 0.1 μm. In this case, the gas flow is governed by Graham's law of diffusion [9], which states that the rate of diffusion of a gas is inversely proportional to the square root of its molecular weight, and it is so because the membrane pore size is the same or smaller than the mean free path of the gas molecules. The third type of gas flow is molecular sieving; this flow takes place when the membrane pore size is of the order of 5 to 20 Å. In this type both the diffusion in gas as well as the adsorption phase take place. The adsorbed species diffuse through the surface of the pores, known as surface diffusion. These three types of microporous membranes are the source of sustenance for the research community working in the membrane gas separation field, but all the commercial gas separation membranes are based on dense polymeric membranes (Figure 1.17, nonporous/dense membrane).

The gas transport through dense membranes is governed by the equation

$$F_x = \frac{\varepsilon_x K_x^G \left(p_{x_o} - p_{x_\alpha}\right)}{\alpha} \tag{1.62}$$

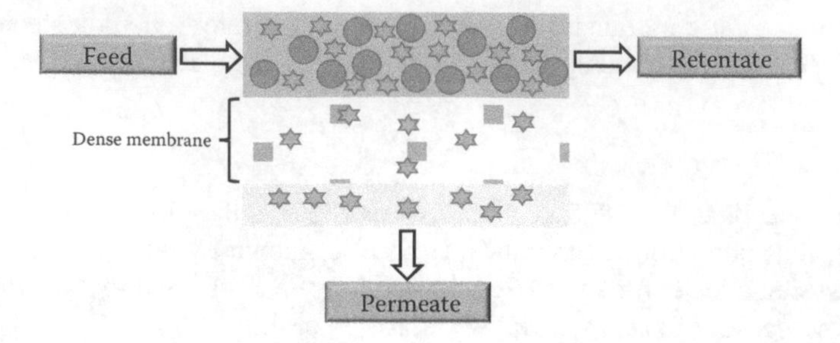

FIGURE 1.16 Schematic representation of membrane gas separation mechanism.

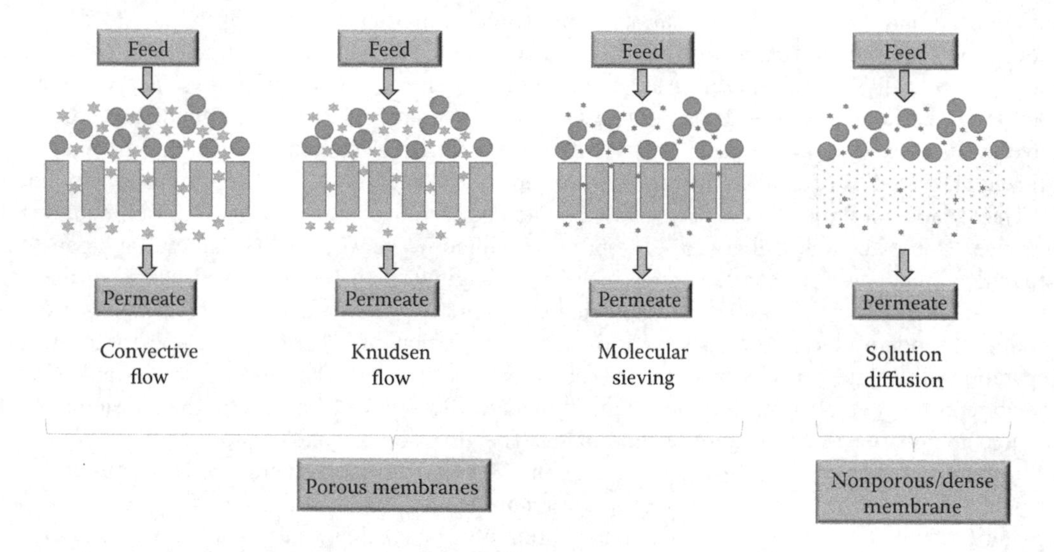

FIGURE 1.17 Schematic representation of gas separation by porous as well as nonporous membranes.

where F_x represents the x component flux (L/m^{2}h), p_{xo} and $p_{x\alpha}$ the partial pressures of x on either side of the membrane, α the membrane thickness, ε_x the permeate diffusion coefficient, and K_x^G the Henry's law [34] sorption coefficient (L/m^3 kPa). Further, Equation 1.62 can be simplified for gas permeation by considering the volume flux as compared to mass flux as

$$F_x = \frac{\varepsilon_x K_x (p_{x_o} - p_{x_\alpha})}{\alpha} \tag{1.63}$$

where F_x represents the volume (molar) flux (L$_x$/m^{2}h) and K_x the sorption coefficient (L$_x$/m^{3}kPa). The product of the permeate diffusion and the sorption coefficient is the membrane permeability, which is the measure of the gas permeation ability of a membrane and can be denoted as P_m. Also, the measure of a membranes ability to separate two gases is known as membrane selectivity. It is the ratio of the permeabilities of the two gases, x and y, ψ_{xy}.

$$\psi_{xy} = \frac{P_x}{P_y} \tag{1.64}$$

TABLE 1.3
Permeabilities of Selective Gases with Some Common Membrane Materials

	Rubbers		Glasses	
Gas	**Natural Rubber at 30°C**	**Silicone Rubber at 25°C**	**Cellulose Acetate at 25°C**	**Polysulfone at 35°C**
Hydrogen	41	550	24	14
Helium	31	300	33	13
Oxygen	23	500	1.6	1.4
Nitrogen	9.4	250	0.33	0.25
Carbon dioxide	153	2700	10	5.6
Methane	30	800	0.36	0.25

In permeability, the diffusion coefficient, ε_x, represents the mobility of individual molecules in the membrane material, and the sorption coefficient, K_x, represents the total number of molecules dissolved in the membrane material. Thus, Equation 1.64 can be rewritten as

$$\psi_{xy} = \left[\frac{\varepsilon_x}{\varepsilon_y}\right]\left[\frac{K_x}{K_y}\right] \tag{1.65}$$

where the ratios $\varepsilon_x/\varepsilon_y$ and K_x/K_y represent the ratio of diffusion coefficients of two gases and the mobility selectivity, and the ratio of the sorption coefficients of the two gases and the sorption or solubility selectivity of the two gases, respectively, which at the end show the relative condensabilities of the two gases.

The mobility selectivity mainly depends upon the glass transition temperature of the membrane material. When below its glass transition temperature, the material is rigid and is known as glassy material. The material above its glass transition temperature is soft and known as rubber. The relative mobility of gases differs in glasses and rubbers. The increase in size of the permeating molecule decreases the diffusion coefficients rapidly in the case of glassy material as compared to rubbers. The sorption or solubility selectivity mainly depends upon the condensability of the permeant. The sorption coefficient increases with an increase in the size of the permeant because bigger molecules condense easily as compared to smaller molecules. The difference between the permeant sorption coefficients in glasses and rubbers is not much as compared to the diffusion coefficients. In broader terms, the mobility selectivity is dominant for glassy materials; the permeability decreases with the increasing size of the permeant. On the other hand, sorption selectivity is dominant for rubbery materials; the permeability increases with the increasing size of the permeant.

Permeabilities of selective gases with some common membrane materials are given in Table 1.3. Due care should be taken when calculating the selectivity of a membrane using Equation 1.65 and the permeabilities given in Table 1.3. The permeabilities given in Table 1.3 are measured with pure gases and the ratio of pure gas permeabilities gives ideal membrane selectivity, which is further an intrinsic property of the membrane or membrane material in particular. In reality this is not the case, as the feed will be a mixture of gases and the gases may or may not interact strongly with the membrane material, which affects the permeability as well as the selectivity of the membranes. Therefore, the actual selectivity may come out to be half or less than the selectivity calculated with pure gases. Work on the membrane selectivity in the mixture of gases is scant. For most of the cases, membrane selectivity was calculated considering pure gases only. Thus, this point should be taken care of while selecting a membrane for a particular gas separation application from its mixture.

STUDY QUESTIONS

1. Give a few examples of membrane processes from everyday life.
2. Cite the advantages, types of transports, and driving forces for membrane processes.
3. Name the important membrane parameters that define the membrane selectivity and permeability.
4. A beaker contains two solutions, S1 with higher salt concentration and S2 with lower salt concentrations, separated by a membrane. What will be the flow direction of the solvent when no pressure is exerted on any of the solution and name the process? If pressure is applied on solution S1, then what will be the flow direction of the solvent and what is this process called? In both cases name the driving force.
5. What is the difference between symmetric and asymmetric membranes, and asymmetric and composite membranes?
6. What will be the osmotic pressure difference between 5 wt% of glucose and an NaCl solution of the same concentration?
7. Jerry watched a video on YouTube regarding water scarcity in the world and thought of starting a water conservation project in his house. Now, he wants to reuse the washbasin water from the kitchen as well as the washroom. What membrane processes should he employ to achieve his goal?
8. Jenny was working on an experiment in the lab. She needs to prepare a mixture of toluene and hexane for her experiment, but instead of using hexane she used heptane. What membrane process she should use to separate toluene from heptane? What other chemicals are required and why?
9. An army research facility working on the development of advanced warfare technology using silicon chips was stationed in a remote area. Workers were using a portable ultrafiltration water treatment plant for getting a clean water supply. They were using the same water as the feed for their research as well as backup batteries. They noticed a downfall in the results of their experiments and the performance of their batteries (even though the batteries were newly installed). Where does the problem lie and what measures should they take?
10. Can membrane processes be employed in the automobile industry. If yes, then how?
11. Elaborate on the transport mechanism of the membrane process that involves both heat and mass transfer.

REFERENCES

1. K. Nath, Membrane separation process, Prentice Hall of India, 2008.
2. B. K. Dutta, *Principles of mass transfer and separation processes*, Prentice Hall of India, 2007.
3. M. Mulder, *Basic principles of membrane technology*, Springer, 2007.
4. H. Darcy, *Les fontaines publiques de la ville de Dijon*, Paris, Dalmont, 1856.
5. S. Loeb and S. Sourirajan, Sea water demineralization by means of an osmotic membrane, *Advances in Chemistry Series*, 38, 117–132, 1962.
6. Richard W. Baker, *Membrane technology and applications*, 2nd ed., John Wiley & Sons, 2004.
7. J. E. Cadotte, Interfacially synthesized reverse osmosis membrane, US Patent 4,277,344, 1981.
8. P. A. Kober, Pervaporation, perdistillation, and percrystallization, *Journal of American Chemical Society*, 39, 944–948, 1917.
9. T. Graham, Liquid diffusion applied to analysis, *Philosophical Transactions*, Royal Society, *London*, 151, 183–224, 1861.
10. L. Cerini, Apparatus for the purification of impure solutions of caustic soda and the like on osmotic principals, US Patent 1,719,754, July 1929; and US Patent 1,815,761, July 1929.
11. W. J. Kolff, H. TH. J. Berk, M. Welle, A. J. W. van der Ley, E. C. van Dijk, and J. van Noordwijk, The artificial kidney: A dialyser with a great area, *Journal of International Medicine*, 117(2), 121–134, 1944.

12. F. Kiil, Development of a parallel flow artificial kidney in plastics, *Acta Chirurgica Scandinavica*, Supplementum 253, 140–142, 1960.
13. F. G. Donnan, Membranpotentiale bei vorhandensein von nicht dialysieren elektrolyte. Ein veitrag zur physikalische chemishen Physiologie, *Zeitschrift für Elektrochemie*, 17, 572, 1911.
14. A. Fick, Über diffusion, *Poggendorff's Annalen der Physik und Chemie*, 94, 59, 1855.
15. P. F. Scholander, Oxygen transport through hemoglobin solutions, *Science*, 131, 585, 1960.
16. S. Schlosser and E. Kossaczky, Comparison of pertraction through liquid membranes and double liquid-liquid extraction, *Journal of Membrane Science*, 6, 83–105, 1980.
17. N. N. Li, Permeation through liquid surfactant membranes, *American Institute of Chemical Engineers Journal*, 17, 459, 1971.
18. C. Antoine, Tensions des vapeurs; nouvelle relation entre les tensions et les températures, *Comptes Rendus des Séances de l'Académie des Sciences*, 107, 681–684; 778–780, 836–837, 1888.
19. W. Ostwald, Elektrische eigenschaften halbdurchl¨assiger scheidewande, *Zeitschrift für Physikalische Chemie*, 6, 71, 1890.
20. E. Manegold and C. Kalauch, Uber Kapillarsysteme, XXII Die Wirksamkeit Verschiedener Reinigungsmethoden (Filtration, Dialyse, Electrolyse und Ihre Kombinationen), *Kolloid Zeitschrift*, 86, 93, 1939.
21. T. R. E. Kressman, Ion exchange resin membranes and impregnated filter papers, *Nature*, 165, 568, 1950.
22. E. A. Murphy, F. J. Paton and J. Ansel, US Patent 2,331,494, 1943.
23. W. Juda and W. A. McRae, Coherent ion-exchange gels and membranes, *Journal of Americal Chemical Society*, 72, 1044, 1950.
24. Permeability of pellicle precipitates, *Journal of the Royal Microscopical Society*, 2, 592, 1879.
25. H. Bechhold, Kolloidstudien mit der filtrationsmethode, *Zeitschrift für Physikalische Chemie*, 60, 257, 1907.
26. A. J. Staverman, The theory of measurement of osmotic pressure, *Recueil des Travaux Chimiques des Pays-Bas*, 70, 344–352, 1951.
27. O. Kedem, Criteria of active transport, in *Membrane transport and metabolism*, edited by A. Kleinzeller and A. Kotyk, 87–93, 1961.
28. R. Schlögl, Stofftransport durch Membranen, *Fortschritte der Physikalischen Chemie*, Brand 9, Darmstadt, SteinkopffD Verlag, 1964.
29. U. Merten, Transport properties of osmotic membranes, in *Desalination by reverse osmosis*, edited by U. Merten, Cambridge, MA, MIT Press, 15–54, 1966.
30. J. E. Cadotte, Reverse osmosis membrane, US Patent 3,926,798, 1975.
31. R. L. Riley, R. L. Fox, C. R. Lyons, C. E. Milstead, M. W. Seroy and M. Tagami, Spiral-wound poly (ether/amide) thin-film composite membrane system, *Desalination*, 19, 113, 1976.
32. K. H. Meyer and W. Strauss, La perméabilité des membranes VI. Sur le passage du courant électrique à travers des membranes sélectives, *Helvetica Chemica Acta*, 23(1), 795–800, 1940.
33. AQUASTAT main database, Food and Agriculture Organization of the United Nations, 2016.
34. W. Henry, Experiments on the quantity of gases absorbed by water, at different temperatures, and under different pressures, *Philosophical Transactions of the Royal Society of London*, 93, 29–42 and 274–276, 1803.

2 Transport Mechanisms and Membrane Separation Processes

2.1 INTRODUCTION

Membranes are selective barriers that permeate feed components from one phase to another. This property of selective permeation of the feed components across the membrane is the reason behind their wide popularity. In this chapter, the membrane transport mechanisms along with various membrane processes are explained in detail. Also, the different membrane modules with different modes of membrane operations are explained. This chapter will help readers to get into the mathematics of the permeation process and will give them a better understanding of the membrane transport mechanisms.

2.2 TRANSPORT MECHANISMS

Permeation across the membrane is generally explained on the basis of two models, shown in Figure 2.1. The first model is the solution–diffusion model and the other is the pore flow model. In the solution–diffusion model, the feed components dissolve in the membrane material and then permeate across the membrane. The permeation of feed components takes place across the membrane due to the effect of concentration gradient, which acts as a driving force for the process. Therefore, the extent of solubilities and the rate of permeation of the feed components determines the degree of separation. On the other hand, in the pore flow model the feed components are permeated through the membrane pores due to the driving force of the process, which is the pressure gradient across the membrane. The feed components flow in a convective flow through the membrane pores. The separation takes place on the basis of the size exclusion principle, where feed components having a size greater than the membrane pores are excluded and the feed components having a size lesser than the membrane pores are permeated across the membrane.

The solution–diffusion model is based on the concentration gradient and diffusion process. Both the diffusion and concentration are interrelated, as diffusion is the transport of molecules from one place to the other in a system based on the concentration gradient. Therefore, the direction of movement of the molecules in diffusion is always ruled by the concentration gradient. This was first explained theoretically and practically by Fick in 1857 [1]. The famous Fick's law of diffusion is based on this concept and is given by

$$F_i = -D_i \frac{dc_i}{dx} \tag{2.1}$$

where F_i represents the transfer rate of component i (Kg/m^2.s), dc_i/dx the concentration gradient of component i, and D_i the diffusion coefficient (m^2/s), the measure of the mobility of individual component molecules of the component i. The minus sign represents the direction of diffusion, that is, along the concentration gradient. The pore flow model is based on the pressure-gradient-driven convective flow of the components through the membrane pores. In 1856, Henry Philibert Gaspard Darcy (June 10, 1803–January 3, 1858) formulated a law based on his experimental results, which is

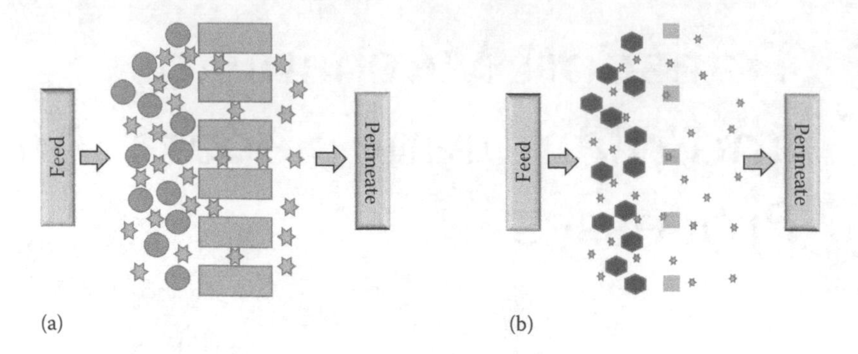

FIGURE 2.1 Membrane permeation models: (a) pore flow and (b) solution–diffusion.

known as Darcy's law. The law governs the flow of a fluid through a porous medium, which is basically the pore flow model. Darcy's law governing the pore flow model is given by the equation [2]

$$F_i = kc_i \frac{dp}{dx} \tag{2.2}$$

where dp/dx represents the pressure gradient, c_i the concentration of i component, and k the intrinsic permeability coefficient of the medium (membrane). The simple diffusion process is very slow as compared to the pressure-driven convective flows across the membranes.

The nonporous or dense membranes follow the solution–diffusion model and Fick's law as the pores in these membranes are very minute. In fact, the pores are just the miniscule spaces present or occurring in between the polymer chains due to the thermal movement of the molecules and appear and disappear continuously as the components are permeating through the membrane. On the contrary, porous membranes follow the pore flow model or Darcy's law, as the pores in these membranes are prominent, fixed, and interconnected. The size of the membrane pores is proportional to the pore flow model characteristics. The membranes or the pores in the pore size range of 0.5 to 1 nm fluctuates between the fleeting (solution–diffusion model) and fixed (pore flow model) pores. On the basis of average pore diameter, the membranes can be classified in following three categories:

- Microfiltration, porous gas separation membranes, and ultrafiltration membranes having pores in the range of 1 to 10000 nm come under one category and follow the pore flow model.
- Reverse osmosis and non-porous gas separation membranes having pores in the range of 0.2 to 0.5 nm come under one category and follow the solution–diffusion model.
- Membranes, such as nanofiltration, that are between the first two categories with pore size in the range of 0.5 to1 nm fall in separate category.

2.2.1 Knudsen Flow

Knudsen flow is a pore-size dependent flow, in which membrane pore size plays a vital role. For instance, if the membrane pore size is very large, then the feed components will be colliding with each other and effective separation will not take place. This type of flow is known as viscous (or Poiseuille) flow. On the other hand, if the membrane pore size is small such that the mean free path of the moving feed components equalizes or increases as compared to the membrane pore size, then the collisions among the feed components reduce and effective separation takes place. This type of flow is known as Knudsen flow. The two types of flow are shown in Figure 2.2.

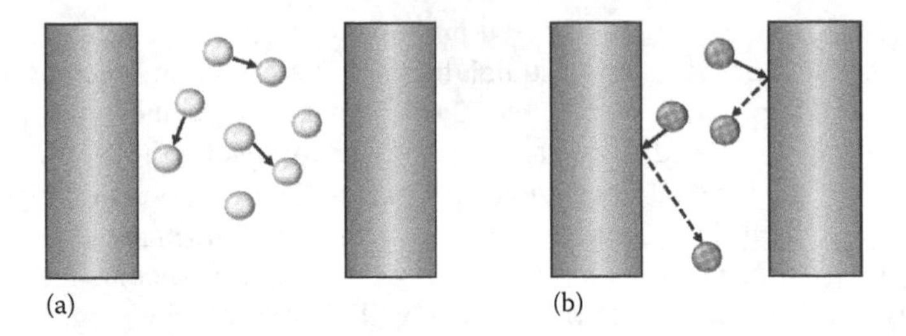

FIGURE 2.2 Schematic representation of (a) viscous flow and (b) Knudsen flow.

The mean free path (λ) for feed components can be defined as the average of the distance covered by a molecule between two collisions. For liquids the Knudsen flow can be neglected, since in liquids the molecules are very close to each other. However, for gases the molecules are free and loosely packed. The mean free path of the gas molecules is pressure and temperature dependent and is given by

$$\lambda = \frac{kT}{\left(\pi\, d_{gas}^2\, P\sqrt{2}\right)} \tag{2.3}$$

where d_{gas} represents the gas molecule diameter, P the pressure, T the temperature, and k the Boltzmann constant. The mean free path of a gas molecule is inversely proportional to pressure and proportional to temperature at constant pressure.

For membranes with large pore size, Knudsen diffusion will play a vital role, like in the case of microfiltration and ultrafiltration membranes. In these membranes, when operating at low pressures the complete transport mechanism is governed by Knudsen flow. In such cases, the flux of the membrane can be given by

$$F = \frac{\pi\, n\, r^2\, D_k\, \Delta P}{R\, T\, l} \tag{2.4}$$

where R represents the gas constant, r the membrane pore radius, D_k the Knudsen diffusion coefficient, n the molar concentration, and l the pore length. The Knudsen diffusion coefficient is equivalent to

$$D_k = 0.66\, r\sqrt{\frac{8\, R\, T}{\pi\, M_w}} \tag{2.5}$$

where T represents the temperature, r the pore radius, and M_w the molecular weight. It can be seen from Equation 2.5 that the membrane flux relies on the square root of the molecular weight, which means that the separation of two molecules is inversely proportional to the ratio of the square root of the molecular weights of the gases.

2.2.2 Viscous Flow

Under the action of a total pressure gradient, convective flow takes place and the exact driving force becomes $x_i \Delta P/RT$. That is, each mixture component is transported through the membrane at a rate that is proportional to its mole fraction and to the pressure gradient. For the Knudsen number $(K_n) << 1$, the continuum assumption is valid and the classical Stokes, Navier-Stokes, or compressible momentum conservation equations can be used in the void space [3,4]. However, integration over the entire membrane structure is far from straightforward due to the presence of the pore structure that is usually quite complicated. Hence, the Darcy equation or some other phenomenological equation is typically used, depending mainly on the porosity level and also on the degree of the equations that describe the boundary conditions. In general, the transport coefficient is, in this case, proportional to P/η, where η is the dynamic viscosity of the fluid. The proportionality constant is the viscous permeability, the value of which is characteristic of the particular membrane or support structure. Viscous flow in a porous membrane is not by itself very efficient regarding selectivity in a mixture separation, since the main hydraulic mechanism of separating the species is the differential driving of the species molecules toward the center or the edges of the pores. Usually, viscous flow is present as a consequence of a pressure drop across the two sides of the membrane, but it occurs in combination with some diffusion mechanism that undertakes the main separation task.

In cases where the mean free path for the feed component is smaller than the membrane pore size, such as

$$\left(\frac{r}{\lambda}\right) > 3 \tag{2.6}$$

then the gas flow will occur primarily via the collisions among the gas molecules. Therefore, as long as the velocity of the gas molecules is zero near the walls of the membrane pore, then the molar flux of the gas molecules can be given by the Hagen-Poiseuille [5] equation:

$$F_v = -\frac{r^2}{8\eta}\frac{P}{RT}\left(\frac{dp}{dx}\right) \tag{2.7}$$

where P represents the transmembrane pressure, R the gas constant, T the temperature, η the dynamic viscosity, and r the membrane pore radius.

Integrating Equation 2.7 for a single membrane pore gives the gas permeation rate as

$$F_r = \frac{\pi r^4 P_{avg}\left(p_{upstream} - p_{downstream}\right)}{8\mu RTl} \tag{2.8}$$

where

$$F_r = \pi r^2 F_v \tag{2.9}$$

and

P_{avg} represent the average transmembrane pressure given by

$$P_{avg} = \frac{\left(p_{upstream} + p_{downstream}\right)}{2} \tag{2.10}$$

where $p_{upstream}$ and $p_{downstream}$ are the upstream and downstream pressures, respectively.

Adding the effects of membrane pore structures to Equation 2.7 gives

$$F_v = -\frac{\eta r^2}{\tau 8\mu}\frac{P}{RT}\left(\frac{dp}{dx}\right) \tag{2.11}$$

or

$$F_v = -\frac{\mu r^2}{\tau 8\mu}\frac{P_{avg}}{RTl}\left(p_{upstream} - p_{downstream}\right) \tag{2.12}$$

where μ represents the membrane porosity and τ the tortuosity.

2.2.3 Surface Flow

It is well known that all gases, to a greater or lesser extent, may get adsorbed on solid surfaces and get transported subject to a driving force such as the surface concentration gradient before they are desorbed back into the gas phase (Figure 2.3). This mechanism of gas transport may become significant under certain conditions of pressure, temperature, and for a specific pore size range. Therefore, gases with a large difference in their molecular masses use this separation mechanism instead of Knudsen flow [6].

This separation mechanism follows four steps during the transport of a gas across the membrane via membrane pores. These four steps are

1. Adsorbing gas molecules diffuse through the bulk to the membrane surface
2. Reversible sorption of the gas molecules on the membrane surface
3. Adsorbed gas molecules transport across the membrane by using surface diffusion via membrane pores
4. Desorption of the adsorbed gas molecules to the permeate bulk from the membrane surface

Surface flow dominates and is important even if the surface area and the adsorbed concentration of the gas molecules is high. Generally, the gas transport is slow through a membrane as compared to the bulk diffusion. This confirms that the porous diffusion is slow as compared to bulk diffusion and surface adsorption. This is known as Fickian diffusion in which the gas diffusion through the membrane pores is slow. On the other hand, non-Fickian diffusion is the one in which the gas diffusion is fast through the membrane pores. The surface flow depends upon the type of gas, membrane surface nature, and operational conditions. The type of gas is important since it reveals the adsorption capacity of the gas molecules and other molecular parameters, such as molecular weight. Likewise, the membrane surface reveals the physicochemical factors and membrane morphology, such as membrane pore diameter and surface roughness. The operational conditions tell about the temperature, pressure, pH, and flow values.

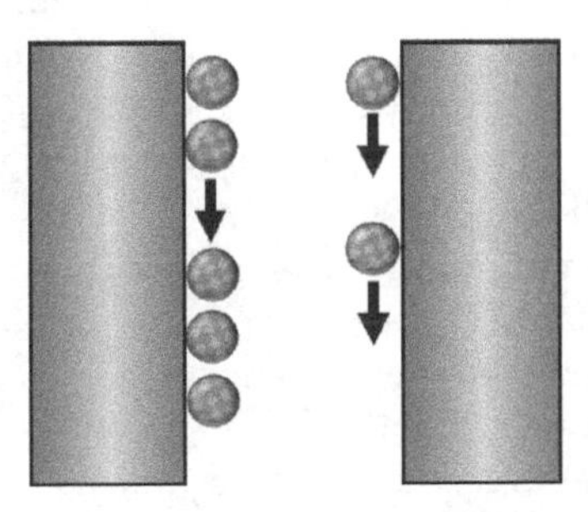

FIGURE 2.3 Schematic representation of surface diffusion.

The surface flow may follow any of the following three models as per the process conditions:

- Hydrodynamic model—In this model, the gas molecules are assumed to be a liquid film under the effect of external factors, such as pressure slides beside the membrane pore.
- Hopping model—In this model, the gas molecules transport through a membrane pore by hopping over the surface.
- Random walk model—In this model, the gas molecules transport by following random steps along the vacant sites present in the pore. The concentration difference is the driving force and guides the gas molecules.

It is always doubtful to confirm the model of surface flow followed by the gas molecules in the transport across a membrane. Also, it is difficult to distinguish the contribution of Knudsen and surface flow in the transport of the gas molecules. Therefore, the gas transport across a membrane via micropores is a result of mutual roles played by Knudsen flow and surface flow.

2.2.4 Capillary Condensation

Capillary condensation is the process in which multilayer adsorption of vapors takes place in the membrane pores up to the point where adsorbed vapors condense to liquid form. There is a reversible equilibrium in this process that maintains the balance between the sorption and desorption processes, where pressure plays an important role [6]. The change in pressure changes the direction of the equilibrium toward sorption (at increased pressure) and desorption (at decreased pressure). All the available membrane sites are occupied by the adsorbing molecules at optimum pressure leading to multilayer adsorption.

The adsorption and condensation process in the membrane pores takes place by two possible ways. First, the molecules adsorb on the whole surface of a pore such that the pressure is higher in the condensing pore as compared to other pores. In this case, the condensation in addition to pressure also depends upon the pore size of the membrane pores. For wider pores, higher pressure is required for the capillary condensation to take place, and vice versa. Also, capillary condensation by this method does not plug the pore and helps in the enhancement of the membrane flux. On the other hand, the second case involves an increase in capillary condensation with increased pressure. Here, the capillary condensation plugs the pore and a channel is formed by the liquid. The condensed liquid forms a curved meniscus on both the permeate as well as the feed side of the liquid–vapor interfaces. The capillary condensation occurs below the saturation pressure of a pure liquid due to the presence of the curved meniscus. In this case, the membrane pore size also matters, since condensation occurs at lower pressures in the case of wider pores as compared to the narrow pores, and vice versa. The condensed liquid flows from the feed side to the permeate side because of hydrostatic pressure differences across the meniscus or capillary suction pressure in the empty pores. The membrane flux decreases with a decrease in the applied pressure, since the flux of condensed molecules is lower than the molecules present in vapor form. However, the membrane selectivity will be very high because the condensed liquid blocks the permeation of uncondensed gas across the membrane pores. Therefore, a higher separation efficiency is achieved.

The role of membrane pores is critical in case of capillary condensation as it is important to have optimum-sized membrane pores for the effective capillary condensation process to take place. The pores larger than the molecular sieves are sufficient for capillary condensation to take place, that is, 2 to 10 nm. As it was explained, membrane pores of smaller size increase selectivity and provide decreased membrane flux, and vice versa. For instance, membrane pores with size less than 2 nm show increased selectivity by a factor of greater than 60. The important factors required to observe the capillary condensation process or phenomenon is the availability of a condensable gas. The pressure and temperature are minimally equivalent to half of the saturation pressure and below the critical temperature of the gas, respectively.

2.2.5 Molecular Sieving

The molecular sieving separation takes place when the membrane pore size is the same as that of the feed (gas) molecules. It is stated that the membrane pore size should not be more than three times the feed molecule size. Generally, the membrane pore size used to be less than 2 nm for molecular sieving.

Temperature plays a vital role in the transport of molecules via molecular sieving. The transport rate of molecules increases with increased temperatures, and vice versa. Maximum separations are obtained at high temperatures in the range of 100°C to 300°C.

Molecular sieving is a simple and effective method for separating gas molecules on the basis of their size. A classic example of molecular sieving is hydrogen separation from other gases.

2.2.6 Driving Forces for Permeation

The mathematics of the separation mechanism in membrane science is based on the thermodynamic branch, which consists of the driving forces, that is, concentration, pressure, temperature, and electrical gradients. These gradients are all related and the master of all these gradients is the chemical potential of the permeant. Therefore, the flux of a membrane, F_i (Kg/m^2 s) of component i, can be given as

$$F_i = -\varnothing_i \frac{d\mu_i}{dx} \tag{2.13}$$

where $d\mu_i/dx$ represents the chemical potential of component i, and $\varnothing_i$ the coefficient of proportionality connecting the chemical potential to the flux. The different driving forces can be expressed in terms of this equation. Many of the membrane processes encompass more than one driving force, therefore, this analogy is advantageous. For example, reverse osmosis, pervaporation, or gas separation includes both concentration as well as pressure gradients as the driving forces for the membrane processes. Considering the concentration and pressure gradients as the only driving forces in a process, then the chemical potential can be given by

$$d\mu_i = R\ T\ d\ln(\varepsilon_i n_i) + \sigma_i\ dp \tag{2.14}$$

where n_i represents the mole fraction (mol/mol) of component i, ε_i the coefficient of activity (mol/mol), p the pressure, and σ_i the molar volume of the component i. The coefficient activity connects the mole fraction to the activity.

In case of incompressible phases, like solid and liquid membranes, there is no effect of pressure on volume. Thus, by integrating Equation 2.14 with respect to concentration and pressure

$$\mu_i = \mu_i^o + R\ T \ln(\varepsilon_i n_i) + \sigma_i(p - p_i^o) \tag{2.15}$$

where μ_i^o represents the chemical potential for pure component i at a reference pressure, p_i^o. For compressible phases like gases, the pressure has a significant effect on the molar volume of the gases. Substituting the gas laws and integrating Equation 2.14 gives

$$\mu_i = \mu_i^o + R\ T \ln(\varepsilon_i n_i) + R\ T \ln\left(\frac{p}{p_{i_{saturation}}}\right) \tag{2.16}$$

To make the chemical potential identical in Equations 2.15 and 2.16, the reference pressure p_i^o is considered as the saturation vapor pressure of component i, $p_{i_{saturation}}$, thus Equation 2.15, for incompressible liquids, can be rewritten as

$$\mu_i = \mu_i^o + R\,T \ln(\varepsilon_i n_i) + \sigma_i(p - p_i^o) \tag{2.17}$$

and Equation 2.16, for compressible gases as

$$\mu_i = \mu_i^o + R\,T \ln(\varepsilon_i n_i) + R\,T \ln\left(\frac{p}{p_{i_{saturation}}}\right) \tag{2.18}$$

There are a number of assumptions to define any permeation model. In the case of membrane transport, there are two main assumptions used: The first one is that the phases on both sides of a membrane are in equilibrium with each other. This means that there is a continuous chemical potential gradient present across the membrane. It must be understood that the rate of transport of the feed components is higher at the membrane interface as compared to the diffusion of feed components through it. This is true for all membrane processes except the few involving chemical reactions (e.g., facilitated transport or gas diffusion in metals), because in these cases the transport rate of the feed components at interfaces is used to be slow. The second one is the applied pressure to a membrane process that is constant throughout the system. Therefore, according to the solution–diffusion model the pressure is uniform and it is only the concentration gradient that governs the chemical potential.

Taking the assumptions into account, Equations 2.13 and 2.14 are combined and assuming σ_i as constant can be rewritten as

$$F_i = -\frac{\varnothing_i\,R\,T}{n_i} \cdot \frac{dn_i}{dx} \tag{2.19}$$

The gradient of component i in Equation 2.19 can be expressed in terms of mole fraction using the concentration term c_i (g/cm^3), which is defined as

$$c_i = M_{w_i} \delta n_i \tag{2.20}$$

where M_{w_i} represents the molecular weight of the component i (g/mol) and δ the molar density (mol/cm^3). Therefore, Equation 2.19 can be rewritten as

$$F_i = -\frac{\varnothing_i\,R\,T}{c_i} \cdot \frac{dc_i}{dx} \tag{2.21}$$

Equation 2.21 resembles Fick's law, and thus can be rewritten as

$$F_i = -D_i \frac{dc_i}{dx} \tag{2.22}$$

where the $\varnothing_i\,RT/c_i$ term is replaced with the diffusion coefficient D_i. Now, integrating Equation 2.22 within the limits of membrane thickness (l) yields

$$F_i = \frac{D_i\left(c_{i_{o(m)}} - c_{i_{l(m)}}\right)}{l} \tag{2.23}$$

where the terms $c_{i_{o(m)}}$ and $c_{i_{l(m)}}$ represent the concentration of i component in the feed, which is in contact with the membrane at the feed interface and permeate interface, respectively.

2.3 MEMBRANE SEPARATION PROCESSES

Membrane processes are characterized on the basis of pore size, type of driving force, mechanism of separation, and various other features. In this section the membrane processes, namely, microfiltration, ultrafiltration, nanofiltration, and reverse osmosis, are discussed while covering the basics of their separation mechanisms, theoretical relations, membrane materials, preparation methods, and applications. The following sections will provide a clear picture of these different membrane processes and help to select the best suitable membrane process with best attributes in terms of membrane process, material, pore size, pressure range, and so on for a particular application.

2.3.1 MICROFILTRATION

Microfiltration membranes are pressure-driven porous membranes with pores in the size range of 100 to 1000 nm. A discussion on the basics of microfiltration membranes is provided in Chapter 1 of this book. They are prepared from both organic (polymers) as well as inorganic (ceramics, glasses, and metals) materials. Various techniques are available for their preparation, such as phase inversion, sintering, track-etching, and stretching. These techniques are explained and discussed in detail in Chapter 3. The membrane flux of a microfiltration membrane can be given by the following relation:

$$F = \alpha \cdot \Delta P \tag{2.24}$$

where F represents the microfiltration membrane flux, α the permeability constant, and ΔP the transmembrane pressure. For a microfiltration membrane system, the Hagen-Poiseuille and Kozeny-Carman equations can be applied [5,7,8]. For instance, if it is assumed that the pores of microfiltration membranes are well connected, symmetrical, and uniform cylindrical pores than the Hagen-Poiseuille relation can be used effectively by replacing α in Equation 2.24 with μr^2 as

$$F = \frac{\mu r^2}{8\eta\tau} \frac{\Delta P}{\Delta l} \tag{2.25}$$

where μ represents the membrane porosity, r the membrane pore radius, η the dynamic viscosity, Δl the membrane thickness, and τ the tortuosity factor. When spherical particles agglomerate, the Kozeny-Carman equation can be applied to the microfiltration membrane system as

$$F = \frac{\mu^3}{\chi\eta A^2} \frac{\Delta P}{\Delta x} \tag{2.26}$$

where χ represents the pore geometry dependent dimensionless constant and A the spherical particles per unit volume surface area.

Equations 2.25 and 2.26 show that the membrane flux depends on the membrane structural features, namely, pore size (r) and porosity (μ). Therefore, it is important to have a microfiltration membrane with narrow pore size distribution and high porosity to get overall effective membrane flux.

Microfiltration membranes prepared from inorganic materials are more suitable for various applications because of their excellent chemical as well as mechanical resistances. The preparation techniques available for the preparation for these membranes allow for the use of a variety of materials for their preparation with better properties. Some of the materials are listed next:

Hydrophobic membranes—Polyvinylidene fluoride, polytetrafluoroethylene, polyethylene, and polypropylene
Hydrophilic membranes—Cellulose, polysulfone, polycarbonate, polyamide, and polyetheretherketone
Ceramic membranes—Alumina, zirconia, and titania

In addition to these, other materials are used for the preparation of glass (e.g., SiO_2) and metal membranes (e.g., stainless steel, silver, tungsten). The successfully prepared membranes are further characterized for their morphological as well as permeation properties by using various characterization techniques, such as Fourier transform infrared spectroscopy, (field) scanning electron microscopy, atomic force microscopy, liquid–liquid displacement porosimetry, hydraulic permeability, and pure water permeation.

In membrane processes fouling is a major problem. This results in the flux decline of a membrane process; therefore, it is important to explore the reasons and development of ideas, strategies, and methods to deal with membrane fouling. The main reason for membrane fouling is the concentration polarization and fouling caused by the deposition of feed components over the membrane surface and inside the pores. This phenomenon of fouling and methods to control it are well discussed in Chapter 7.

Microfiltration is used for various industrial applications. The basic function of a microfiltration membrane is to retain components from a liquid feed, that is, microfiltration is simply used as a sieving process. Dead-end filtration is the most famous mode of use, but on the industrial scale, due to the problem of fouling, it is replaced by cross-flow filtration. Microfiltration is widely used in the pharma, food, and biotechnology sectors for sterilization of a feed, where the feed is heat sensitive and cannot be sterilized by using heat-based sterilization methods. Nowadays, microfiltration membranes are used for various advanced applications and are suitable for the developments in the fields of health care, food, textile, and wastewater treatment. The various advanced and developing applications of both polymeric as well as ceramic membranes are discussed in detail in Chapters 5 and 6.

2.3.2 Ultrafiltration

Ultrafiltration membranes are also pressure-driven membranes with pore size in the range of 1 to 100 nm. The basics of ultrafiltration membranes are also discussed in Chapter 1. Ultrafiltration membranes are used to remove macromolecules and colloids from the feed. Their basic principle of separation is similar to microfiltration membranes. Ultrafiltration membranes are usually asymmetric membranes, which means they have a dense (selective) top layer and porous (supporting) sublayer. Due to the dense nature of the top layer, it exerts much resistance on the feed and feed components.

The ultrafiltration membrane flux, similar to microfiltration membranes, can be given by the following relation:

$$F = \alpha \cdot \Delta P \tag{2.27}$$

where F represents the membrane flux, α the permeability constant dependent on pore size and porosity of the membrane, and ΔP the transmembrane pressure.

Phase inversion is the most popular method for the preparation of ultrafiltration membranes by using polymeric materials, such as cellulose, polysulfone, polyvinylidene fluoride, polyimide, and polyacrylonitrile. Ceramic and polymeric ceramic composite ultrafiltration membranes are also prepared. The most common ceramic materials are alumina and zirconia. The sol-gel technique is the commonly used preparation technique for ceramic ultrafiltration membranes. The sintering method is used to prepare the supporting layers for ceramic ultrafiltration membranes. For nanofiltration and reverse osmosis, these ceramic ultrafiltration membranes are used as sublayers.

Ultrafiltration membranes are characterized by using the same characterization techniques used to characterize microfiltration membranes. Molecular weight cut-off is also a commonly used characterization technique to characterize ultrafiltration membranes (for more details refer to Section 4.2.2.3 of Chapter 4). Ultrafiltration membranes also suffer from the problem of fouling as microfiltration membranes. The feed components used to poison the membrane pores and surface, which results in an overall decline in membrane selectivity and flux. The intrinsic properties, mainly the material type (hydrophobic or hydrophilic) used to prepare the membranes also influence the amount of fouling. It is now well established that membranes prepared with hydrophobic materials are more fouled as compared to membranes prepared with hydrophilic materials due to the hydrophobic nature of the feed components. Therefore, it is important to take care of this aspect of membranes in addition to chemical and mechanical resistances during membrane development so as to reduce fouling as well as performance of membranes in harsh conditions of pH, temperature, pressure, and chemicals. These advancements will further increase the number of applications where membranes can be successfully employed. The development of membrane modules and process designs are also important for the enhancement in performance as well as antifouling nature of the membranes.

Ultrafiltration membranes are the most widely used membranes among other membranes for various applications ranging from wastewater treatment to biotechnology. Ultrafiltration membranes are very effective and efficient in purifying drinking water; protein and fruit juice concentration; sterilization of food, pharmaceutical, and biotechnology products; wastewater processing and treatment; hemodialysis, blood oxygenation, and drug delivery in the health care field. Recent developments further make ultrafiltration membranes eligible to be employed to novel applications, such as smart textile and advanced military gear. Therefore, ultrafiltration membranes are a very important category of membranes vital for the future developments and advancements in the field of membrane science.

2.3.3 Nanofiltration

Nanofiltration membranes are the membranes with pore size in the range of 0.5 to 5 nm. Generally, nanofiltration membranes are used to separate low molecular weight solutes or feed components from a feed, such as sugars (glucose and sucrose). Nanofiltration membranes are denser, which exert very high hydrodynamic resistance as compared to micro- and ultrafiltration membranes. These membranes are, therefore, considered as intermediate between the porous micro- and ultrafiltration membranes and nonporous (dense) gas separation membranes. Due to high hydrodynamic resistance, they need high pressures to permeate a feed across them.

The nanofiltration membrane flux can be given by the following relation:

$$F = \alpha\,(\Delta P - \Delta\pi) \tag{2.28}$$

where F represents the membrane flux, α the permeability coefficient, ΔP the transmembrane pressure, and $\Delta\pi$ the transmembrane osmotic pressure. In practicality, some of the solutes also permeate across the membrane. Therefore, the transmembrane osmotic pressure is not $\Delta\pi$ but $\delta\Delta\pi$, where δ represents the solute reflection coefficient. In case of $\delta < 1$, Equation 2.28 becomes

$$F = \alpha\,(\Delta P - \delta\;\Delta\pi) \tag{2.29}$$

The solute flux is given by the following relation:

$$F = \beta \cdot \Delta c_s \tag{2.30}$$

where β represents the solute permeability coefficient and Δc_s the difference across the membrane between the solute concentrations.

It can be seen from Equation 2.28 that the membrane pure water flux increases linearly with increasing applied pressure. On the other hand, the solute flux is independent of the applied pressure and barely showed any effect of pressure on its values. Solute permeability totally depends upon the solute concentration difference across the membrane. Usually, as explained earlier, very high pressures are used in a nanofiltration membrane process as compared to micro- and ultrafiltration membranes. The applied pressure with nanofiltration membranes is in the range of 1000 kPa to 2000 kPa.

In the case of nanofiltration membranes, the choice of material is very crucial as it directly affects the performance of the membrane, unlike the micro- and ultrafiltration membranes. The material chosen for nanofiltration membranes should have a high permeability coefficient and a low solute permeability coefficient to achieve a satisfactory selectivity. Thus, it can be said that for nanofiltration membranes, unlike micro- or ultrafiltration membranes, the intrinsic properties of the membranes play a major role in its separation performance and the pores play a sleek role in the overall performance of the nanofiltration membranes. Therefore, nanofiltration membrane materials should be chosen very carefully.

The membrane selectivity as well as flux defines the overall performance of a membrane, therefore, for a better membrane these two values should be high. Now, because the flux of nanofiltration membranes depends on their intrinsic properties, it can be enhanced either by choosing a better membrane material or reducing the membrane thickness. Since the membrane flux and membrane thickness are inversely proportional to each other, asymmetric or composite membranes are better for nanofiltration membranes due to a thin top layer, which defines the overall membrane flux. For asymmetric membranes, the material for the top as well as the sub- (support) layer is the same, but for composite membranes the materials are different for top layer and sublayer. In these membranes, the porous sublayer works just as a support for the thin top layer. Also, these membranes can be prepared by any material due to the preparation technique used, that is, phase inversion, but the feed type should be considered before choosing the material. For example, for hydrophobic feed, hydrophilic membrane materials should be chosen, and vice versa. Cellulose, polyamides, polysulfone, polyvinylidene fluoride, polyimides, and polybenzimidazoles are used for the preparation of hydrophilic, hydrophobic or membranes with properties of both hydrophilic as well as hydrophobic membranes. Phase inversion, dip coating, interfacial polymerization, and plasma polymerization methods are some of the commonly used preparation methods for the preparation of asymmetric or composite nanofiltration membranes.

Nanofiltration membranes are used for a wide range of applications, such as ionic separations (separation of monovalent [Na^+ or Cl^-] or divalent [Ca^{2+}] ions); removal of micropollutants, such as dyes, sugars, herbicides, pesticides, and insecticides; and in food, health care, biotechnology, and pharmaceutical industries. Nanofiltration is preferred over the reverse osmosis membrane process, where the solute molecular weight as well as concentration is low, due to the high flux values of nanofiltration membranes. Therefore, it is better to employ nanofiltration membranes, which save both energy and time.

2.3.4 Reverse Osmosis

Reverse osmosis is the membrane process that uses membranes of very small pore sizes and is used for the separation of very small feed components, therefore, the pressure used is very high. It is well known that if a selective permeable membrane is kept in between pure water and water having salt content, then the water will start permeating from the pure water side to the water with salt content side. This process is known as osmosis. On the other hand, if pressure is applied to the water with salt content side, then the pure water flux will stop and this pressure is known as osmotic pressure (π). If the pressure is further increased, then the water from water with salt content side will start to flow toward the pure water side, and this process is known as reverse osmosis.

Low molecular weight solutes can be separated with the reverse osmosis membrane process. The basic details of reverse osmosis are provided in Chapter 1, and here, basically the transport mechanism of the reverse osmosis membrane process is given. This process utilizes pressure as the driving force for the separation of the feed components. In the case of completely efficient membranes, the solute flux can be neglected, since complete separation of the solutes will take place. Also, if the solvent flux is compared to the solute flux of the process, the amount of solvent flux is very less. Therefore, the total flux of a reverse osmosis membrane process is given by

$$F_{total} = F_w + F_s \approx F_w \tag{2.31}$$

Generally, water (w) and salts (s) are used as the solvent and solute in the reverse osmosis process, respectively. Therefore, F_w is the water flux and F_s the salt flux. Initially, the pressures on both sides of the membrane are equal, therefore the chemical potentials on both sides will also be equal and can be given by

$$\mu_{i_o} = \mu_{i_{o(m)}} \tag{2.32}$$

where μ_{i_o} represents the chemical potential at the feed–membrane interface and $\mu_{i_{o(m)}}$ at membrane–permeate interface.

Equating the chemical potentials at this interface yields the expression

$$c_{i_{o(m)}} = K_i \cdot c_{i_o} \tag{2.33}$$

In reverse osmosis a pressure difference is always present between the membrane permeate interphase and the permeate solution from p_o to p_l, respectively. Now, by equating the chemical potentials across this interface gives

$$\mu_{i_l} = \mu_{i_{l(m)}} \tag{2.34}$$

where μ_{i_l} represents the chemical potential at the membrane permeate solution and $\mu_{i_{l(m)}}$ at the membrane interface.

Substituting Equation 2.34 in Equation 2.17 yields

$$\mu_i^o + RT\ln\left(\varepsilon_{i_l} n_{i_l}\right) + \sigma_i\left(p_l - p_{i_{saturation}}\right) = \mu_i^o + RT\ln\left(\varepsilon_{i_{l(m)}} n_{i_{l(m)}}\right) + \sigma_i\left(p_o - pi_{saturation}\right) \tag{2.35}$$

which reduces to

$$\ln\left(\varepsilon_{i_l}\, n_{i_l}\right) = \ln\left(\varepsilon_{i_{l(m)}}\, n_{i_{l(m)}}\right) + \frac{\sigma_i\left(p_o - p_l\right)}{R\,T} \tag{2.36}$$

The sorption coefficient can be defined as

$$K_i = \frac{\varepsilon_{i_o}\ \delta_m}{\varepsilon_{i_{o(m)}}\ \delta_o} \tag{2.37}$$

Substituting and rearranging Equation 2.36 for the sorption coefficient, K_i, with Equations 2.20 and 2.37 yields

$$c_{i_{l(m)}} = K_i \cdot c_{i_l} \cdot e^{\left(\frac{-\sigma_i\left(p_o - p_l\right)}{R\,T}\right)} \tag{2.38}$$

Now, substituting Equations 2.33 and 2.38 in Equation 2.23 yields the expression for the water and salt flux across the reverse osmosis membrane in terms of the pressure and concentration gradient:

$$F_i = \frac{D_i\, K_i}{l}\left[c_{i_o} - c_{i_l} \cdot e^{\left(\frac{-\sigma_i\,(p_o - p_l)}{RT}\right)}\right] \tag{2.39}$$

2.3.5 Liquid Membranes

Feed components are of various natures with different solubilities and diffusion coefficients. Since the transport across a membrane depends on the flux of the membrane, the properties of the solutes and the membrane play a vital role in the membrane transport. The main obstructions of membrane transport are the membrane thickness, viscosity of the liquids, and geometry of the system. The higher the membrane thickness and viscosity of the liquids, the higher will be the resistance for the membrane transport, and vice versa. If a steady-state system is considered for liquid membranes, then the flux of a species i across a membrane of thickness l, can be given by Fick's first law [1] of diffusion as

$$F_i = -D_i \frac{di}{dl} \tag{2.40}$$

where D_i represents the diffusion coefficient.

In the case of thin membranes, the high rate of diffusions can be maintained, which results in high fluxes. The high diffusion can be explained on the basis of Fick's second law of diffusion as

$$\frac{di}{dt} = D_i \frac{d^2 i}{dl^2} \tag{2.41}$$

For steady-state diffusion through the membranes, Equation 2.40 can be written as

$$F_i = -k_i([i_2 - i_1]) = k_i([i_1 - i_2]) \tag{2.42}$$

where k_i represents the individual mass transfer coefficient dependent on the membrane thickness as

$$k_i = \frac{D_i}{l} \tag{2.43}$$

The individual mass transfer coefficient for the interdiffusion in the membrane pores can be given as

$$k_i^m = \frac{D_i \mu_m}{l_m \tau_m} \tag{2.44}$$

where μ_m, τ_m, and l_m represents the membrane porosity, membrane tortuosity, and membrane thickness, respectively.

The different diffusion steps (as shown in Figure 2.4) of a liquid membrane process are given next:

- Diffusion through membrane-feed interface:

$$F^F = k^F\left(\left[i^F\right] - \left[i^{F1}\right]\right) \tag{2.45}$$

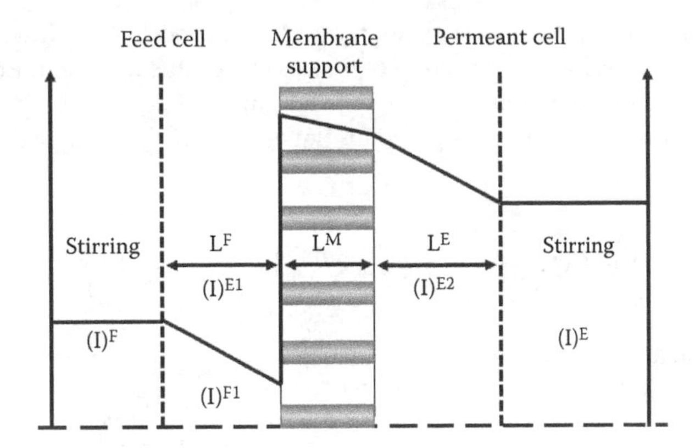

FIGURE 2.4 Schematic representation of different diffusion steps through a membrane.

This diffusion through the membrane–feed interface is present in all types of liquid membrane systems (i.e., bulk liquid membrane [BLM], supported liquid membrane (SLM), and emulsion liquid membrane [ELM]).

- Diffusion through the support membrane pores in SLM:

$$F^{FM} = k^{FM}\left(\left[i^{E1}\right] - \left[i^{E2}\right]\right) \tag{2.46}$$

This type of diffusion is present for SLM or BLM with hydrophobic membrane support. For BLM systems with hydrophilic membrane supports, it can be given as

$$F^{FM} = k^{FM}\left(\left[i^{F1}\right] - \left[i^{F2}\right]\right) \tag{2.47}$$

- Diffusion through the membrane–permeate interface:

$$F^{FE} = k^{FE}\left(\left[i^{E2}\right] - \left[i^{E}\right]\right) \tag{2.48}$$

And for the ELM it can be given as

$$F^{FE} = k^{FE}\left(\left[i^{E1}\right] - \left[i^{E2}\right]\right) \tag{2.49}$$

In the case of a BLM system with hydrophilic or ion-exchange membrane supports and for BLM without any support, it can be given as

$$F^{FE} = k^{FE}\left(\left[i^{E1}\right] - \left[i^{E}\right]\right) \tag{2.50}$$

Thus, from the total of individual resistances and bulk phase concentrations, the overall mass transfer rate can be calculated.

2.3.6 Gas Separation

In a gas separation membrane process, a higher pressure (p_o) is applied on the feed side, and the pressure (p_l) on the permeate side is lower than the feed side. Thus, the gases are separated by virtue of the pressure driving force in the gas separation membrane process [3]. Now, following the same procedure as reverse osmosis for the derivation of the gas transport expression, Equation 2.18 may be considered for the incompressible gases. The chemical potential equated at the gas and membrane interface yields the same expression as is given by Equation 2.32. Now, substituting Equation 2.18 into Equation 2.32 yields

$$\mu_i^o + R\,T \ln\left(\varepsilon_{i_o}\, n_{i_o}\right) + R\,T \ln\left(\frac{p_o}{p_{i_{saturation}}}\right) = \mu_i^o + R\,T \ln\left(\varepsilon_{i_{o(m)}}\, n_{i_{o(m)}}\right) + \sigma_i\left(p_o - p_{i_{saturation}}\right) \quad (2.51)$$

Rearranging of Equation 2.51 yields

$$n_{i_{o(m)}} = \frac{\varepsilon_{i_o}}{\varepsilon_{i_{o(m)}}} \cdot \frac{p_o}{p_{i_{saturation}}} \cdot n_{i_o} \cdot e^{\left(-\frac{\sigma_i\left(p_o - p_{i_{saturation}}\right)}{R\,T}\right)} \quad (2.52)$$

The exponential term for gases is close to one even in the case of very high pressures. Therefore, Equation 2.52 reduces to

$$n_{i_{o(m)}} = \frac{\varepsilon_{i_o}}{\varepsilon_{i_{o(m)}}} \cdot \frac{p_o}{p_{i_{saturation}}} \cdot n_{i_o} \quad (2.53)$$

The term $n_{i_o} p_o$ represents the partial pressure, p_{i_o}, of the i component in the feed gas. Therefore, Equation 2.53 simplifies to

$$n_{i_{o(m)}} = \frac{\varepsilon_{i_o}}{\varepsilon_{i_{o(m)}}} \cdot \frac{p_{i_o}}{p_{i_{saturation}}} \quad (2.54)$$

or, in terms of molar concentration, as

$$c_{i_{o(m)}} = M_{w_i} \cdot \sigma_m \cdot \frac{\varepsilon_{i_o}}{\varepsilon_{i_{o(m)}}} \cdot \frac{p_{i_o}}{p_{i_{saturation}}} \quad (2.55)$$

The sorption coefficient for gas sorption, K_i, can be defined as

$$K_i = M_{w_i} \cdot \sigma_m \cdot \frac{\varepsilon_{i_o}}{\varepsilon_{i_{o(m)}}} \cdot \frac{\delta_m}{p_{i_{saturation}}} \quad (2.56)$$

At the feed–membrane interface, the concentration of the i component can be given as

$$c_{i_{o(m)}} = K_i \cdot p_{i_o} \quad (2.57)$$

Similarly, at the membrane–permeate interface, the concentration of i component can be given as

$$c_{i_{l(m)}} = K_i \cdot p_{i_l} \quad (2.58)$$

Combining Equations. 2.57 and 2.58 in Equation 2.23 results in

$$F_i = \frac{D_i K_i\left(p_{i_o} - p_{i_l}\right)}{l} \quad (2.59)$$

The term D_iK_i is commonly referred to as the gas permeability coefficient, α_i; thus Equation 2.59 simplifies to

$$F_i = \frac{\alpha_i (p_{i_o} - p_{i_l})}{l} \tag{2.60}$$

The expression denoted by Equation 2.60 is widely used to precisely predict the gas separation membrane process properties.

2.3.7 Pervaporation

In the pervaporation membrane process a liquid feed is permeated through the membrane and the permeant is collected in vapor form. The driving force for this process is the partial pressures of the feed components. A vacuum at laboratory scale or temperature difference at industrial scale on the permeate side of the membrane is maintained so as to keep the driving force active and the process running [4]. The chemical potential is equilibrated with that at the feed–membrane interface at constant pressure to derive the pervaporation membrane process flux expression. Therefore, in this case Equation 2.17 gives

$$\mu_i^o + R\,T\,\ln(\varepsilon_{i_o}\, n_{i_o}) + \sigma_i(p_o - p_{i_{saturation}}) = \mu_i^o + R\,T \ln(\varepsilon_{i_{o(m)}}\, n_{i_{o(m)}}) + \sigma_i(p_o - p_{i_{saturation}}) \tag{2.61}$$

which further gives the concentration expression at the feed side interface,

$$c_{i_{o(m)}} = \frac{\sigma_{i_o}\,\delta_m}{\sigma_{i_{o(m)}}\,\delta_o} \cdot c_{i_o} = K_i^L \cdot c_{i_o} \tag{2.62}$$

where K_i^L represents the feed phase sorption coefficient and L denotes the liquid phase.

The pressure (p_o) drops to (p_l) from the membrane phase to the vapor phase. Therefore, the corresponding expression for the chemical potential of each phase is given by

$$\mu_i^o + R\,T\,\ln(\varepsilon_{i_l}\, n_{i_l}) + R\,T\,\ln\left(\frac{p_l}{p_{i_{saturation}}}\right) = \mu_i^o + R\,T \ln(\varepsilon_{i_{l(m)}}\, n_i) + \sigma_i(p_o - p_{i_{saturation}}) \tag{2.63}$$

Rearranging Equation 2.63 yields

$$n_{i_{l(m)}} = \frac{\varepsilon_{i_l}}{\varepsilon_{i_{l(m)}}} \cdot \frac{p_l}{p_{i_{saturation}}} \cdot n_{i_l} \cdot e^{\left(\frac{-\sigma_i\left(p_o - p_{i_{saturation}}\right)}{R\,T}\right)} \tag{2.64}$$

Since the condition of this case is similar to the gas separation, the behavior of the system will be same. Thus, the exponential term being close to unity is omitted. Hence, the permeate side concentration can be given by

$$n_{i_{l(m)}} = \frac{\varepsilon_{i_l}}{\varepsilon_{i_{l(m)}}} \cdot \frac{p_l}{p_{i_{saturation}}} \cdot n_{i_l} \tag{2.64}$$

The term $p_l n_{i_l}$ represents the partial pressure, p_{i_l}; hence, Equation 2.64 can be written as

$$n_{i_{l(m)}} = \frac{\varepsilon_{i_l}}{\varepsilon_{i_{l(m)}}} \cdot \frac{p_{i_l}}{p_{i_{saturation}}} \tag{2.65}$$

Substituting for the concentration term from Equation 2.10 into Equation 2.65 gives

$$c_{i_{l(m)}} = M_{w_i}\delta_m \cdot \frac{\varepsilon_{i_l}}{\varepsilon_{i_{l(m)}}} \cdot \frac{p_{i_l}}{p_{i_{saturation}}} = K_i^G \cdot p_{i_l} \tag{2.66}$$

where K_i^G represents the gas-phase sorption coefficient and G denotes the gas phase, which is defined in the gas separation section by Equation 2.56.

Now, to achieve the final expression for the membrane flux in the pervaporation membrane process, Equations 2.57 and 2.62 are substituted in Equation 2.23, which gives

$$F_i = \frac{D_i\left(K_i^L c_{i_o} - K_i^G p_{i_l}\right)}{l} \tag{2.67}$$

Nevertheless, two separate liquid and gas phase sorption coefficients are present in the equation, therefore it is important to derive an expression that can handle both phase separation coefficients conveniently. Thus, a hypothetical vapor is considered to be in equilibrium with the feed liquid. This vapor–liquid equilibrium will yield an expression:

$$\mu_i^o + R\,T\,\ln\left(\varepsilon_i^L\, n_i^L\right) + \sigma_i\left(p - p_{i_{saturation}}\right) = \mu_i^o + R\,T\ln\left(\varepsilon_i^G\, n_i^G\right) + R\,T\,\ln\left(\frac{p}{p_{i_{saturation}}}\right) \tag{2.68}$$

The liquid and vapor phases are shown by the L and G superscripts. Following the same procedure from Equations 2.63 to 2.66, Equation 2.68 yields

$$n_i^L = \frac{\varepsilon_i^G}{\varepsilon_i^L} \cdot \frac{p_i}{p_{i_{saturation}}} \tag{2.69}$$

Substituting with Equation 2.20 yields the expression in terms of concentration as

$$c_i^L = M_{w_i}\delta \cdot \frac{\varepsilon_i^G}{\varepsilon_i^L} \cdot \frac{p_i}{p_{i_{saturation}}} = \frac{K_i^G}{K_i^L} \cdot p_i \tag{2.70}$$

This expression connects the concentration of the i component in the feed phase (liquid), c_i^L to the partial vapor pressure, p_i, component i in equilibrium with the liquid. By substituting Equation 2.70 with Equation 2.68 gives

$$F_i = \frac{D_i K_i^G\left(p_{i_o} - p_{i_l}\right)}{l} \tag{2.71}$$

where p_{i_o} and p_{i_l} represent the partial vapor pressures of the i component present on each side of the membrane. Equation 2.62 can also be written as

$$F_i = \frac{P_i^G}{l}\left(p_{i_o} - p_{i_l}\right) \tag{2.72}$$

Equation 2.72 determines the driving force of the pervaporation membrane process as the vapor pressure difference present across the membrane.

2.3.8 Dialysis

The dialysis membrane process transport can also be explained on the basis of the solution–diffusion model, since only concentration gradients are involved [3]. The dialysis is used to separate liquid feeds of different compositions and, due to the concentration gradient, the feed components flow across the membrane.

Now, substituting the chemical potential expression for incompressible fluids from Equation 2.17 in Equation 2.32 gives

$$\mu_i^o + R\,T \ln\left(\varepsilon_{i_o}\, n_{i_o}\right) + \sigma_i\left(p_o - p_{i_{saturation}}\right) = \mu_i^o + R\,T \ln\left(\varepsilon_{i_{o(m)}}\, n_{i_{o(m)}}\right) + \sigma_{i_{o(m)}}\left(p_o - p_{i_{saturation}}\right) \quad (2.73)$$

which further yields

$$\ln\left(\varepsilon_{i_o}\, n_{i_o}\right) = \ln\left(\varepsilon_{i_{o(m)}}\, n_{i_{o(m)}}\right) \quad (2.74)$$

Therefore,

$$n_{i_{o(m)}} = \frac{\varepsilon_{i_o}}{\varepsilon_{i_{o(m)}}} \cdot n_{i_o} \quad (2.75)$$

or, from Equation 2.20,

$$c_{i_{o(m)}} = \frac{\varepsilon_{i_o}\,\delta_m}{\varepsilon_{i_{o(m)}}\,\delta_o} \cdot c_{i_o} \quad (2.76)$$

Hence, the sorption coefficient is defined as

$$K_i = \frac{\varepsilon_{i_o}\,\delta_m}{\varepsilon_{i_{o(m)}}\,\delta_o} \quad (2.77)$$

Thus, Equation 2.76 reduces to

$$c_{i_{o(m)}} = K_i \cdot c_{i_o} \quad (2.78)$$

Similarly, on the permeate side an equivalent expression is obtained as

$$c_{i_{l(m)}} = K_i \cdot c_{i_l} \quad (2.79)$$

By substituting Equations 2.78 and 2.79 in Equation 2.23, the expression for permeation in the dialysis membrane can be obtained as

$$F_i = \frac{D_i\,K_i}{l}\left(c_{i_o} - c_{i_l}\right) = \frac{\beta_i}{l}\left(c_{i_o} - c_{i_l}\right) \quad (2.80)$$

where D_iK_i represents the permeability coefficient, β_i, of the membrane process. Generally, the permeability coefficient is taken as constant for pure material. It is assumed to be dependent on the permeant and membrane material only, but this is not true since the nature of the solvent used in the feed phase is also important. Therefore, from Equations 2.80 and 2.77, the permeability constant, β_i, comes out to be

$$\beta_i = \frac{D_i \, \varepsilon_i}{\varepsilon_{i(m)}} \cdot \frac{\delta_m}{\delta_o} \tag{2.81}$$

Here, the presence of the term ε_i represents the functionality of permeability coefficient with respect to the solvent used in the feed phase.

2.3.9 Membrane Distillation

Membrane distillation is an isothermal membrane separation process with vapor pressure as the driving force (discussed in Chapter 1). Membrane distillation, depending on the type of configuration used, is mainly of four types: (1) vacuum, (2) sweeping gas, (3) direct, and (4) air gap. This membrane process has applications in the desalination, water reuse, food, and pharmaceutical industries. The availability of immense possibilities for the integration of membrane distillation with other processes makes it a process of great interest for industrial scale. Instead of having so many advantages associated with membrane distillation, it is scarcely successful on the industrial scale due to the limitations of membrane and membrane modules, wetting of membrane pores, and low and decay of permeate flux, including the process energy and cost uncertainties.

Membrane distillation is mainly advantageous in the case of a liquid feed with water as the main constituent. The liquid feed remains in contact with the membrane and vapors permeate through the membrane on the other side (permeate side). It is important that the membrane pores are not wet by the liquid feed, and thus mostly hydrophobic membranes are entertained for the membrane distillation process. The hydrophobicity of the membranes helps in keeping the membrane pores dry. The membrane pores may get wet if the transmembrane pressure is higher than the liquid feed pressure. The various modes of membrane distillation stated earlier maintain the driving force by applying different possibilities available, which are explained next:

- A vacuum is applied on the permeate side of the membrane distillation process. This vacuum gives rise to the driving force of separation in the process, because the pressure on the permeate side will be now lower than on the feed side. The volatile feed components raise the vapor pressure and thus separation takes place by virtue of the pressure difference driving force. The permeate is condensed after coming out of the permeate side of the membrane module. The vacuum membrane distillation utilizes this scheme.
- In the case of the *sweeping gas membrane distillation*, an inert gas at a temperature lower than the feed temperature is used to create the driving force and to separate the feed components from the feed. The vaporized feed components are carried by the cool inert gas across the membrane where they are condensed. A limitation of this configuration is that the temperature of the sweeping gas increases with time due to the heat transfer from the feed side to the permeate side across the membrane. To circumvent this limitation a new configuration is used, which is known as thc *thermostatic sweeping gas membrane distillation*. In this configuration, the sweep gas temperature is maintained by keeping the permeate sidewall temperature below that of the membrane module. This eliminates the problem associated with the previous configuration.
- An air gap is introduced between the membrane and the permeate surface in the *air gap membrane distillation*. The vaporized feed components come across the membrane, and after passing the air gap condense on the permeate surface and are collected from the permeate outlet.

- The feed and permeate side liquids are kept at different temperatures and circulated tangentially, while the permeate side liquid is in direct contact with the membrane. This results into the first configuration, the *direct contact membrane distillation*, of membrane distillation mentioned earlier. The temperature difference between the two liquids give rise to the driving force of the process, which is the vapor pressure difference. Due to this driving force, the volatile feed components vaporize and pass through the membrane in their vapor form. On the permeate side, since the temperature is lower, the vapors condense and are collected from the permeate outlet. In another version of this configuration of membrane distillation, the *gap liquid direct contact membrane distillation*, a gap is kept between the membrane and permeate liquid by keeping distilled water in between the two.

2.3.9.1 Transport in Membrane Distillation

Membrane distillation simultaneously deals with both heat and mass transfer. The membrane pores are responsible for the mass transfer, and pores together with the membrane matrix are responsible for the heat transfer across the membrane. There is a temperature polarization effect in the membrane distillation due to the difference in the temperatures at the membrane surface and the bulk phases of the solution. The transport mechanism of membrane distillation is explained by considering the direct contact membrane distillation in the following sections.

2.3.9.1.1 Heat Transfer

The heat transfer in direct contact membrane distillation occurs in three regions, covering the membrane–feed phase interface, membrane, and membrane–permeate phase interface, as shown in Figure 2.5. The first one is the convectional heat transfer in the membrane–feed boundary, $Q^F_{\text{Convection}}$, and heat transfer across the feed boundary layer due to the mass transfer, $Q^F_{\text{Mass transfer}}$. The second is the synergistic heat transfer due to the combination of conductive heat transfer across the membrane, $Q^M_{\text{Conduction}}$, and water vapor flow through the membrane pores, $Q^M_{\text{Mass transfer}}$. The last is the heat transfer due to convection at membrane–permeate boundary, $Q^P_{\text{Convection}}$, and mass transfer based heat transfer across the permeate phase, $Q^P_{\text{Mass transfer}}$. The said heat transfers can be expressed mathematically as:

- Heat transfer across the feed phase:

$$Q^F = Q^F_{Convection} + Q^F_{Mass\ transfer} = h^F\left(T^{BF} - T^{MF}\right) + F_m H^F_L\left[\frac{T^{BF} + T^{MF}}{2}\right] \tag{2.82}$$

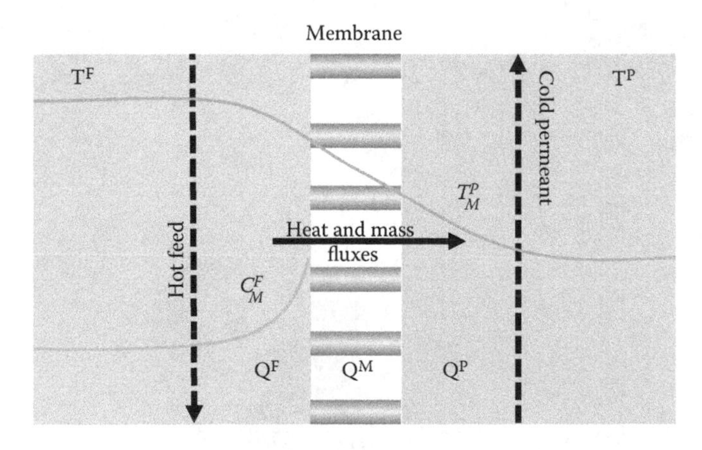

FIGURE 2.5 Schematic representation of heat and mass transfer in direct contact membrane distillation process.

- Heat transfer across the membrane:

$$Q^M = Q^M_{Conduction} + Q^M_{Mass\ transfer} = h^M\left(T^{MF} - T^{MP}\right) + F_m H^V \tag{2.83}$$

- Heat transfer across the permeate phase:

$$Q^P = Q^P_{Convection} + Q^P_{Mass\ transfer} = h^P\left(T^{MP} - T^{BP}\right) + F_m H^P_L\left[\frac{T^{MP} + T^{BP}}{2}\right] \tag{2.84}$$

where h^F and h^P represent the feed and permeate boundary heat transfer coefficients, respectively, and h^M is the heat transfer coefficient of the membrane. F_m is the permeate flux, H^V the vapor enthalpy, and H^F_L and H^P_L the feed and permeate phase enthalpies, respectively. T^{BF} and T^{BP} are the average bulk feed and permeate temperatures, respectively. T^{MF} and T^{MP} are the temperatures at membrane–feed and membrane–permeate interfaces, respectively. The h^M can be calculated from the thermal conductivities of the membrane polymer (σ_M) and air inside the membrane pores (σ_G) as

$$h^M = \frac{\sigma_G\,\mu + \sigma_M(1-\mu)}{l} \tag{2.85}$$

where μ and l represent the membrane porosity and thickness, respectively.

The ratio of the heat transferred through membrane pores and the total heat transferred through the membrane is known as the evaporation efficiency, E, and is given as

$$E = \frac{Q^M_{Mass\ transfer}}{Q^M_{Mass\ transfer} + Q^M_{Conduction}} = \frac{F_m H^V}{F_m H^V + h^M\left(T^{MF} - T^{MP}\right)} \tag{2.86}$$

The convectional mode of heat transfer is important for the feed and permeate phases of membrane distillation. The mass transfer–based heat transfer across the feed and permeate phase increases with an increase in the feed temperature. A significant increase can be seen in heat transfer through the feed phase as compared to the permeate phase at higher feed temperatures.

Since the vapor enthalpy $\left(H^V\right)$ is equal to the latent heat of vaporization $\left(\Delta H^V\right)$, Equations 2.82 to 2.84 can be rearranged as

$$Q^F = h^F\left(T^{BF} - T^{MF}\right) \tag{2.87}$$

$$Q^M = h^M\left(T^{MF} - T^{MP}\right) + F_m\,\Delta H^V \tag{2.88}$$

$$Q^P = h^P\left(T^{MP} - T^{BP}\right) \tag{2.89}$$

In case of a steady state,

$$Q = Q^F = Q^M = Q^P \tag{2.90}$$

Combining Equations 2.87 to 2.90 gives

$$Q = \left(\frac{1}{h^F} + \frac{1}{h^M + \dfrac{F_m\,\Delta H^V}{T^{MF} - T^{MP}}} + \frac{1}{h^P}\right)^{-1}\left(T^{BF} - T^{BP}\right) \tag{2.91}$$

Therefore, the overall heat transfer for direct contact membrane distillation can be written as

$$U = \left(\frac{1}{h^F} + \frac{1}{h^M + \dfrac{F_w \, \Delta H^V}{T^{MF} - T^{MP}}} + \frac{1}{h^P} \right)^{-1} \tag{2.92}$$

2.3.9.1.2 Mass Transfer

The mass transfer in the membrane distillation is also explained by considering the direct contact membrane distillation. A linear relationship between the mass flux (F_m) and the vapor pressure gradient through the membrane distillation coefficient (β_m) is assumed and the mathematical expression for the same can be given as

$$F_m = \beta_m \left(p^F - p^P \right) \tag{2.93}$$

where p^F and p^P represent the partial pressures of the feed and permeate phases, respectively.

There are a number of mechanisms that are proposed to explain the transport of gases and vapors across porous membranes. Some of the commonly used models are the Knudsen, viscous, diffusion, and a combination of these. The Knudsen number (K_n) is important because it provides the details of the mechanism of operation of the process under given experimental conditions. It can be defined as the ratio of the mean free path (λ) of the permeated feed components to the pore size (diameter, d) of the membrane, that is, $K_n = \lambda/d$.

Similar to heat transfer, mass transfer also takes place in three regions. They are the Knudsen, continuum (or diffusion), and the transition region (or a combination of the Knudsen/diffusion region). As previously discussed, in the case of Knudsen flow, the mean free path of a feed component is larger compared to the membrane pore size (i.e., $K_n > 1$ or $r < 0.5\lambda$, where r is the membrane pore radius). The collision occurs between the feed component and the membrane pore wall rather than solely between the feed components. Therefore, Knudsen flow will be the mechanism of vapor transfer through the membrane pores. Thus, the total permeability (β_m^K) of the membrane can be given as

$$\beta_m^K = \frac{2 \, \mu \, r}{3 \, \tau \, l} \left(\frac{8 \, M_w}{\pi \, R \, T} \right) \tag{2.94}$$

where μ, r, τ, and l represent the membrane porosity, pore radius, pore tortuosity, and membrane thickness, respectively. M_w represents the molecular weight of the feed. R and T represent the gas constant and absolute temperature, respectively.

In the case of air trapped in the membrane pores, the feed component diffusion will be the choice to explain the mass transport through the continuum region. In this case $K_n < 0.01$ (i.e., $r > 50\lambda$) and, therefore, the total permeability (β_m^D) can be given as

$$\beta_m^D = \frac{\mu \, P \, D}{p^a \, \tau \, l} \left(\frac{M_w}{R \, T} \right) \tag{2.95}$$

where P, p^a, and D represent the total pressure, partial air pressure in the membrane pores, and diffusion coefficient, respectively. PD can be calculated by using the following relation:

$$PD = \left(1.895 \times 10^{-5} \right) \cdot T^{2.072} \tag{2.96}$$

Last, in case of the transition region, where $0.01 < K_n < 1$ (i.e., $0.5\lambda < r < 50\lambda$), the feed components mostly collide with each other, and the transport mechanism is a combination of both Knudsen and diffusion. The total permeability (β_m^C) in this case can be given as

$$\beta_m^C = \left[\frac{3\,\tau\,l}{2\,\mu\,r}\left(\frac{\pi\,R\,T}{8\,M_w}\right)^{0.5} + \frac{\tau\,l\,p^a}{\mu\,P\,D}\,\frac{R\,T}{M}\right]^{-1} \tag{2.97}$$

2.3.10 Membrane Contactors

Membrane contactors are the membrane systems where a membrane does not control the rate of permeation of feed components and works just as an interface between the feed and permeate phases. Membrane contactors are generally used in shell and tube configurations with microporous hollow fiber membranes. There are different types of membrane contactors that are differentiated on the basis of phases involved, such as liquid–liquid (immiscible liquids) and gas–liquid (feed in gas phase and permeate in liquid phase). In the case of miscible liquids (liquid–liquid) the process is known as membrane distillation. These different types of membrane contactors are shown in Figure 2.6. The membrane contactors are better than the conventional liquid–gas absorbers or liquid–liquid extractors, and the most important advantage of membrane contactors is the impressive contactor area they provide to the two phases. It is 10 times more than conventional towers of the same size provide. A great example of this is the membrane blood oxygenator, which made open-heart surgeries possible by bringing the required amount of blood to a convenient level. Another important advantage of membrane contactors is that the two phases are always separated and never come in contact with each other. This reduces the channeling or flooding due to different large flows. Also, the counterflows can be used to their best efficiency to separate or concentrate the products or feed components. Therefore, a small amount of extractant can be used for the extraction of components from a very large feed volume, thus making the process economical and efficient. Due to the noncontact of the two phases, liquids of the same densities can also be used unlike with conventional contactors. The disadvantages associated with membrane contactors are mainly due to the membranes. The rate of transport depends upon the membrane properties and usually results in slower separation rates. The fouling of membranes is another important disadvantage, which further makes the rate of transport slower. Last, the membranes in use are not capable of facing high temperatures,

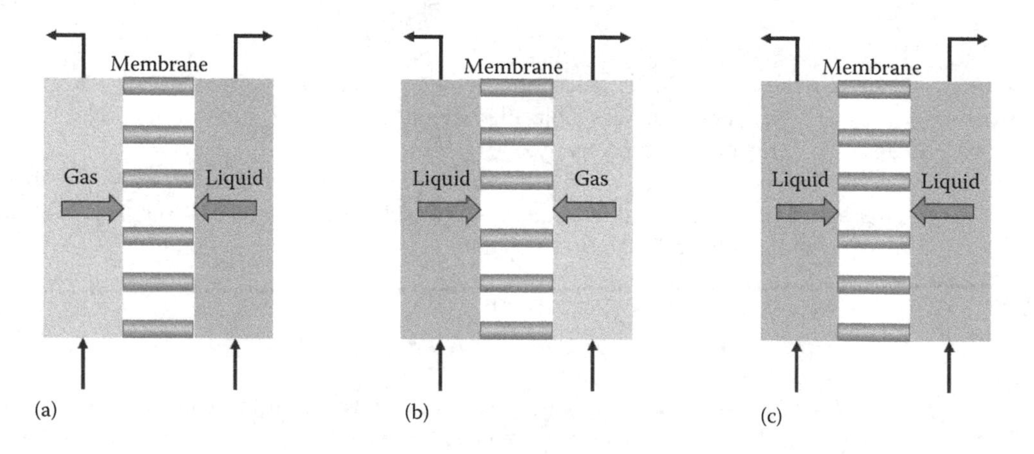

FIGURE 2.6 Schematic representation of membrane contactors: (a) gas–liquid, (b) liquid–gas, and (c) liquid–liquid.

pressures, and harsh chemical conditions, therefore, this limits the use of membrane contactors in many industrial separation processes. Thus, there is the need to develop membranes with better temperature, pressure, and chemical resistance.

2.3.10.1 Types of Membrane Contactors

2.3.10.1.1 Gas–Liquid

Gas–liquid membrane contactors are the most widely used membrane contactors for the removal or transport of gases from liquids. In this type of membrane contactor, a gas phase is present on one side of the membrane and on the other side a liquid. The feed components, due to the presence of a partial pressure gradient, diffuses from the feed phase to the permeate phase. The blood oxygenator is the best example of a gas–liquid membrane contactor widely used worldwide, where pure oxygen or air flows on one side of the membrane and on the other side flows the blood. The pure oxygen diffuses to the blood from the gas phase and the carbon dioxide flows from the liquid (blood) phase to the gas phase due to the presence of the partial pressure gradient between the two species.

Porous as well as nonporous (dense) membranes are used for these membrane contactors. The primary function of the membranes is to act as a barrier. The hydrophilic and hydrophobic nature of the membranes decides whether a liquid or a gas will wet the membrane pores. For hydrophilic membranes, it is the liquid that wets the pores. On the other hand, for hydrophobic membranes gas phase wets the pores.

2.3.10.1.2 Liquid–Liquid

Liquid–liquid membrane contactors are the membrane contactors where two liquids are employed: one as the feed phase and the other as the permeate phase on two sides of the membrane. The components from the feed phase liquid transmit to the permeate phase liquid via the membrane. With liquid–liquid membrane contactors, there will be different conditions of wetting or nonwetting of the membrane pores based on the type of liquids used as the feed and permeate phase. For example, in the case of an organic solvent, the aqueous phase and a hydrophobic membrane, then the membrane pores will be filled by the organic solvent. For a hydrophilic membrane the case will be opposite and the membrane pores will be filled with the aqueous solvent. These conditions privilege the phases to maintain their distinct profiles, since there will be no intermixing of the two phases due to their contrasting features. The interface for the exchange will be formed on the side of the membrane that consists of the nonwetting liquid. Liquid–liquid membrane contactors are applied for the removal of pollutants, bioproducts, pesticides, and volatile organic compounds.

2.3.11 Membrane Reactors

A membrane reactor is the combination of a membrane and reactor, where the role of a membrane is in product separation, reactant distribution, or catalyst support. This combination helps in achieving better control, efficiency, and economy of a process. Depending upon the application, different types of membranes and membrane reactors in different modes are used. Membrane reactors are used in various chemical, pharmaceutical, and biotechnology industries for different applications, like production and separation of a valuable product in the pharmaceutical or biotechnology industry, separation and reaction in the chemical industry, and purifying feeds in the case of environmental security.

2.3.11.1 Principle

In membrane reactors, the production as well as separation takes place simultaneously, unlike in conventional production processes where both these unit operations take place separately [3]. The synthesized products are separated in the production unit itself and thus lower the load of a downstream process. Therefore, a membrane reactor is symbiotic between a membrane and a reactor, where the membrane plays the role of a separator or reaction interface and the reactor provides the required operational conditions. This symbiosis between the membrane and a reactor makes the whole process simple, efficient, safe, and economical.

Membrane reactors as discussed use different kinds of membranes depending upon the application. Organic (polymeric) membranes are used in biochemical processes due to their inability to withstand high temperatures and harsh chemicals. Thus, they are used in the production of products by using immobilized enzymes, biogas from waste, and environmental safety by cleaning the polluted feeds; whereas inorganic membranes, due to their properties and ability to handle high temperatures and harsh chemicals, are used in catalytic membrane reactors. The membrane reactors have immense potential to be used in a number of applications. However, the limitation comes from the membranes. Therefore, it is important to have membranes with the desired properties of separation, selectivity, efficiency, and toughness.

There are various ways in which the membranes can be used in a membrane reactor for effective and efficient operation of a membrane reactor:

- Product extractor—The removal of a single product shifts the reaction equilibrium to the product side in thermodynamically equilibrium reactions; therefore, when membrane reactors are employed for such reactions the membrane acts as the product extractor and increases the product conversion efficiency. In other cases, the production of unwanted or secondary products is inhibited by the membrane acting as an extractor for an intermediate product of the process. This increases the selectivity and efficiency of the process.
- Reactant distributor—A reactant required for a particular reaction or process is added to the process with a controlled rate by virtue of the membrane reactors. The reactant is added to the feed side of the membrane and released to the reaction zone in a controlled fashion. This limits the production of secondary products or side reactions of the process and thus increases the selectivity and efficiency of the process. In this process, an impure reactant can also be used. This makes the process more economical.
- Active contactor—The reactants come in contact with the catalyst in a uniform and controlled manner via the membrane. The reactants are diffused through the membrane from one or simultaneously from both sides. This method provides more active catalyst sites to the reactants and thus increases the efficiency of the catalyst and in particular the process.

2.3.11.2 Catalysts in Membrane Reactors

Catalysts are the heart of the chemical reactions taking place in a membrane reactor. The catalysts can be incorporated into the membranes in four different ways, as shown in Figure 2.7.

- Catalysts layered on the membrane surface—In this case, the membrane surface is packed with a layer of catalysts, physically separated from the membrane. Thus, the membrane does not act directly in the catalytic reaction but as a product extractor or reactant distributor in the process.

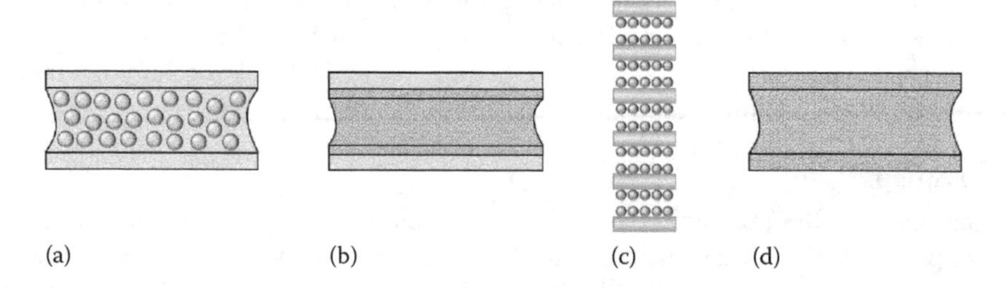

FIGURE 2.7 Schematic representation of catalyst presence in the membranes: (a) catalysts layered on the membrane surface, (b) catalysts coated on the membrane surface, (c) catalysts dispersed in membrane, and (d) catalytic membranes.

This method is widely used due to the ease of its operation. The membrane selectivity and catalyst activity can all be controlled separately, since both the catalyst layer as well as the membrane are physically separated. The catalyst layer can be placed on either side of a membrane, but on the porous side if the catalyst is reactive to the selective layer of the membrane.

- Catalysts coated on the membrane surface—In this case, the membrane surface is coated with catalysts. The catalyst layer is porous in nature and, unlike the layered case, is integrated with the membrane surface. A catalyst that is reactive to the selective layer of the membrane should be used in the layered configuration as discussed in the case of catalyst layered on the membrane surface.
- Catalysts dispersed in membrane—In this case, the catalysts will be dispersed in the membranes and is known as the membrane catalyst. The membrane pores are the sites of catalyst dispersion, and the reactant or product diffuses through these reactive zones and access the catalysts. The transmembrane pressure and uniform pores make it easier for the reactants to access the catalysts. This will not only improve the access of the reactants to the catalysts but also improve the overall efficiency of the process. The membrane thickness, structure of pores, porosity, and amount and site of catalysts, if optimized correctly, and then the overall activity of the catalysts and efficiency of the process, would be 10-fold that of catalysts used in a pellet form.
- Catalytic membranes—In this case, the membrane itself is both a catalyst as well as a separator. The membrane's intrinsic properties are the reason for these catalytic properties, which directly depend upon the type of material used. Many membrane materials are catalytic in nature or materials with catalytic properties that can be used for the preparation of such membranes. For example, titanium dioxide (TiO_2) is a catalytic (photocatalytic) material, which if used for the preparation of membranes, can impart photocatalytic properties to the membranes. In this case, the problem of catalyst loss is rare with high regeneration capacities. Therefore, these membranes when used in a membrane reactor make the process efficient and economical.

Membranes containing catalysts in or on the surface are known as catalytic membranes. The maintenance or holding of the catalytic property is of prime importance for these membranes as compared to permselectivity. As discussed, the optimization of the catalyst composition, loading, activity, stability, and membrane pore structure are very important for catalytic membranes.

2.3.11.3 Configuration of Membrane Reactors

The configuration of a membrane reactor mainly depends on the membrane configuration and the respective application of use. The two main configurations used for membrane reactors are tubular and flat sheet.

2.3.11.3.1 Tubular Membrane Reactor

These reactors are tubular in shape, as shown in Figure 2.8. They are similar in configuration to the shell and tube heat exchanger as they also consist of a shell and tube. The feeds are given in either a

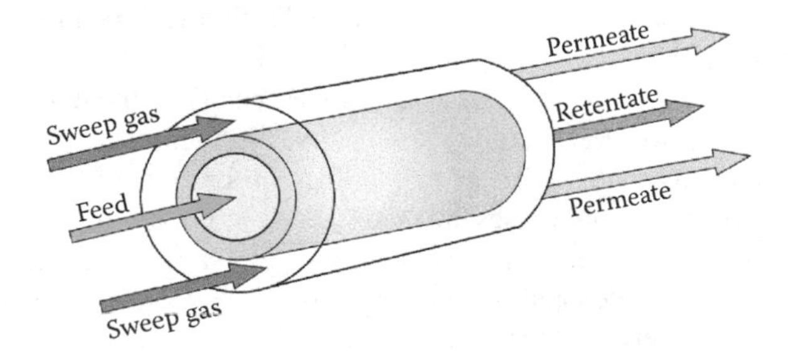

FIGURE 2.8 Schematic representation of a tubular membrane reactor.

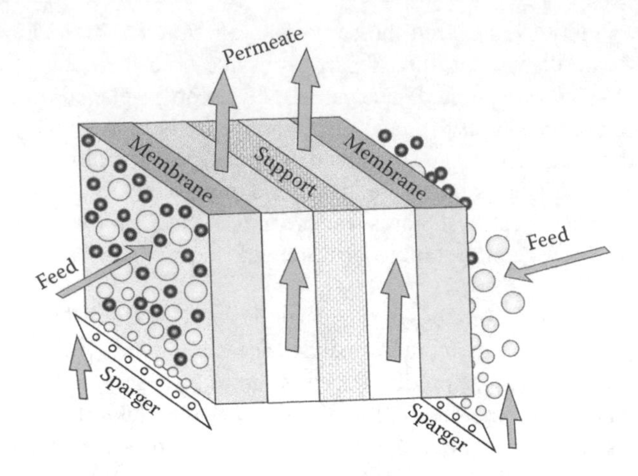

FIGURE 2.9 Schematic representation of a flat sheet membrane reactor.

cocurrent or countercurrent manner and the catalyst can be placed on either the tube side or shell side. The tubular type of membrane reactors is widely used on the laboratory and industrial scale. The hollow fiber membranes also follow the tubular membrane reactor configuration.

2.3.11.3.2 Flat Sheet Membrane Reactor

The flat sheet membrane reactors are of a flat sheet or disk type configuration as shown in Figure 2.9. As discussed, the membrane reactor configuration depends upon the membrane configuration; here the membrane is flat or disk shaped. The catalyst in layered or coated form is used in this configuration, and catalytic membranes are also used. The membrane reactors with a flat sheet configuration are mainly used on the laboratory scale because of their ease in making and need for fewer materials, namely, membrane and reactor.

Membrane reactors used in different types of configurations are membrane microreactors, catalyst fluidized bed membrane reactors, and electrolytic membrane reactors. These reactors vary slightly from the two important configurations, that is, tubular and flat sheet.

2.3.11.4 Types of Membrane Reactors

Membrane reactors can be classified on the basis of the membrane properties or reactor type. A widely used membrane reactor is the packed bed membrane reactor. In this type of membrane reactor, the membrane and catalyst in the layered or coating configuration are used, where the membrane and catalysts are not in contact with each other. The membrane plays the role of a product extractor or reactant distributor. These reactors can be used in different reaction and production combinations, for example, two simultaneous reactions taking place on both sides of the membrane, where the product from one side of the membrane works as a reactant for the other side of the membrane, or vice versa. The packed bed reactor is called a fluidized bed membrane reactor, when the catalyst is present in a fluidized state.

The other type of membrane reactor is the one that uses catalytic membranes and is known as a catalytic membrane reactor. In these membrane reactors, the membrane has the dual function of catalysis and separation or distribution. These membranes are prepared from catalytic materials and thus are intrinsically catalytic. In some cases the membranes are not required to be permselective and act as supports or providers of active sites; they are known as catalytic nonpermselective membrane reactors. In this type of reactor, the catalysts are coated or deposited on the membrane surface or in the membrane pores, respectively, and the reactants can be fed from either side of the membranes. In the case of membrane reactors, where the membranes are not directly involved in catalysis but act as a support and have the basic function of separation or distribution of products or reactants, respectively, are known as inert membrane reactors. In the cases where the membrane used is of <1 mm length, they are known as membrane microreactors. When electrical power is given to a membrane reactor with the help of an external electrical circuit, these are known as electrical membrane reactors. The electricity required for such membrane reactors can also be generated in situ by virtue of the chemical reactions taking place in the membrane reactor.

2.4 MEMBRANE MODULES

Membranes with a large area are required on the industrial or commercial scale. Therefore, they have to be assembled in units that are compact and at the same time spacious, so as to fulfill the area requirements of the membranes. These membrane assemblies are known as membrane modules. For the most basic membrane module, there will be an inlet for the feed and outlets for the permeate and retentate, as is shown in Figure 2.10. The feed enters the membrane module via the feed inlet at a specified temperature, pressure, and flow rate conditions. The feed stream then further separates into permeate and retentate streams depending upon the feed composition and membrane properties. The feed components that are allowed by the membrane to be passed through form the permeate stream and the ones that are not form the retentate stream.

There are many module configurations available for the hollow/tubular and flat sheet/disk membranes. Plate-and-frame and spiral-wound modules are for flat sheet membranes, and tubular (approximate membrane diameter <10 mm), hollow fiber (approximate membrane diameter <0.5 mm), and capillary (approximate membrane diameter 0.5–10 mm) modules fit hollow membranes [3,9].

The membrane processes, as per the process requirements, employ more than one module out of various designs in different configurations. There are also provisions made for recirculation of the retentate or permeate through the modules, if the process demands so. The module type, its arrangement in the process system, and the overall design of the process system are all based on the absolute engineering parameters and on the goal to achieve maximum efficiency with economical design or membrane process. The main parameters are the type of feed and its composition, scale, compactness, operation, cleaning, and maintenance of the system, and possible replacement of the fouled membranes. This important information regarding membrane modules is mostly present in patents, which are secured by the membrane companies and thus individuals like academicians and researchers do not have access to them. Therefore, many in the membrane community ignore the importance and areas of further development in membrane modules. In the following sections, different membrane modules will be discussed with their principle of working and design.

2.4.1 PLATE-AND-FRAME MODULE

The plate-and-frame module assembles flat sheet membranes in a casket form. The plate-and-frame module is shown schematically in Figure 2.11. In this module, more than one membrane can be

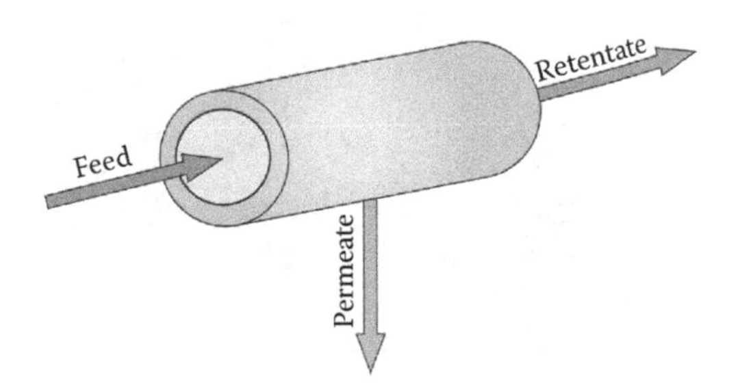

FIGURE 2.10 Schematic representation of simple membrane module.

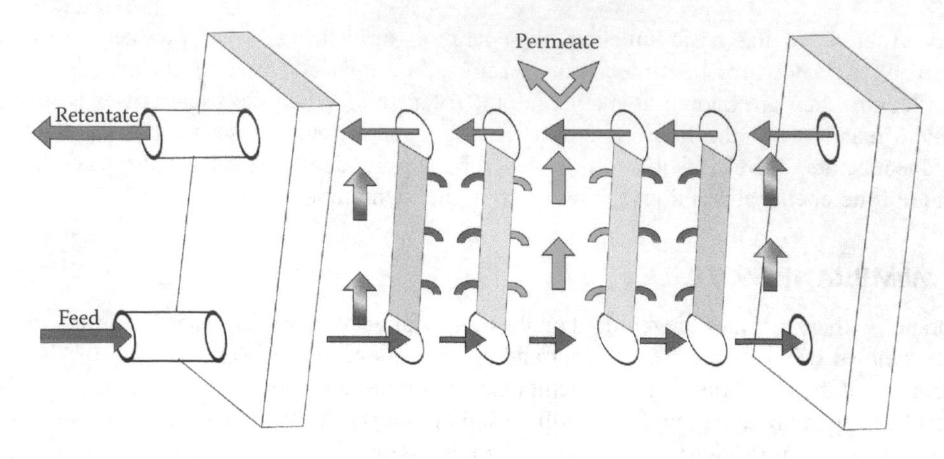

FIGURE 2.11 Schematic representation of a plate-and-frame membrane module.

accommodated in a stacked form over one another. The feed and permeate side of the membranes face each other; it gives an impression of an individual compartment made up of a set of membranes. Spacers are placed in between these compartments and the number of the compartments or membrane sets in the module are sealed with closing rings. Lastly, the module is enclosed between plates to form a plate-and-frame stack. This finally results in the plate-and-frame module. The normal packing density that can be attained in the plate-and-frame module is in the range of 100 to 400 m^2/m^3. The feed is introduced in the plate-and-frame module from the feed side of the module, which travels across the membrane to the permeate channel and finally enters a common permeate channel, taking the permeant out of the module. Generally, the plate-and-frame module is used for small-scale applications due to its expensiveness and leakage problems. Plate-and-frame modules are rarely used nowadays; spiral-wound modules are more commonly used. Electrodialysis and pervaporation membrane processes are some of the processes that are still entertaining the plate-and-frame modules in high numbers. Membrane processes like ultrafiltration and reverse osmosis employ plate-and-frame modules where feed induces high fouling.

2.4.2 Spiral-Wound Module

Spiral-wound modules are similar to plate-and-frame modules, but swathed around a central permeant collection tube, as presented in Figure 2.12. These modules are now the most widely used

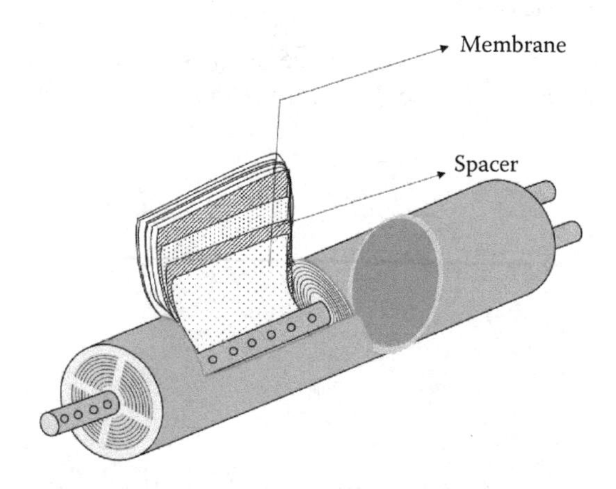

FIGURE 2.12 Schematic representation of spiral-wound membrane module.

membrane modules for flat sheet membranes. In this membrane module, the membrane with a permeate side spacer is glued from three sides in the form of an envelope and then swathed around a porous central collection tube. The feed flows parallel to this central porous tube and axially to the module. Then it exits the module via this collection tube. This module provides a packing density of around 300 to 1000 m^2/m^3. Usually, a number of spiral-wound modules are assembled together in a single pressure vessel so as to be economical and efficient at the same time. If a single module is used to encompass a large area membrane, then there will be a huge pressure drop across the collection tube due to the large path the permeate has to follow through. The length of the collection tube or overall module depends upon the diameter of the module. Generally, in industries, modules with 0.2 m diameter and 1 m length are used on a standard basis.

2.4.3 Tubular Module

Tubular membranes, unlike hollow fibers and capillaries, are not self-supporting. Therefore, these membranes are kept inside a porous metallic, ceramic, or polymeric tube. Inside this tube, the number of tubular membranes that can be packed is not limited and as many as possible membranes can be packed. Figure 2.13 represents the schematics of the tubular membrane module. In this module, the feed flows through the packed tubular membranes and the permeate is collected via the porous supporting tube. The tubular module is widely used for housing ceramic membranes with a packing density of 300 m^2/m^3.

2.4.4 Capillary Module

In a capillary module, a huge number of capillaries are assembled together. Since the capillaries are self-supporting, there is no need for a housing for them to be assembled as a module. The schematics of a capillary module are shown in Figure 2.14. When assembling the capillaries in the module, both the ends of the module are sealed with sealing agents, such as silicone rubber, epoxy resins, or polyurethanes.

Capillary modules are used in two ways: First, the feed flows through the lumen of the capillary and the permeate outside the capillary. Second, the feed flows from outside the capillaries and permeates flow through their lumen. The process parameters, such as membrane material, trans-membrane pressure, and pressure drop across the membrane, play an important role in deciding which method to be used. In the case of porous membranes, the capillaries carry a pore size gradient across them, which also helps in the selection of a suitable configuration. This selection is made on the basis of the location of the smallest pores on the capillaries that are inside or outside. Also, based

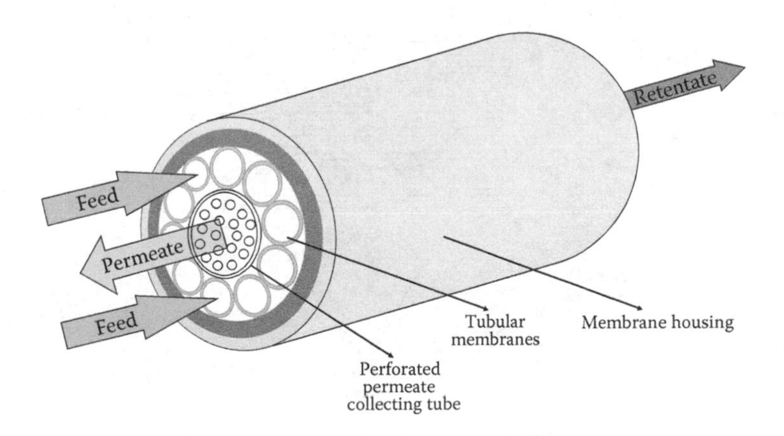

FIGURE 2.13 Schematic representation of tubular membrane module.

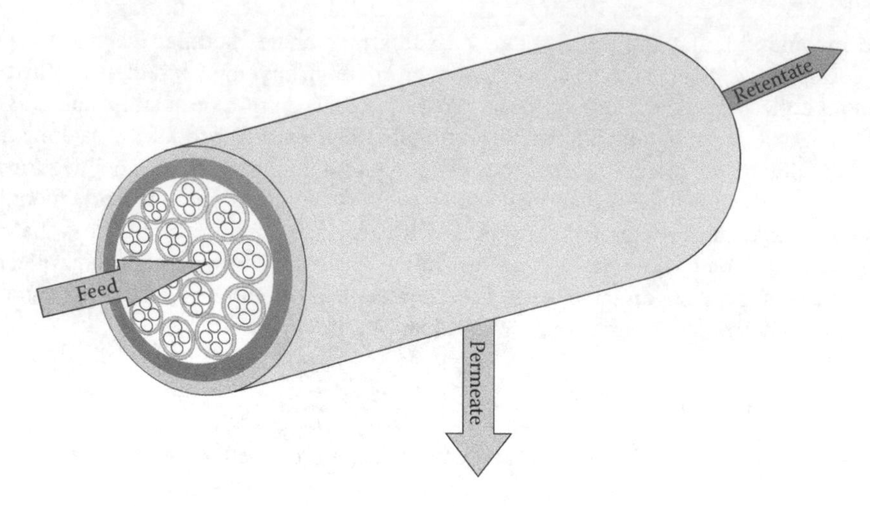

FIGURE 2.14 Schematic representation of capillary membrane module.

upon the method chosen the skin of the asymmetric capillaries are inside or outside. For a capillary module, a packing density of 600 to 1200 m^2/m^3 is possible.

2.4.5 Hollow Fiber Module

Hollow fiber modules and capillary modules are conceptwise similar with the only difference in their dimensions. The hollow fiber module resembles the reverse osmosis membrane module in shape and design. Similar to a capillary module, the feed can flow through the lumen of the hollow fiber or outside the fiber. The schematics of a hollow fiber module are shown in Figure 2.15. In a hollow fiber module, maximum packing density can be achieved that is equivalent to 30,000 m^2/m^3. This module is used in cases where the feed is quite clean, for example, in gas separation and pervaporation, and for desalination after an active pretreatment of the feed.

A hollow fiber module is used in a configuration where feed flows outside of the fiber and permeate inside its lumen, so as to increase the total membrane area and to reduce the overall pressure loss. Whereas for pervaporation it is used in a reverse configuration, that is, the feed flows in the lumen of the fiber and the permeate outside the fiber to circumvent the effect of permeate pressure increase inside

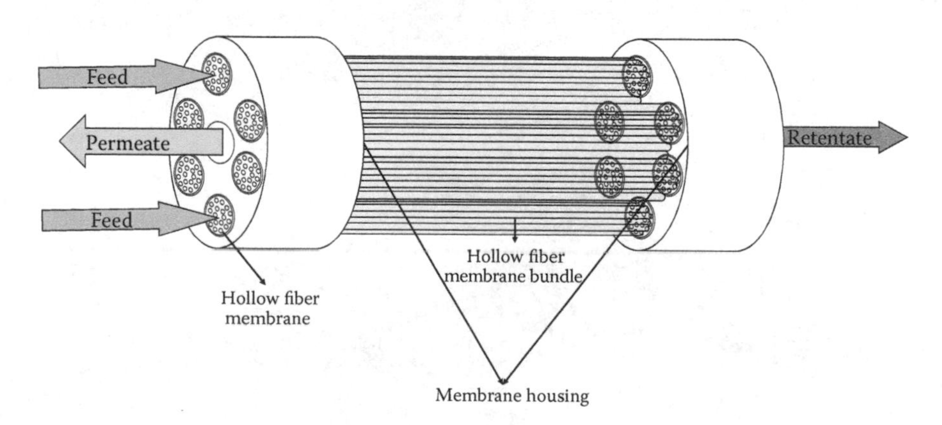

FIGURE 2.15 Schematic representation of hollow fiber membrane module.

the fibers. The advantage of the first configuration is that it provides high membrane surface area, and the best thing about the second configuration is that the thin selective top layer is well secured.

2.5 MODE OF MEMBRANE OPERATIONS

2.5.1 Batch

Membrane separation processes, generally, use the batch mode of operation. In this mode a limited amount of feed is given to the membrane, which is stored in a membrane cell before permeation. The process runs until the feed is not stopped or finishes and feed has to be refilled in the membrane cell for futher runs. It is similar to a batch reactor in operation. This mode of operation is good for the membrane separation process, since in this the membranes get time for their cleaning during the refilling step. Therefore, it enhances the membrane process efficiency and membrane life, and reduces cost and fouling of the membranes. It is widely used in industries related to food, pharmaceuticals, and biotechnology. The schematic of a batch membrane process is shown in Figure 2.16a.

2.5.2 Semibatch

The semibatch mode of operation is quite similar to the batch mode but with some modifications to allow addition/removal of feed or retentate/permeate or recirculation of the retentate/permeate to the membrane setup. This improves the selectivity and control of the membrane process. This mode also helps in reducing the burden on the downstream processing of the permeate and the retentate, since the permeate is recirculated many times in the process. The semibatch process is also widely used in process industries. This mode of operation also shows a reduced amount of membrane fouling and good overall process efficiency. The graphical representation of a semibatch membrane operation is presented in Figure 2.16b.

2.5.3 Continuous

The continuous mode of membrane operation is the one in which the addition of feed and the removal of permeate goes on continuously. In this mode of membrane operation, the speed, capacity, and efficiency to handle a large amount of feed are high. This mode reduces the time of operation but

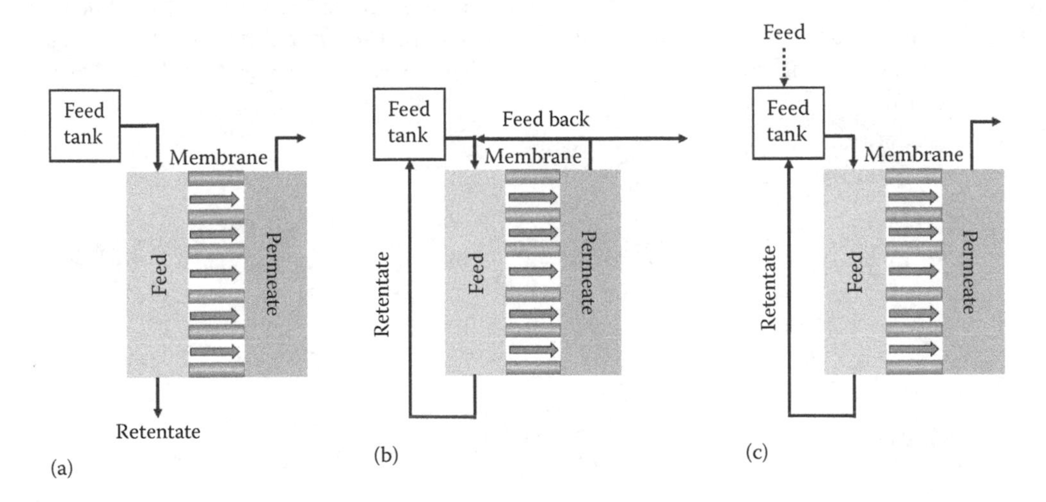

FIGURE 2.16 Schematic representation of (a) batch, (b) semibatch, and (c) continuous membrane separation process.

increases the chances of membrane fouling due to its nature of continuous operation and thus reduces overall efficiency of the process. The continuous mode of membrane operation is schematically shown in Figure 2.16c.

2.5.4 Other Modes

The other important modes of operation are dead-end filtration and cross flow. Membrane processes throughout the world utilize any one of these modes of membrane operation. Dead-end filtration is mostly used with the batch membrane operation and the cross flow with the continuous mode of membrane operation. Both have their own advantages and disadvantages. With dead-end filtration the recovery as well as the fouling is high. The membrane fouling makes the flux decline with time. The good recovery also makes it a choice to be used for better results. On the other hand, cross flow faces less of a fouling problem but has less recovery efficiency. Control over membrane fouling makes it a choice for use. A membrane process system with a combination of both dead-end filtration and cross flow will be very effective and beneficial in terms of use and operation. Both modes can be used in single or multiple pass configurations, where the feed or permeate can be circulated for a single time through the membrane process system or recirculated several times as shown in Figure 2.17a.

2.5.4.1 Dead-End Filtration Mode

In dead-end filtration mode the feed is given perpendicular to the membrane in the membrane process system, as shown in Figure 2.17b. This mode is good for the batch membrane operation. The efficiency of this mode is very high as compared to cross flow. The only disadvantage with this mode is membrane fouling; over time the membrane gets fouled and the efficiency of the process decreases. The membrane needs to be cleaned many times during the operation, which adds to the wear and tear cost. This makes the process a bit costly and unsuitable for longer periods of use. Therefore, the cross-flow mode is preferred for longer duration membrane processes.

2.5.4.2 Cross-Flow Mode

In cross-flow mode the feed is given parallel to the membrane in a membrane process system, as is shown in Figure 2.17b. The cross-flow mode is famous for its low membrane fouling profile. It reduces concentration polarization and membrane fouling to a great extent. A proper membrane module with optimized cross-flow feed velocity makes it the best available mode. The optimized feed velocity and membrane module help in achieving better mass transport and low membrane fouling in a membrane process. The cross-flow mode can be used in different configurations like cocurrent, countercurrent, cross flow, and perfect mixing, as shown in Figure 2.18. Out of these,

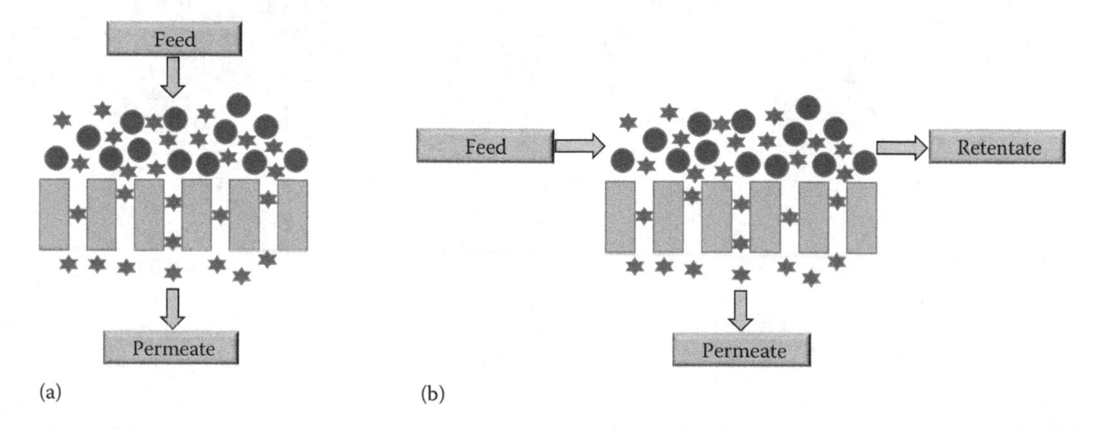

FIGURE 2.17 Schematic representation of (a) dead-end and (b) cross-flow membrane separation modes.

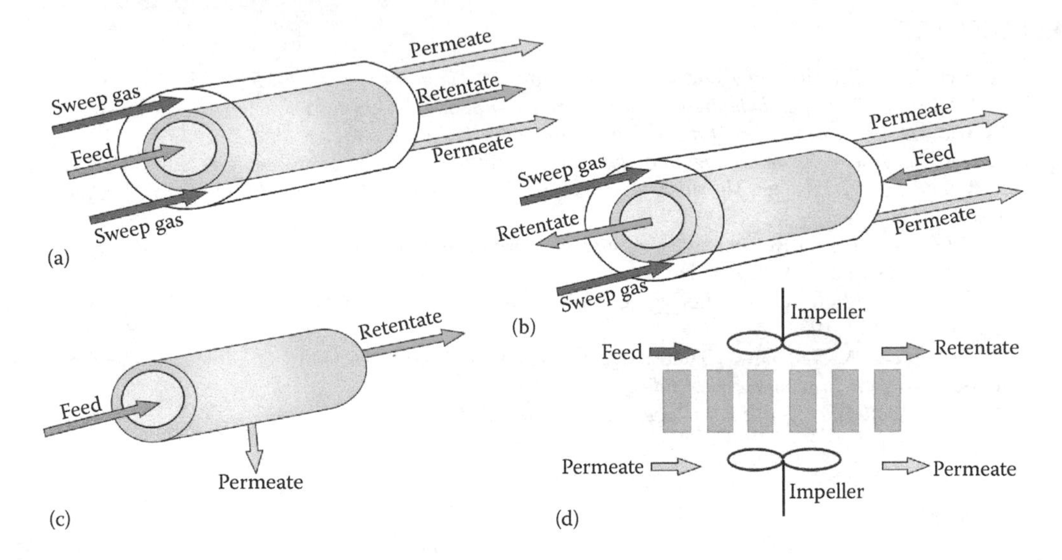

FIGURE 2.18 Schematic representation of different configurations of cross-flow membrane separation mode: (a) cocurrent, (b) countercurrent, (c) cross flow, and (d) perfect mixing.

countercurrent is the best in terms of results and is followed by cross flow, cocurrent flow, and perfect mixing, respectively.

STUDY QUESTIONS

1. Explain the basic models governing the transport mechanisms in the membrane processes.
2. What are the different types of flows available in membrane processes? How can they be helpful in the selection of a membrane for a particular application?
3. There are some assumptions made in models defining various parameters of a membrane process. Do these assumptions project the real-time facts of a membrane process? If not, then why are these assumptions are important?
4. An enzyme is produced using microorganisms. Sam wants to isolate and use this enzyme in its native form free from microorganisms for a catalytic reaction. Can he solve this task by using membrane processes? If yes, then how he should proceed to achieve his goal?
5. Is it possible to use nanofiltration instead of reverse osmosis? If yes, then what are the parameters that decide it and what is the advantage?
6. Zeenat wants to use salty water as a boiler feed water. What membrane process and membrane material should she use to get purified water with high output and in acceptable limits to be used as boiler feed water? Why is the membrane material important in the process?
7. Why it is advised to use asymmetric or composite membranes in case of nanofiltration and reverse osmosis membrane processes?
8. Elaborate the membrane process that can be successfully used for the separation of liquids with same densities.
9. Is it possible to use a single bioreactor as a reactor, distributor, and contactor? If yes, then explain with an example.
10. Jonathan wants to install a reverse osmosis plant for his home, but he has little space. Conventional reverse osmosis plants are huge and cannot be accommodated in a house. How can he solve this problem?
11. Is it better to use a batch process or continuous process? Explain with reasons.

REFERENCES

1. A. Fick, Über diffusion, *Poggendorff's Annalen der Physik und Chemie*, 94, 59, 1855.
2. H. Darcy, *Les fontaines publiques de la ville de Dijon*, Paris, Dalmont, 1856.
3. M. Mulder, *Basic principles of membrane technology*, Springer, 2007.
4. Richard W. Baker, *Membrane technology and applications*, 2nd ed., John Wiley & Sons, 2004.
5. S. P. Sutera and R. Skalak, The history of Poiseuille's law, *Annual Review of Fluid Mechanics*, 25, 1–19, 1993.
6. Vitaly Gitis and Gadi Rothenberg, *Ceramic membranes: New opportunities and practical applications*, Weinheim, Germany, Wiley-VCH, 2016.
7. P. C. Carman, Fluid flow through granular beds, *Transactions of the Institution of Chemical Engineers London*, 15, 150, 1937.
8. O. Kedem and A. Katchalsky, Thermodynamic analysis of the permeability of biological membranes to non-electrolytes, *Biochimica et biophysica Acta*, 1958.
9. K. Nath, *Membrane separation process*, Prentice Hall of India, 2008.

3 Preparation Techniques

3.1 INTRODUCTION

In Chapters 1 and 2, various membrane processes and their characteristics are discussed. These characteristics vary with the properties of the membrane precursor materials and the methods used to prepare the membranes. The fast growth in the strategies and methods for the design and fabrication of membranes with required properties of selectivity, permeability, functionalities, and physicochemical nature established them in the commercial membrane market. Functions of membrane precursors and various fabrication methods are discussed here.

3.2 MEMBRANE PRECURSORS AND THEIR ROLES

The choice of material is very important for membrane preparation as it defines various properties like pore size, porosity, strength, and permeability thickness. The chemical and physical characteristics of these materials are responsible for these membrane properties. The chemical properties of the material show how a particular material will behave under different chemical conditions such as pH, presence of other compounds, temperature, and pressure. The physical properties of the membrane material, such as density, melting point, compressibility, and glass transition temperature, represent its strength, durability, and flexibility under different physical conditions such as stress, temperature, and pressure.

In the membrane preparation process it is important that the constituent material retains its characteristic features and the desired membrane structure. Therefore, the membrane material, especially the polymeric materials, should have good chemical, mechanical, and thermal resistance with a degree of flexibility. For polymeric membranes, thermoplastics and cellulose are used on a large scale as the precursors. Theoretically, every available polymer can be used as a membrane material, but the need for specific chemical and physical properties limits their number. Nowadays, use of noncellulosic membrane materials are given importance as compared to cellulose-based materials. This is because the noncellulosic materials are more durable, less susceptible to biodegradation, and chemically and thermally more resistant as compared to the cellulosic material. Commercial membranes are prepared from membrane materials of all types based on the type of application, for example, membranes are prepared from completely hydrophilic material (cellulose acetate), hydrophobic material (polypropylene, polyethylene), and from materials that are neither hydrophilic nor hydrophobic (polysulfone, polyethersulfone, polyvinylidene fluoride, and polyacrylonitrile). For hydrophobic materials, different strategies are used to make them hydrophilic and of less fouling nature. The main methods used are surface coating, grafting, and blending. On the other hand, ceramic membrane materials are of inorganic nature like α-alumina, silica, zirconia, kaolin, and titania. It is already known that ceramic membranes are superior to polymeric membranes for operational stability and resistance to thermal, chemical, and mechanical stresses. Generally, the thickness of a ceramic membrane is in the range of 2 to 5 mm or more depending on the requirements. In the case of composite or asymmetric ceramic membranes, the coated layer measures up to 10 to 100 μm. It is also important to note that the principal cost of membrane materials should be as low as possible, since it will add to the final cost of a membrane and in particular to the membrane separation process.

For polymeric membranes, polymers such as polysulfone, poly(vinylidene fluoride), and cellulose are used as base polymers for the fabrication of polymeric membranes using various preparation methods, such as phase inversion. These polymers induce chemical and mechanical stability to the prepared membranes. Similarly, materials such as polymers, nanoparticles, carbon nanotubes, and modified or responsive materials are used as additives so as to improve the membrane properties such as hydrophilicity and antifouling nature. On the other hand, polymers like poly(ethylene glycol) and poly(vinyl alcohol) are used as pore formers, which induce pore formation in the membranes. In the case of ceramic membranes, materials such as zirconia, kaolin, and quartz are used as base materials for the fabrication of a ceramic membrane. Materials, such as sodium metasilicates are used as binders, which enhance the mechanical strength. Similarly, boric acid and sodium carbonate are used for improving the dispersion of the inorganic precursors used for the fabrication of ceramic membranes. Therefore, the choice of material for the membrane precursor is very important, as the attributes of the fabricated membranes directly depend on the properties of these materials. Thus, membrane materials should be chosen accordingly, mainly based upon the demands and requirements of the membrane application.

3.3 POLYMERIC MEMBRANES PREPARATION METHODS

The membrane is the heart of any membrane-based process. It controls the process and the process parameters, therefore, it is important to select a membrane wisely. A membrane is the key to the efficiency and results of a membrane process, and thus it is important to have a specific membrane for a specific application. Membrane properties, such as structure and function, vary. A membrane acquires all these properties and functions from its preparation method. Thus, the membrane preparation methods are also as important as the membranes are and need to be selected accordingly. Membranes with specific properties and functions can be made by using specific membrane preparation methods. There are also various options available for the membranes' porosity (porous and nonporous) and structure (symmetric, asymmetric, and composite), and they may be charged or uncharged. All these options and properties make membranes and membrane processes suitable for almost all types of applications related to liquid and gaseous feeds. In the case of polymeric membranes, membranes can be made into any shape, including flat, tubular, and hollow fiber. There are a number of materials that can be used for the preparation of a particular membrane, but, as said before, the selection of these membrane materials, membrane types, and preparation methods all depend on the type of application for which these membranes are going to be used. The various preparation methods for the preparation of polymeric membranes are explained in this section.

3.3.1 Phase Inversion Method

The phase inversion technique is widely used in polymeric membrane preparation. It is the easiest, most secure, and time tested method to prepare various porous polymeric membranes for different applications. The steps involved in the phase inversion process are schematically shown in Figure 3.1.

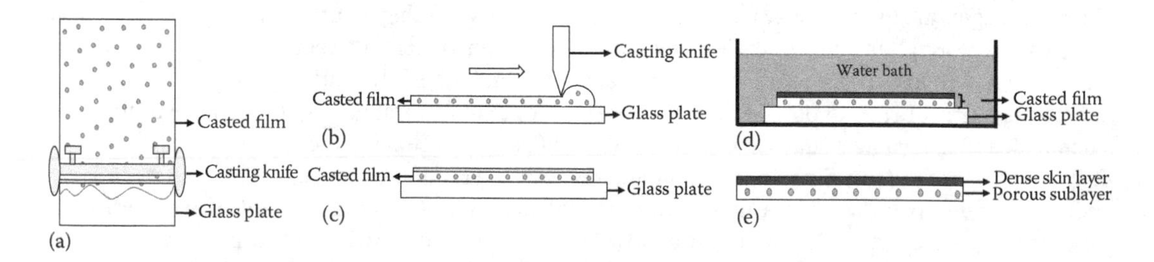

FIGURE 3.1 Schematic representation of the steps involved in a phase inversion process: (a) casting of the polymer solution over a glass plate with the help of a casting glass, (b) casted film, (c) evaporation/drying of the casted film, (d) immersion of the casted film into a coagulation bath containing the nonsolvent, and (e) prepared asymmetric polymeric membrane with a top dense skin layer and a porous sublayer.

The method works by controlling the separation state of the two phases. The one with the concentrated phase, after the phase separation, is solidified immediately and results in the formation of a membrane. The structure and function of the prepared membrane depends upon the changes taken place during the time of phase separation and solidification. In general, phase inversion separation of a homogeneous polymeric solution takes place in a polymer-rich continuous phase and a polymer lean phase due to the thermodynamic instability caused by of the presence of external effects. The driving forces commonly used for the phase inversion or phase separation of the polymeric solutions are temperature (thermally induced phase separation), nonsolvent (nonsolvent-induced phase separation), evaporation (drying-induced phase separation), and nonsolvent vapor (vapor-induced phase separation). These phase inversion methods are explained in the following sections.

3.3.1.1 Thermally Induced Phase Separation

Liquid–liquid demixing plays a prominent role during membrane preparation by this method. It is known that with decreasing temperature the solvent quality also decreases. The solvent is homogeneous at high temperatures, but separates into two phases—the polymer rich and polymer lean phases—at lower temperatures. Therefore, if induced at higher temperatures, the phase separation will not result in a porous structure but occurs at lower temperatures. The two separated phases are characterized by upper and lower critical solution temperatures. The boundary between the two is known as binodal and for polydisperse polymers it is known as the cloud point curve. The gap of liquid–liquid demixing is further subdivided into spinodal demixing, and regions of nucleation and growth. The transition of decomposition between binodal and spinodal is gradual and not immediate. A binary polymer solvent system phase diagram is shown in Figure 3.2, where temperature is plotted against the polymer concentration and explains the thermally induced phase separation process.

3.3.1.2 Nonsolvent-Induced Phase Separation

Nonsolvent-induced phase separation is a method where two phase separation processes take place for the membrane preparation, as shown in Figure 3.3. Here, liquid–liquid demixing and solid–liquid demixing take place. Liquid–liquid demixing helps in the formation of membrane porous structures and solid–liquid demixing occurs in crystalline membrane parts. In this method, the casted thin film is immersed in a nonsolvent bath and precipitation takes place due to the phase separation with

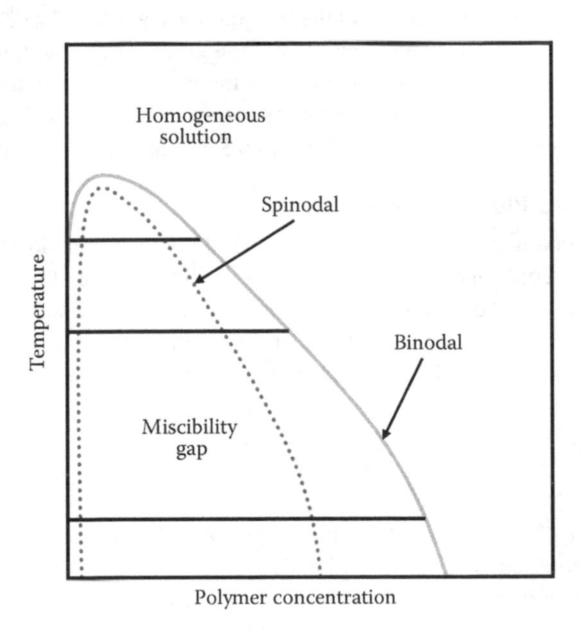

FIGURE 3.2 A binary phase diagram showing liquid–liquid demixing gap of a polymer solution.

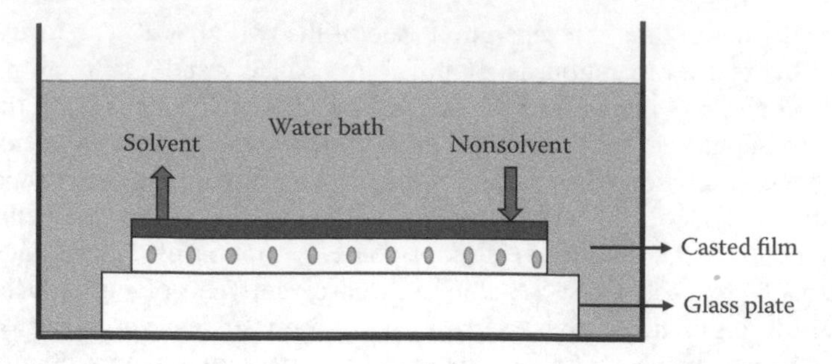

FIGURE 3.3 Schematic representation of nonsolvent-induced phase separation process.

diffusional exchange of solvent and nonsolvent. To understand this process of phase inversion, it is important to understand the thermodynamics of the method. The mass transfer in this method also plays a great role and is also important to understand. In this method liquid–liquid demixing results in two types of morphologies in the synthesized membranes. First is the formation of membranes with a dense skin layer and porous sublayer, and the second type is the formation of membranes with a porous skin layer due to instantaneous demixing. The study of ternary systems will help in understanding the changes in the membrane morphology due to solid–liquid demixing. The solid–liquid demixing also depends upon the type of polymers used for the membrane preparation. Polymers such as cellulose acetate, cellulose nitrate, and chitosan are crystallized by the solid–liquid demixing to a great extent and other polymers to a lesser extent. These different crystallizations of the polymers impart different properties to the prepared membranes and also affect the existing membrane properties. Therefore, to get a membrane with desired properties, choice of polymer or membrane material plays an important role.

3.3.1.3 Drying-Induced Phase Separation

The drying-induced phase separation method involves a volatile solvent for the dissolution of the polymer and a less or nonvolatile nonsolvent. The evaporation of the solvent induces phase separation of the polymer due to its decreasing solubility in the evaporating solvent. The polymer-rich phase forms the solid matrix upon solidification and the formed pores are filled with the nonsolvent. In this method a coagulation bath is not used and thus the complexities present with the use of a coagulation bath can be avoided. The rate of evaporation controls the morphology of the prepared membranes. A dense layer is obtained with fast evaporation rates and thin casting films. The nonsolvent penetrates the skin layers' weak points during its formation and this results into the formation of macrovoids.

3.3.1.4 Vapor-Induced Phase Separation

Vapor-induced phase separation is the process in which the phase separation of a polymer solution initiates when the nonsolvent vapors penetrate the casted films. This method is a combination of the dry and wet casting process, where the casted films are exposed to the nonsolvent vapor for a specific time period and then immersed in a coagulation bath for the completion of the process. Membranes with a variety of morphologies can be prepared by controlling the penetration of nonsolvent vapors into the casting films.

3.3.2 Stretching

Stretching is the method of membrane preparation where a thin extruded film or foil of crystalline or semicrystalline polymeric material is stretched perpendicular to the extrusion direction, as shown in Figure 3.4. This stretching imparts a mechanical stress on the thin film that results in the formation of pores with sizes in the range of 0.1 to 3.0 μm. Normally, this method results in highly porous membranes. Polymeric materials of crystalline or semicrystalline nature are only suitable for the

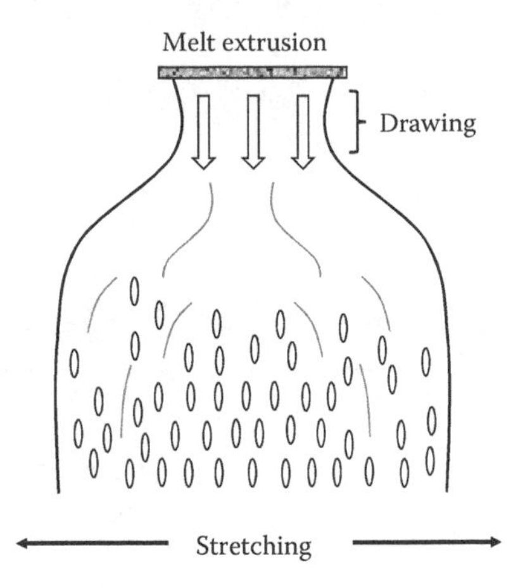

FIGURE 3.4 Schematic representation of stretching method of membrane preparation.

preparation of membranes by this method. Generally, microporous membranes are prepared by this method.

3.3.3 Deep Coating Method

Deep coating is also an anisotropic composite membrane preparation method like the interfacial polymerization method. In this method, a thin selective layer is solution coated over a microporous membrane support and is shown in Figure 3.5. A volatile water-insoluble solvent is used to prepare a polymer solution that is coated over the microporous membrane support. Usually, the polymer solution is casted over the water surface in a water bath between two rods and these rods are then moved apart so as to spread the polymer film over the water surface. This polymer film is then taken up by the microporous membrane support surface. In this particular method, it is difficult to transfer a very thin and fragile layer over the membrane support surface from the water surface. Therefore, to make the method successful the membrane support is moved under the spread polymer thin film. This same method is utilized to coat hollow fiber membranes for gas separation applications. Nowadays, similar to the interfacial polymerization method, the polymer solution is directly coated over a clean and defect-free microporous membrane support surface having fine pores. This helps in preventing the polymer solution from entering the microporous membrane support pores. The coated layer has a thickness around 50 to 100 μm, which upon evaporation of the solvent reduces to 0.5 to 2.0 μm. The success of this method depends on the preparation method and polymer solution.

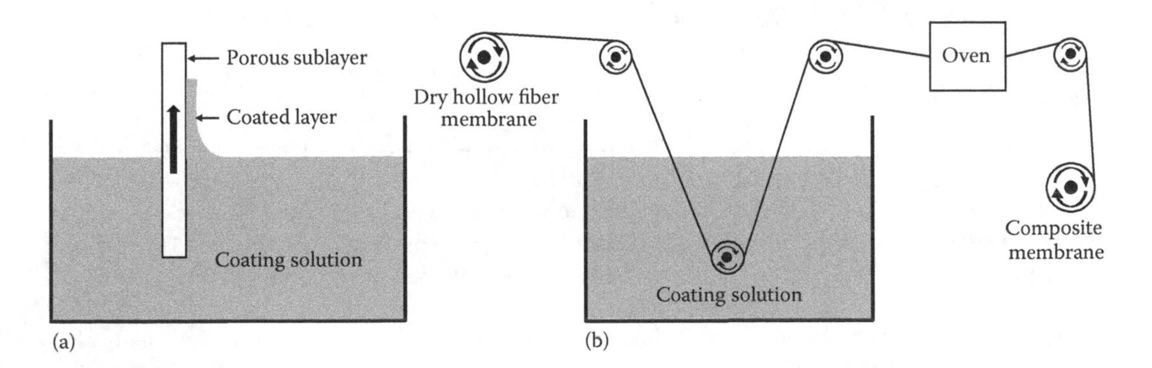

FIGURE 3.5 Schematic representation of dip casting method for (a) flat and (b) hollow fiber membranes.

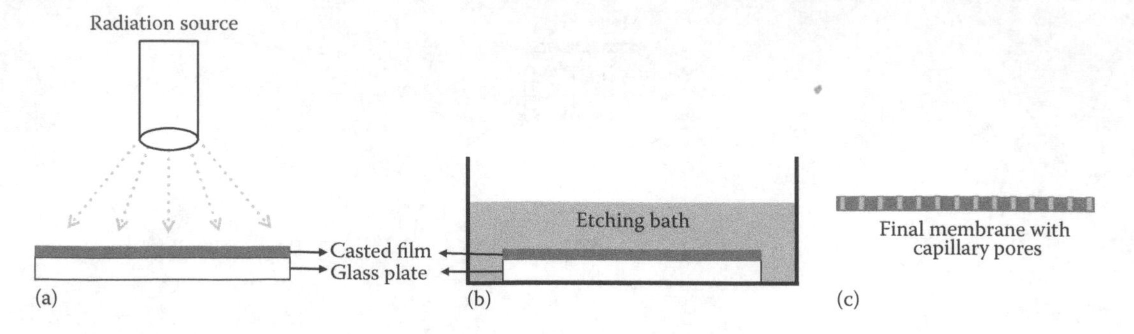

FIGURE 3.6 Schematic representation of track-etching method: (a) impregnation of the film with radiation, (b) radiated film immersed in the etching bath, and (c) final membrane.

3.3.4 Track-Etching Method

The track-etching method is suitable for getting cylindrical pores with uniformity in a membrane. In this method, particle radiation is applied perpendicular to a polymer film, which results in the formation of tracks due to the damage of the film material at the points of particle impact. The method is represented by Figure 3.6. This radiated film is further immersed in an acidic or alkaline bath, so that the tracks result into cylindrical pores of uniform geometry due to the erosion of the film material from the formed tracks. This method results in membranes with pore sizes in the range of 0.02 to 10 μm. The porosity of the prepared membranes is generally low. The material is chosen on the basis of the energy of the particles and thickness of the film. If the thickness of the film is high than higher energy can be used, and vice versa, which results in higher porosity, or vice versa. The porosity in this method depends on the time of irradiation of the films and the pore diameter on the etching time. A longer irradiation time causes a more porous membrane and for that a thicker film is required. Similarly, longer etching time means pores with larger diameters.

3.3.5 Template Leaching

The template leaching method is similar to track etching in operation. In both there is the leaching or removal of the film material. The template leaching method can be used for the preparation of porous glass membranes. In this method, a homogeneous glass melt consisting of three components is cooled to room temperature and thus the system separates into two different phases. One of the phases mainly consists of SiO_2, which is not soluble in an acid or base, but the second SiO_2 lean phase is soluble in an acid or base system. Therefore, when the cooled two-phase glass is immersed in an acid or base, then the soluble phase dissolves and leach from the system. This results in the formation of pores in the glass. A wide range of pores can be obtained on the basis of the immersion time and the materials used. The minimum pore diameter obtained is about 5 nm.

3.3.6 Interfacial Polymerization Method

Interfacial polymerization is a method involving the deposition of a polymer thin layer over a microporous support to prepare an anisotropic membrane, as shown in Figure 3.7. This method was first developed by John Cadotte at North Star Research. He used the polymer polyethylenimine and chemical toluene-2,4-diisocyanate [1]. Mulder [2] and Baker [3] explained the preparation method as the deposition of a reactive prepolymer, for example, polyamine, in the pores of a microporous membrane support, usually a polysulfone ultrafiltration membrane. This amine-containing membrane support is then immersed in a water-immiscible solvent bath containing a reactant, like diacid chloride. The reaction between the amine and acid chloride takes place at the interface of the two immiscible solutions, which results in a very thin membrane layer

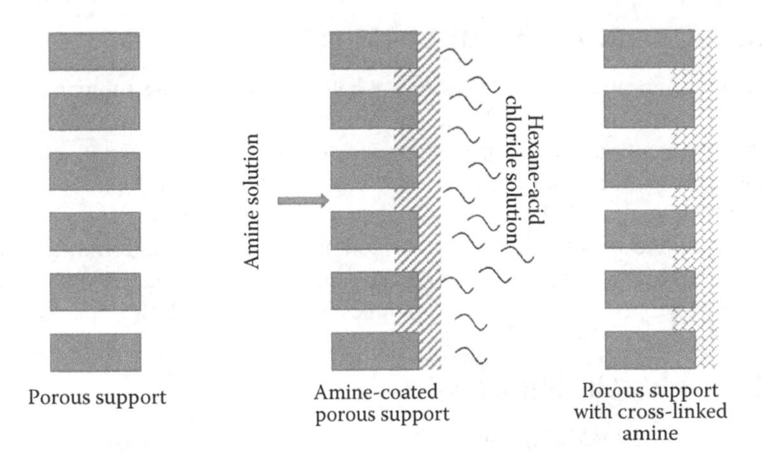

FIGURE 3.7 Schematic representation of interfacial polymerization method. A porous support is treated with amine solution and the formed amine film is reacted with a multivalent cross-linking agent (dissolved in an organic solvent, such as hexane), which results in the formation of a very thin film of polymer over the support surface.

over the membrane support. The thin layer of membranes prepared by this method is very dense and highly cross-linked. The very low thickness of the polymer layer gives high permeability, and high cross-linking imparts high selectivity to the prepared membranes. The membrane layer has to be finely porous to withstand the extreme pressures applied but at the same time must have high porosity to have good permeability. The interfacial polymerization method is easy to follow at the laboratory scale but difficult to take to an industrial or large scale, due to the complexities involved in the development of equipment for the production of the membranes. The interfacial polymerization method is widely used for the production of reverse osmosis membranes. The membranes prepared by this method are not used for gas separation because the membrane layer consists of a hydrogel, which when dry forms a glassy polymer. This glassy polymer in turn fills the membrane support pores and block the membrane, though the selectivity of such membranes are high but have reduced permeabilities for gases.

3.3.7 Plasma Polymerization

In the plasma polymerization method a radio frequency of 2 to 20 MHz is used to generate plasma by introducing inert gases in a chamber at pressures of 50 to 100 mTorr. Then a monomer vapor is introduced (the pressure should not exceed 200 to 300 mTorr) to the chamber. The present chamber conditions are maintained for at least 1 to 10 minutes. In the mean time, a thin polymer film will be deposited onto the membrane sample placed in the plasma field. Initially, plasma polymerization was utilized to induce electrical insulations and protective coatings to devices, but nowadays many membrane enthusiasts are using this method to prepare selective membranes. In the 1970s to 1980s, membrane researchers started using this method for the preparation of composite membranes. Yasuda was the first to develop polysulfone reverse osmosis composite membranes using this method [4]. The polymerization process is totally different in this method as compared to conventional polymerization. The polymerization is a complex process involving ionized molecules and radicals. In this method, the membrane samples get highly polymerized and, as a result, a thin layer is deposited over the membrane samples. The thin layer may contain radicals that continue reading even after the process is completed. It is very difficult to predict the extent of polymerization and the characteristics of the deposited thin film. The percent polymerization and characteristics of the thin film deposited totally depend upon whether the procedure is followed genuinely. Mostly, the resultant layers deposited are very thin and selective in nature, though the permeate flux is still low in plasma polymerized membranes.

3.4 POLYMERIC TUBULAR MEMBRANES

Polymeric tubular membranes are casted with the help of a supporting tubular material (a tube or pipe), since the polymeric tubular membranes are not self-supporting. The casting solution is passed through these tubular materials with a specific applied pressure through a manhole of a casting bob, which is present in the tubular material, as shown in Figure 3.8. The movement of this tubular material coats the inner side of itself with the polymeric casting solution. This coated tubular material is then immersed into a coagulation bath, where, due to the immersion precipitation of the casting solution, the formation of a polymeric tubular membrane takes place. The polymeric tubular membranes are different from the polymeric hollow fiber membranes in their method of preparation as well as use.

3.5 MIXED MATRIX MEMBRANES

Mixed matrix membranes (MMMs) consist of a continuous polymer phase and a dispersed phase, which usually consists of inorganic particles, such as zeolite, nanoparticles, metal-organic frameworks, carbon molecular sieves, carbon nanotubes, or catalysts, as shown in Figure 3.9. The zeal to have a membrane with properties of both a polymeric as well as ceramic membrane resulted in the development of MMMs. The polymeric membranes are widely famous for their toughness and ease of preparation, but limited because of their compromise between permeability and selectivity. On the

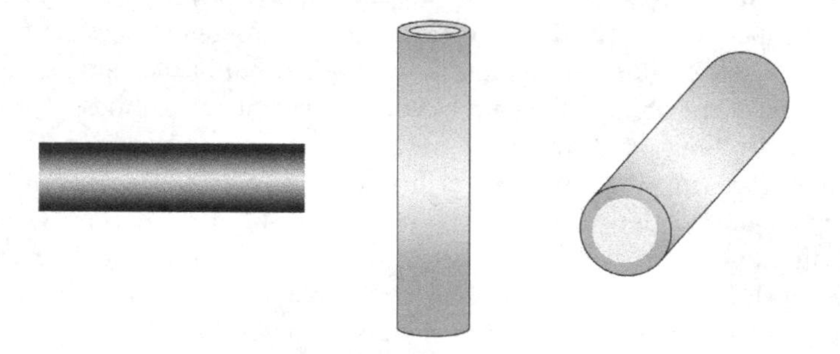

FIGURE 3.8 Schematic representation of polymeric tubular membranes.

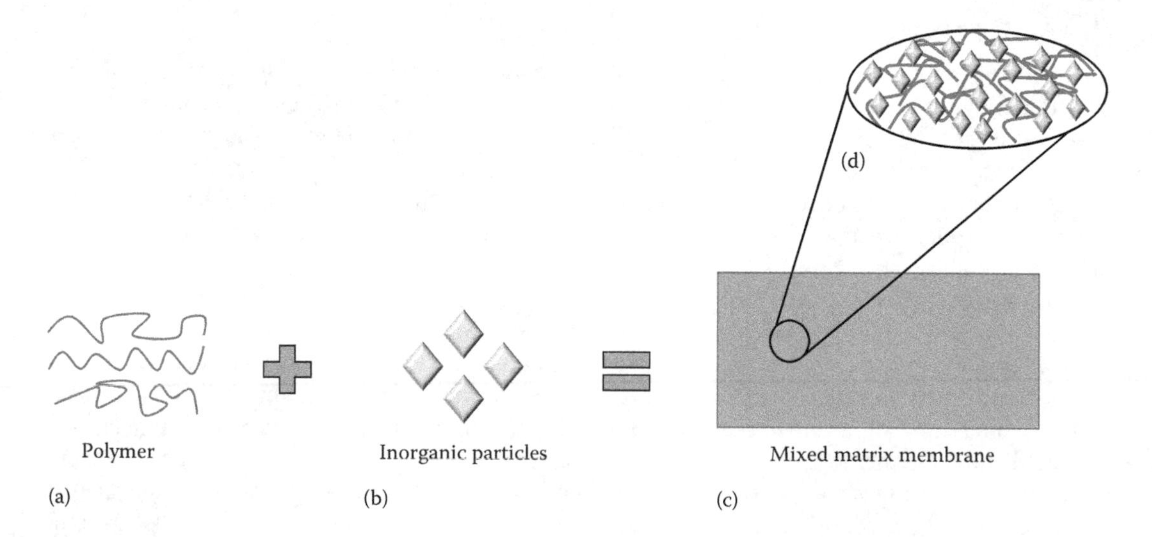

FIGURE 3.9 Mixed matrix membrane schematic representation: (a) polymer and (b) inorganic material (carbon nanotube, zeolites, etc.) used for the membrane preparation, (c) final mixed matrix membrane, and (d) specific magnified area of the mixed matrix membrane showing its content (polymer and inorganic material).

other hand, inorganic or ceramic membranes have great separation capabilities, but their preparation process is tough and costly. However, MMMs have properties of both types of membranes. A good inorganic filler in MMMs not only enhances the selectivity of the separation process but will not compromise the permeability capability of the membrane process.

The permeability of a MMM can be given as

$$P_{MMM} = P_{cp} \frac{P_{dp} + 2P_{cp} - 2\varnothing_{dp}\left(P_{cp} - P_{dp}\right)}{P_{dp} + 2_{cp} + 2\varnothing_{dp}\left(P_{cp} - P_{dp}\right)} \quad (3.1)$$

where P_{cp} represents the continuous phase permeability, P_{dp} the dispersive phase permeability, and $ø_{dp}$ the dispersed phase volume fraction. It is to be noted here that Equation 3.1 in the limits of $ø_{dp} = 0$ or 1 goes to the fitting value of P assuming that the dispersing phase is uniformly distributed through the continuous polymer phase and the two phases are in complete interfacial contact.

It is important to have materials of better compatibilities in both phases. It will negate the efficiency and performance of the membrane if the materials used are not compatible with each other. For example, compatibility of the materials used is important in for permeability, selectivity, and thus performance of the MMMs. In addition to the compatibilities of the membrane material, the processing of the materials to be casted in a membrane should also be better, otherwise it will be difficult to get a better performing membrane. The particle size of the fillers also plays an important role in the performance of the MMMs, since it governs the membrane thickness as well as the interfacial contact between the two phases. The distribution of these filler particles is also an important factor. The filler should be uniformly distributed in the continuous phase of the membrane. Equation 3.1 also demands uniform distribution of the fillers as a key assumption. The filler material may agglomerate if not uniformly distributed, and thus reduces the efficiency and performance of the prepared MMMs. This agglomeration of filler particles increases the gap between the continuous as well as the dispersive phases and thus reduces the interfacial contact between the two phases, which is important for an efficient membrane. The interfacial contact between the two phases should always be optimum, which will not result in any gap but still allow access to the filler surface or pore structure.

The future directions for MMMs should address membrane processing and performance. The main points to work out are the incomplete adhesion of the filler particles (dispersive phase) in the polymer matrix (continuous phase). There are some methods used to achieve full adhesion between the two phases, such as using nonvolatile liquids (e.g., ionic liquids), but there is still room for improvement. The particle size and particle distribution are the other important areas that need the attention of researchers. Since the membrane thickness plays a vital role in the membrane process, particles in the nanometer size range are required, and it is not easy to prepare particles on this size scale with accuracy. The shape of the prepared particles will also change when reducing their size and is not acceptable in most cases as it results in a change in particle properties, which in return affects both the membrane preparation and performance. The other area that can be explored with MMMs is the reactive separation, where a membrane with catalytic properties catalyzes and separates a product efficiently. In this area the important things to note are the membrane preparation and economic cost. In this case particles with large size and membranes with higher thickness can be used, since both of these will have a positive effect on the catalytic membrane process.

The MMMs are mostly used in gas separation applications, such as the separation of carbon dioxide and methane, pervaporation, and ionic and liquid separations.

3.6 CERAMIC MEMBRANE PREPARATION METHODS

Ceramic membranes are prepared from different materials, including zeolite, alumina, zirconia, and titania. Ceramic membranes are famous for their long stability under harsh conditions of temperature, pressure, chemical, and mechanical. Their resistance against biodegradation, easy cleaning, and catalytic activities are some of the other traits that make them eligible for a wide area of applications.

Altogether, ceramic membranes also have some disadvantages like high cost, low permeability, difficult to assemble in a membrane module, and ceramic membranes are also brittle. Ceramic membranes are generally prepared by using the following steps:

- Powder preparation (chemical powder preparations, milling)
- Shaping (pressing, extrusion, slip- or tape-casting, spinning)
- Temperature treatment (drying, calcination, and sintering)
- Layer deposition (sol-gel, chemical vapor deposition, powder suspension layers, membrane material layers)
- Functionalization (catalytic property activation or hydrophobization of membranes)

3.6.1 Sol-Gel Method

The sol-gel method is a widely used method for the preparation of ceramic membranes. This method helps in the preparation of ultrafiltration or dense membranes for various liquid and gas-based applications. The method utilizes two routes based on the preparation of the coating colloidal solution, namely, colloidal sol route and polymeric sol route. Both processes are shown in Figure 3.10. The colloidal sol route utilizes the peptization process to prepare a stable colloidal solution. In this process, precipitate of a metal alkoxide is used. This metal alkoxide is dissolved in an alcohol and hydrolyzed in the presence of water or acid. As a result, the metal alkoxide precipitates and this precipitate is kept at a high temperature for quite a good duration. This precipitate thus results in a stable colloidal solution. This colloidal solution is further used for the coating of a microporous ceramic membrane support. The drying of this coating is important because if not done properly cracks will develop in the coating. The formation of cracks can be decreased by using a binder in the coating solution. The coated ceramic membrane support is then sintered as a final step at 500°C to 800°C. The different steps involved in the process are precipitation, peptization, and sintering.

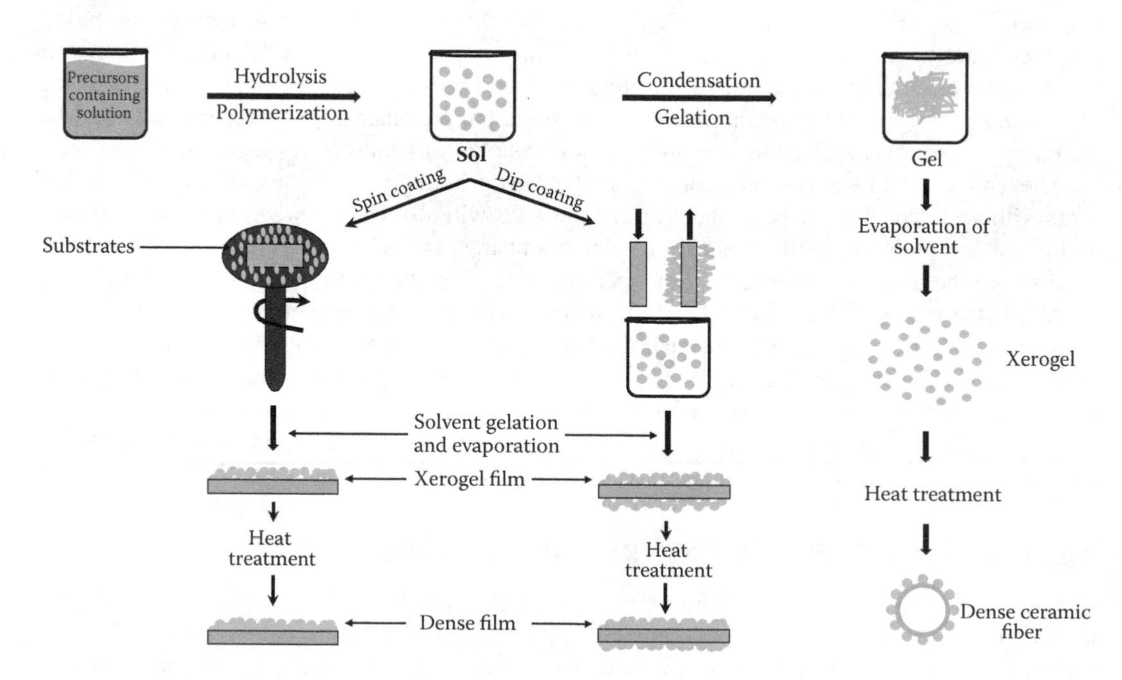

FIGURE 3.10 Schematic representation of the sol-gel method.

For the polymeric sol route, the hydrolysis of the metal alkoxide is partially hydrolyzed in the presence of an alcohol by using the least amount of water. The alkoxide active hydroxyl groups react to result in an inorganic polymer, which is coated on the microporous ceramic membrane support. The different steps involved in the polymer sol route are hydrolysis, polymerization, and cross-linking.

The sol-gel method can be used to prepare ceramic membranes with a wide range of properties based on the type of material and coating procedure used. The membrane pore size is well controlled in the technique and thus membranes with various pore size range can be prepared. The prepared membranes, thus, can be used in various applications ranging from liquid feeds to gases.

3.6.2 Dip Coating

Dip coating is the ceramic membrane preparation technique where a layer of a material is coated over a support by dipping the substrate into the suspension of the coating material. Ceramic membranes with pore sizes of the order of 100 nm can be produced using this method. The important factors that influence this method are the viscosity of the coating solution and the dipping time or speed. The process cycle of dipping, drying, and calcination can be fixed as per the requirements.

3.6.3 Pressing

Pressing is the method of flat ceramic membrane production. It is widely used at the laboratory scale and in fundamental research with ceramic membranes. Disk shapes are formulated from the ceramic membrane precursor materials (powders) and are compacted by applying high pressures, which results in the formation of the ceramic membrane after the sintering process. Pressures of the orders 20 to 100 MPa are employed by using a press machine. This method is used to explore new ceramic membrane materials and ceramic membranes as it is simple, fast, and precise.

3.6.4 Extrusion

Extrusion is the process of ceramic membrane paste flow by the virtue of compaction and shapes due to exiting through a nozzle. This method is widely used for the production of porous ceramic tubes on a large scale. In this method, the shape, size, and porosity depend on the die used for the fabrication of the ceramic membrane. The shaping of the ceramic membrane paste is only possible if it complies closely with the properties of a clay. This is ensured by adding organic compounds to the inorganic powder. The rheological properties of the ceramic membrane paste are important and thus great care should be taken regarding the ceramic grain size, type, and ratios of the organic additives. Also, the binder, plasticizer, and solvent used should be removed/evaporated completely so as to keep the final membrane in its desired shape.

Extrusion is quite similar to the fiber spinning process, but they differ in that that in extrusion the ceramic membrane paste must exhibit a plastic behavior, meaning it behaves as a rigid solid at lower stresses and softens or deforms at higher stresses. However, with spinning, a suspension is the precursor and a stable shape is the final product obtained in a coagulation bath after passing through a spinneret. The other difference between the two processes lies in the symmetry of the product formed from the two processes. The product from the extrusion process is uniform having a homogeneous cross-sectional structure and in spinning a product with an asymmetrical structure is obtained.

3.6.5 Slip Casting

The slip casting method is the most commonly used ceramic membrane preparation method. In this method, a powder suspension is poured in to a porous mold. The solvent of the well-mixed suspension thus diffuses through the pores of the mold forming a gel layer over the porous surface of the mold by precipitation of the particles. The capillary suction process of the porous substrate helps in concentrating the suspension particles at the substrate–suspension boundary. The membranes are then dried and sintered. The membranes prepared by this method are highly permeable, but the wall thickness is difficult to control in this method. This method is also time consuming due to the preparation of inorganic membranes. The gel layer thickness of the membrane prepared by this method can be given by the following relation:

$$L_G = k \cdot \frac{2\,\gamma\,\cos\alpha\,\Delta P_G}{\rho} t^{0.5} + L_A \tag{3.2}$$

where k represents a constant, γ the surface tension (liquid–vapor), α the contact angle between the liquid and solid interface, ΔP_G the pressure drop across the gel layer, ρ the liquid viscosity, t the casting time, and L_A the adhering film thickness.

The total membrane layer thickness can be calculated by the following relation:

$$L = \frac{0.94\,(\rho\,v)^{2/3}}{\gamma^{1/6}\,(\delta\,g)^{1/2}} \tag{3.3}$$

where L represents the total membrane layer thickness, v the withdrawal speed, δ the liquid specific mass, and g the gravity. It can be seen that from Equation 3.3 that thicker films can be obtained from a viscous suspension and lower withdrawal speed.

3.6.6 Tape Casting

Tape casting is also a widely used inorganic or ceramic membrane preparation method. In this method the obtained membranes are of less thickness as compared to the slip casting method. This method consists of two steps: slurry preparation and shaping. The slurry consists of a liquid-dispersed inorganic powder along with binders and plasticizers. The solvent for the slurry can be water or any organic liquid. The time taken for drying depends upon the type of solvent used. With water, the drying process is a bit longer as compared to organic liquids like alcohols or aromatic hydrocarbons. By using a mixture of solvents, the drying time can be controlled. The most important property of the solvent is its capability to dissolve all the membrane constituents. The volatile solvents are suitable for thin membrane films and less volatile solvents for thick membrane films. In addition to solvents, deflocculants are used to keep the membrane constituents stable in suspension. Their basic way of working is based on charges and steric hindrance. They are adsorbed on the membrane constituents' surface making them stable. Their effect increases with decreasing size of the membrane constituents. Phosphate ester, glyceryl trioleate, and fish oil are some of the commonly used deflocculants. Milling and sonication methods are used for the homogenization of the slurry. The viscosity of the slurry, amount of deflocculants, and time of application of these procedures are some of the parameters that affect the process efficiency of milling and sonication. Binders and plasticizers are also added to the slurry for imparting better mechanical strength to the membranes. They are usually long-chain polymers, and in addition to mechanical strength help in achieving better flexibility. The viscosity of the slurry and solvent type determines the type of binder to be used. The most common binders are acrylic compounds, polyvinyl acetate, polystyrene, and polyvinyl alcohol. The most common examples of plasticizers are polyethylene glycol and alkyl phthalate. The compatibility of both binders and plasticizers is important to obtain a membrane with good properties. Generally, they are used in couples, such as acrylic compounds with alkyl phthalate. The other important factor to note is the ratio of plasticizers to binder; it should not be more than 2. The deflocculant amount should be sufficient for complete adsorption and not more. The amount of the solvent and organic-to-inorganic powder weight ratio should be as low as much possible (0.05–0.15). The viscosity of the slurry is also an important factor; it should be in the range of 1000 to 5000 mPa s.

Shaping in tape casting is done either by moving the doctor blade while keeping the carrier stationary, or vice versa. In laboratories, a glass plate is used as a stationary carrier and the doctor blade is moved. After casting, the glass plate is placed in an oven for complete drying. With higher productions, the carrier is moved and the doctor blade is stationary. The resultant thickness of the tape depends upon the inorganic powder content, slurry viscosity, casting speed, and, most important, the doctor blade height. To reproduce the same membrane with same characteristic features, these factors should be optimized.

3.6.7 Chemical Vapor Deposition

The chemical vapor deposition (CVD) method involves deposition of a thin substrate layer over the ceramic membrane support. Chemical reactions in a gaseous medium are used to deposit layers with the same or different compositions at an elevated temperature. The method is shown in Figure 3.11. A simple CVD system consists of a chamber for metering gases having carrier and reactive roles, a reaction chamber, and a system for safe disposal of the exhaust gases. In this method, the gas mixture, which is usually a combination of hydrogen, nitrogen, or argon with reactive gases (e.g., metal halides and hydrocarbons), enters the reaction chamber, which is heated to the reaction temperature. Here, the reaction takes place and a thin layer of the substrate is deposited over the ceramic membrane support. The surface coatings can be done in various ways, including thermal decomposition, oxidation, hydrolysis, or coreduction. There are various types of CVD systems available, including plasma assisted, moderate temperature, and laser. In this method of ceramic membrane preparation, many reactive, toxic, and hazardous materials develop. These hazardous materials from a CVD system should be treated well before their release into the atmosphere. It is also important to note that a CVD system operates at subatmospheric pressures and high temperatures. Therefore, the pumping system used should be carefully monitored for the hot, reactive, and corrosive gases, as it may lead to a disaster.

3.6.8 Sintering

Sintering is usually the final step in ceramic membrane production. It is a widely studied and used method in the field of ceramic membranes. There are three main steps in a sintering process—presintering, thermolysis, and final sintering—explained in the following subsections.

3.6.8.1 Presintering

The presintering is the preheating step of the sintering process. The material goes through changes like vaporization of water present on the membrane precursor surface or from the inorganic phases, which contains the water of crystallization. In this step, importance is given to the fact that the

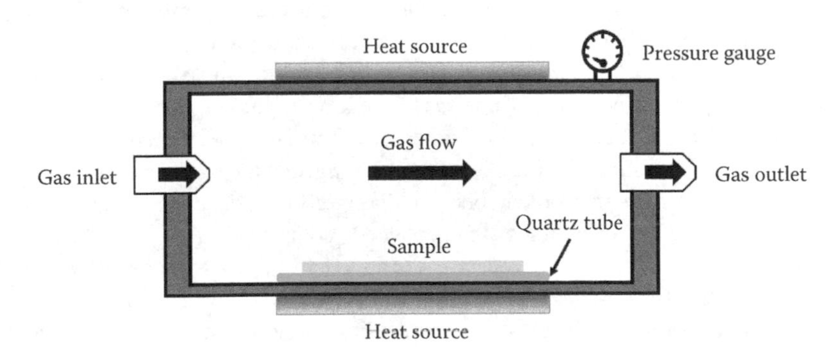

FIGURE 3.11 Schematic representation of chemical vapor deposition method.

membrane precursor should not develop cracks or fractures, as the water vaporization process imparts uneven stress on the membrane precursor. In general, presintering is said to be a step of removal of water of any type from the membrane precursor, for example, liquid present after the forming and drying of the membrane precursor, or moisture taken up by the membrane precursor after its transport and setting period. The membrane precursor is free from water after attaining temperature more than 200°C in the presintering step of the sintering process.

3.6.8.2 Thermolysis

Thermolysis is the step of the sintering process where the removal of organic components, such as binder, plasticizer, or dispersant, takes place. The presence of these organic components in the membrane precursor gives rise to membrane defects, which hamper the membrane performance. In thermolysis, controlled heating is important to control the formation of membrane defects. Therefore, a good binder and controlled heating are both important for the successful completion of the thermolysis process and to obtain a defect-free membrane precursor. The complete removal of the binder from the membrane precursor depends upon many factors like its composition and gas flow. The important factors that impact the process of thermolysis are the amount of binder, membrane precursor size and shape, thermochemistry of the binder and additives, rate of heating, and gases used. The burnout and vaporization of various organic components give rise to an internal gas pressure, which should be released effectively. For higher binder concentrations, it is slow; and it is fast when fewer binders are used. This is because with less binder concentration, the pore channels are sufficiently opened, and this is not the case with high concentration of binder. Therefore, it is said that the type and amount of binder used affects the thermolysis step the most.

3.6.8.3 Final Sintering

The final stage of sintering consists of three steps: the initial stage, the intermediate stage, and the final stage. The membrane precursor and the particles present have different features and movements at each stage, including full densification, grain coarsening, and closing of pores. The temperature regulation depends upon the membrane precursor and especially on the material type. At lower temperatures, fast densification with low grain growth takes place in the membrane precursor, and at higher temperatures it is reverse, that is, fast grain growth and low densification. The main mode of transport in this step is surface diffusion, which results in smoothing of the membrane surface, joining of particle, and rounding of pores. This results in the final ceramic membrane product.

3.7 CERAMIC HOLLOW FIBER MEMBRANES

Ceramic hollow fiber membranes (shown in Figure 3.12) are very new to the field of membrane science. Generally, polymeric hollow fiber membranes are widely used, because there was a lack in the development of a preparation method for the ceramic hollow fiber membranes. These membranes are very much needed in the applications where harsh chemical and thermal conditions prevail and polymeric hollow fiber membranes cannot be used because of their swelling and morphological changes that make them unfit for use. Hollow fiber membranes are also important because of their immense surface area per unit volume, which results in a high packing density compared to flat or tubular membranes. The first ceramic hollow fiber membranes were prepared by using the phase inversion method, which consists of three steps: (1) spinning suspension preparation, (2) spinning of the ceramic hollow fiber membranes precursors, and (3) sintering. This method is not an easy one and many factors influence the performance of the produced membranes.

The first step in the phase inversion process is the successful dispersion of the required materials in a solvent. The dispersion has three primary goals: deagglomeration of the particles, complete coating of the particles with the dispersant, and avoiding agglomeration of the particles by using charge and steric effects. The suspension is degassed after the addition and mixing of binders and plasticizers into the suspension. The binders should be such that they should increase the spinnability

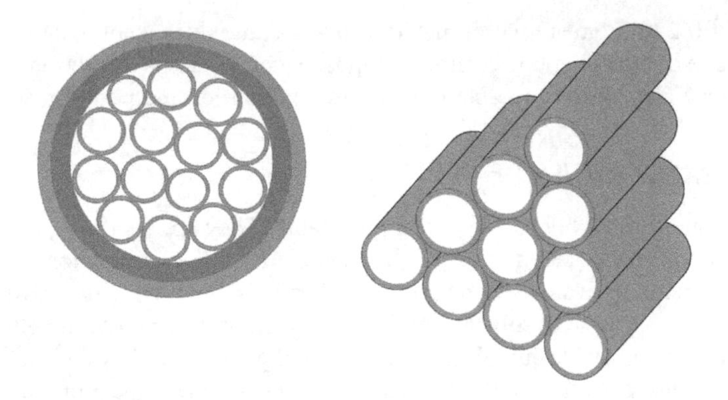

FIGURE 3.12 Schematic representation of ceramic hollow fiber membranes.

and invertability of the suspension for the successful production of the ceramic hollow fiber membranes. Degassing is important to avoid membrane defects due to the presence of air bubbles. Degassing is done by keeping the suspension in air or in vacuum for a specific period of time. The next step in the process is the spinning of the suspension for the production of ceramic hollow fiber membranes. A spinning apparatus is used for the spinning of the suspension. The suspension passes through a tube in the orifice spinneret and the rate of flow is controlled by the nitrogen gas pressure or a gear pump. The suspension is directly extruded into a coagulation bath, where phase inversion takes place and then the formed tube is passed through the washing bath and then dried. The last step in the process is sintering, where the dried ceramic hollow fibers go through three steps of the sintering process, namely, presintering, thermolysis, and final sintering. These steps are explained in the preceding sections. The ceramic hollow fiber membranes are ready for use after the successful completion of the sintering process. Nowadays, ceramic hollow fiber membranes are successfully used for gas separation and production applications.

3.8 POLYMERIC CERAMIC COMPOSITE MEMBRANES

Polymeric ceramic composite membranes (shown in Figure 3.13) are the polymer-coated porous ceramic support prepared by using various methods, including dip, spray, or spin coating, grafting, vapor deposition, or self-assembly. These composite membranes are superior in their robustness, permeability, selectivity, and antifouling nature. Polymeric ceramic composite membranes are mostly used in ultrafiltration, desalination, pervaporation, and gas separation applications. Generally, polymers such as cellulose acetate, polysulfone, polyvinyl acetate, polyvinyl pyrrolidone, polydimethylsiloxane are used for the top thin skin layer of these membranes. A ceramic support with

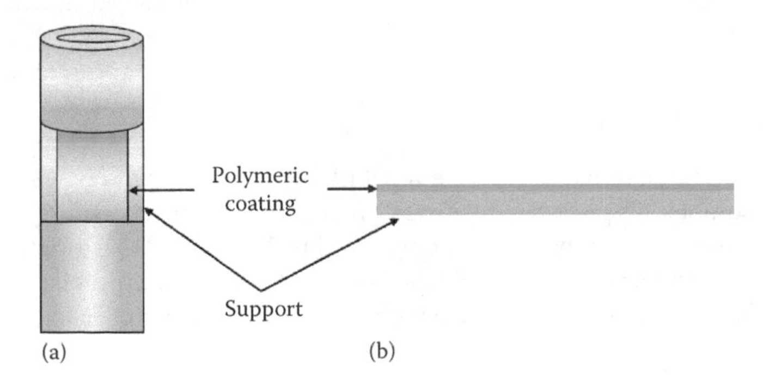

FIGURE 3.13 Schematic representation of polymer ceramic composite membrane: (a) tubular and (b) flat.

pore size range of 0.2 to 2 μm is used of materials like metal oxides, kaolin, alumina, and zirconia. The main advantage of the polymeric ceramic composite membranes is that the each coated layer can be optimized independently so as to obtain the desired properties and performance.

3.9 ZEOLITE MEMBRANES

Polymeric membranes are highly selective and permeable. They are also cheap to prepare and use. There are well-established preparation techniques available for the preparation of polymeric membranes. Polymeric membranes are used in various applications, but their main drawback is resistance to high temperature and pressure, solvents, and corrosive environment. On the other hand, ceramic membranes are robust and stable at high temperatures and pressures. They can also withstand harsh chemicals and solvents. They have a rigid structure with better mechanical properties. Due to these reasons they have longer life expectancies than the polymeric membranes. Zeolite membranes are the latest of the inorganic membranes. Zeolites are crystalline microporous aluminosilicate materials with uniformly distributed pores. This enables membranes to have shape-size-selective catalysis or separation. Zeolite fulfills the requirements of a better inorganic membrane material since it has properties like stability at high temperatures, can be acidic or basic, and have hydrophilic or organophilic nature. The properties of the zeolites can be made according to the demands of a specific application by ion exchange, isomorphous substitution, dealumination or realumination, or by addition of catalytically active sites by using alkali metals or metal oxides, transition metal ions, or complexes. The unique properties of zeolites, including molecular sieving, selective sorption, catalytic activity, with high thermal and pressure stability, make zeolites suitable for the separation as well as catalytically active membranes. Suzuki in the year 1987 [5] was the first person to synthesize zeolite membranes. Thereafter, zeolite membranes saw tremendous growth and interest from membrane enthusiasts.

Zeolite membranes over the flat or tubular supports were extensively used in gas separation, pervaporation, and catalytic reactors. Zeolite membranes have shown selectivities similar to the membranes following the Knudsen flow, which means that the membranes contain intercrystalline pores. Therefore, the quality of a zeolite membrane can be assessed on the basis of intercrystalline porosity, orientation of the crystal in accordance with the membrane layer, crystal size, thickness of the zeolite layer, and its uniformity. Zeolite membranes can be prepared by using various methods, including dip or spin coating, sputtering, chemical vapor deposition, or laser ablation. The commonly used methods for the preparation of gas separation zeolite membranes are in situ crystallization, zeolite–polymer composite, and secondary growth. In the crystallization method, a porous ceramic membrane support is coated with a thin layer of zeolite material to form a supported zeolite membrane. In the zeolite–polymer composite method, a composite is formed by adding zeolite material to a polymer matrix and this zeolite–polymer composite is coated over a porous ceramic membrane support to form a zeolite membrane. In contrast, in the recently developed secondary growth method, colloidal zeolite material is deposited over a porous ceramic membrane support, which is followed by hydrothermal synthesis.

3.10 GLASS MEMBRANES

Glass membranes are nothing but porous glass. Glass is formed of two phases: one the hot acid, water, or alcohol-soluble, alkali-rich borosilicate phase; and second, the silica phase. The porous glasses are phase-separated alkali borosilicates produced by leaching of alkali-rich borate phase. The pore size of a glass membrane ranges from 0.3 to 1000 nm. The pore size of a glass membrane mainly depends upon the glass constituents, phase separation time and temperature, and leaching conditions.

The porous glasses are better in terms of thermal and chemical resistance or stability, optical transparency, and better access to the active sites present inside the pores. Glass membranes, due to their better properties, can be used as gas detectors, catalyst support, and gas separation.

Glass membranes have been used for gas separation for the last 40-plus years, mainly by using commercially available Vycor glass material. The problem associated with these membranes is the pore size, which has huge effects for gas separation application. Therefore, membranes with less pore size are required for better gas separation applications. This can be achieved by optimizing the phase separation and leaching conditions, in addition to the membrane modifications after or during synthesis. The selectivity of the glass membranes was also not good, since the Knudsen ratio of a typical glass membrane is not sufficient. Therefore, processes for surface modifications of the glass membranes were introduced to increase selectivities of the glass membranes. The surface of the glass membranes was modified with organic functional groups by attaching these groups to the membrane surface via covalent bonds. It is important to consider here that the functional groups should have selectivities as required for the separation application.

Glass membranes are prepared from a glass, for example, with a glass having composition of SiO_2 (70%), B_2O_3 (23%), and Na_2O (7%). The glass is heated up to 500°C to 600°C for 72 h to acquire the required phase separation. This heat treatment leads to the separation of the glass material into two phases: sodium-rich borate phase, which is acid-soluble, and a silicate phase, which is insoluble in an acid. The glass was then immediately cooled to a temperature below 500°C to restrict any further phase separation to occur. The leaching process is the next step in the process of glass membrane preparation, where the glass is immersed in a hot (~90°C) acid (HCl) bath (for 3 h) to dissolve the acid-soluble sodium-rich borate phase. This will lead to the formation of pores in the glass, and a porous silica framework will remain which forms the glass membrane. This method can be used to prepare glass membranes of pore sizes up to 120 nm. The effect of heating on the glass or in the formation of glass membranes is that the longer heat treatments at higher temperatures will lead to increased phase separation, which results in pores with larger pore sizes and vice versa. This will also result in dissolution of silica phase in the borate phase, which further results in the filling of membrane pores with silica species.

3.11 DENSE MEMBRANES

Dense membranes are the membranes that are almost impermeable and allow very specific compounds to permeate. Membranes made up of metals like palladium, silver, and their alloys come under this category of membranes. Alloys are better since they are less brittle as compared to the metals. Membranes made up from these metals and alloys are impermeable to almost all compounds, but allow to permeate gases like hydrogen and oxygen. Palladium has very high affinity for hydrogen separation. The hydrogen gas permeates through the membranes as a hydrogen atom and not in the form of molecular hydrogen. The hydrogen changes to the atomic form at the membrane surface and changes to the molecular form at the other end (permeate side) of the membrane and desorbs from it.

These membranes are associated with the problem of very low flux. Therefore, to eliminate this problem, composite membranes are used with a dense top layer and a porous sublayer or support. In these membranes, a very thin layer of the metal or alloy is deposited over the support. The usual thickness of these dense layers lies in the range of 15 to 300 nm and are able to retain entities smaller than 200 to 1000 Da. A similar strategy can be used for immobilized liquid membranes, where inorganic materials can be used for adding specific separation properties to the membranes. For example, molten salts can be used for the separation of oxygen, carbon dioxide, or ammonia by adding the molten salts into the porous inorganic membranes.

3.12 MEMBRANE COST

The industrial and commercial competitiveness of any product depends on its cost; the same goes for polymeric as well as ceramic membranes. The main cost factors considered for the optimization or fixing the cost of a membrane are

- Capital costs, which includes the expenses related to the membranes, membrane framework, compression and other equipment
- Operation and maintenance costs
- Energy consumption costs

In any membrane process, capital cost is the main constituent of the total cost. Therefore, the ceramic membrane modules cost 10 times more than polymeric membrane modules. In terms of capital cost investments, ceramic membranes are a bit costly, but if their operational life, stability, and performance are considered, then they are on par with the polymeric membranes, which are a bit low on life span and stability scale. Ceramic membranes are famous for their stability and performance under harsh operational conditions. Polymeric membranes are prone to fouling and are not able to withstand harsh operational conditions. Therefore, the average life span of a polymeric membrane module is in the range of 3 to 5 years; on the other hand, ceramic membrane modules can last up to more than 10 to 20 years. This reduces the operation and maintenance costs for ceramic membranes, and thus, it can be said that it is economical to use ceramic membranes on an industrial scale as compared to polymeric membranes. There is still a need for low-cost membrane materials and membranes with low operation and maintenance costs, and researchers all over the world are trying their best to develop the best and most economical membrane separation processes. Nowadays, membranes with positive properties of both ceramic and polymeric membranes are extensively explored. There are techniques for membrane preparations, as discussed in this chapter, by which membranes having such properties can be prepared. Membrane costs can be further reduced by using low-cost membrane material and development of membranes with low fouling and stable nature. Therefore, by taking in account all these factors, membrane costs can be reduced to nominal.

STUDY QUESTIONS

1. What are the basic things required for a good membrane material? Name some polymeric as well as ceramic membrane materials with their positive properties.
2. Nowadays, cellulosic material is replaced by noncellulosic material due to its plausible biodegradation and less chemical and mechanical resistance. So, can cellulosic material be used to prepare biodegradable membranes?
3. Why is phase inversion a popular and widely used preparation method for polymeric membranes? What are the differences between its various types? Are the membranes prepared from different phase inversion methods different in morphology?
4. Josh got an order from an industry for polymeric membranes, but with a limited time period. According to you, which method should he follow to prepare the polymeric membranes in the least time possible and why?
5. Monika got a written request from her supervisor to prepare microfiltration membranes for an upcoming project. Monika successfully prepared microfiltration membranes and updated her supervisor, but then found out that her supervisor mistakenly wrote "microfiltration" membrane instead of "nanofiltration." What can Monika do to rectify the problem?
6. In the case of plasma polymerization, how can you use the percent polymerization in your favor for the separation of low molecular weight solutes?
7. Randeep's friend Adil has developed a new magnetic powder with particles in the submicron range and good dispersion properties. This material is capable of imparting magnetic

properties to the membranes, if used properly. How should this powder be used to prepare polymeric as well as ceramic membranes?

8. Cynthia prepared ceramic membranes by using the sol-gel method, but at the end found cracks on the ceramic membranes. What might be the plausible reasons?
9. Laxmi was preparing ceramic membranes by using the extrusion method, but she was not successful. Then, she asked Shaad to help her. Later, she got the ceramic membranes prepared successfully. What might have been the problem and what advice could Shaad have given her?
10. Kamal needs a thin ceramic membrane for separating proteins from a solution. Which method should he use to prepare a thin ceramic membrane? What other materials and things he will need and should be taken care of?
11. What important role does sintering play in ceramic membrane preparation? What are the things to be considered?
12. Nitesh wants to set up a low-cost water treatment plant in his community. Should he go for a polymeric membrane-based water treatment plant or ceramic membrane and why? To cut the preparation cost, which materials and methods should he use to prepare the chosen membrane?

REFERENCES

1. J. E. Cadotte, Reverse osmosis membrane, US Patent 3,926,798, 1975.
2. M. Mulder, *Basic principles of membrane technology*, Springer, 2007.
3. Richard W. Baker, *Membrane technology and applications*, 2nd ed., John Wiley & Sons, 2004.
4. H. K. Yasuda, *Plasma polymerization*, 1st ed., Elsevier, 1985.
5. H. Suzuki, Composite membrane having a surface layer of an ultrathin film of cage-shaped zeolite and processes for production thereof, US Patent 4,699,892, 1987.

4 Characterization Techniques

4.1 INTRODUCTION

The structure of a membrane is as important as its application. Since all the virtues of a membrane mainly depend on the morphology, it is necessary to analyze its structure. The characterization of the structure and membrane in particular helps to get better performance out of a membrane indirectly. Indirectly, because the performance of a membrane is directly proportional to its structure. Different applications have wide ranges of membrane requirements that mainly depend upon their structure. Therefore, membrane characterization techniques provide the best solution for applying specific membranes for exact applications. These techniques also help in identifying appropriate membrane structures for specific applications.

The characterization techniques are also important so as to make sure the membranes are defect free and have proper structural and functional properties intact, since a single defect can lead to a total loss of membrane performance. The characterization techniques are also the tools to make ensure consistency in the membranes produced over batches. The membranes should be checked for the presence of any defect or to establish a specific protocol for the preparation of particular membranes. The characterization techniques are also important in relating the membrane parameters of pore size, pore size distribution, porosity, and permeability to the area of application stated. Since different membrane applications require different membrane properties, characterization of membranes makes it easier for the user to select the most appropriate membrane for a specific application.

The characterization of membranes reveals their structural and morphological properties. Characterization of membranes is the first step after the development of membranes. Different techniques are used to characterize the prepared membranes in order to know different parameters. The characterization techniques vary from membrane to membrane with application involved. These are varying for porous and nonporous membranes. These two membrane types are shown in Figure 4.1. The pores found in different membranes are characterized on the basis of pore size (diameter) as macropores (<50 nm), mesopores (2 nm–50 nm), or micropores (<2 nm) [1]. The first two pore types are representative of microfiltration and ultrafiltration membranes. The last one outlines membranes for nanofiltration, reverse osmosis, and gas separation application. Usually, the dense or gas separation membranes lacks the presence of any fixed pores and thus come under the nonporous category. Other than the porous and nonporous nature, the type of material, solvent/nonsolvent system, fillers, and so forth used for the preparation of membranes also define the performance as well as characteristic parameters. The physical parameters like pH, temperature, and operating pressure are also crucial for ascertaining the characterization techniques to be employed.

An important aspect to notice in membrane processes based upon their characterization is that the membrane performance differs significantly (practically) from the theoretical conventions. The reason behind this difference in a membrane's performance is the presence of concentration polarization and fouling. Also, to fit the data to various physical equations (Poiseuille and Kozeny-Carman) [1–3] during the measurements of pore size and pore size distribution some assumptions are made regarding the membrane pores (geometry), which are only representative and absent in reality. Therefore, these contradictions in pore geometries also add to the difference in the membrane performance, for instance, the pore size and pore size distribution of a particular membrane, as shown in Figure 4.2. The figure contains pore size and pore size distribution of polysulfone membranes M1, M2, and M3, modified with poly(ethylene glycol) (PEG) of different molecular weights 400 Da, 2000 Da, and 6000 Da, respectively. It can be seen from the figure that the use of a high molecular weight additive increases the pore size of a membrane.

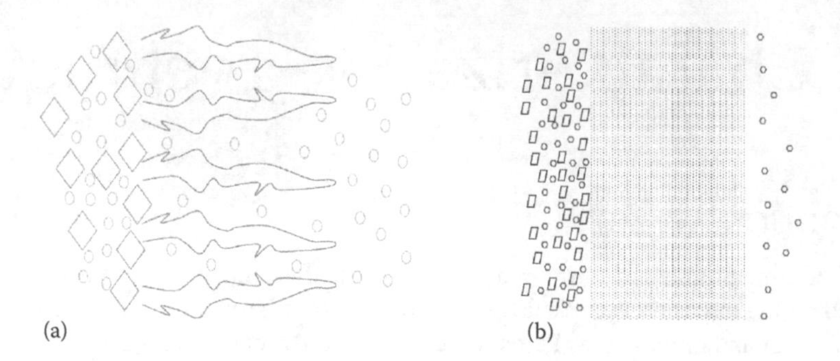

FIGURE 4.1 Schematic representation of (a) porous and (b) nonporous membranes.

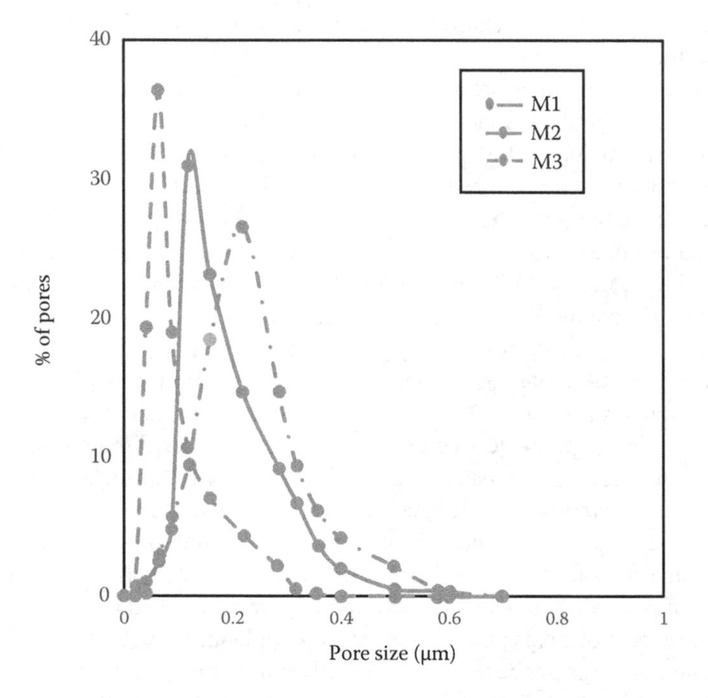

FIGURE 4.2 Pore size and pore size distribution of PEG modified polysulfone membranes.

In this chapter, the membrane characterization techniques are basically presented under two categories: first, based on morphological aspects of the membranes and microscopy techniques, and, second, based on the permeation factors and the permeation techniques. Though sometimes it is difficult to relate the two as the pore size and shape are inconsistent in nature and in some characterization techniques several assumptions are made regarding the two. Figure 4.3 is a representation of the fact regarding the pore size and shape (actual and model). Therefore, conclusive interpretations ought to be made after employing a combination of morphological and permeation characterization techniques. Other techniques are also available to characterize the membrane materials and other physical properties of the membranes. Techniques like x-ray photoelectron spectroscopy, thermogravimetric analysis, x-ray diffraction, and Fourier transform infrared spectroscopy are also covered in this chapter. These techniques help in analyzing the components present in the membranes or the components used in manufacturing of the membranes. These techniques also reveal almost everything about the physical properties of the membranes like stability, performance, different conditions of pH, temperature, and pressure.

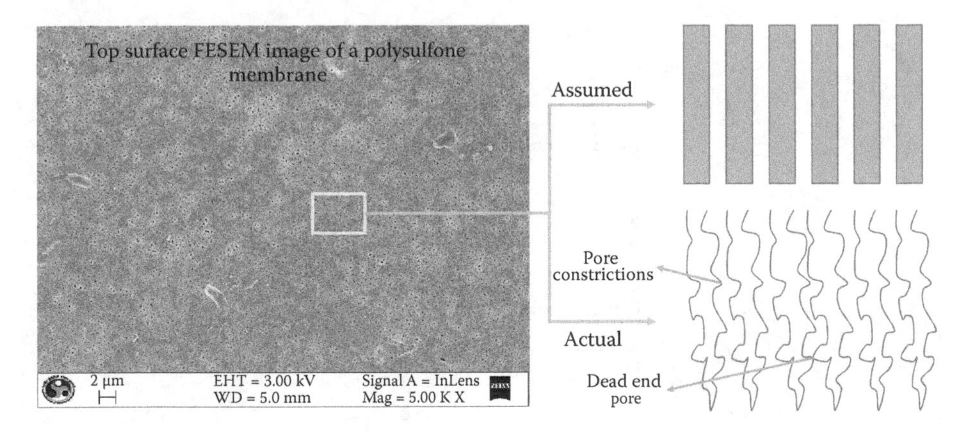

FIGURE 4.3 Representation of actual and assumed pores.

4.2 MORPHOLOGICAL ASPECTS

This section will discuss characterization techniques related to the structure and morphology of the membranes. The characterization techniques under this section are divided into two subsections: microscopy and permeation techniques. The microscopy techniques include scanning electron microscopy, field emission scanning electron microscopy, confocal laser scanning microscopy, and atomic force microscopy. The permeation techniques include characterization of the membranes by gas permeation, water permeation, liquid–liquid displacement porosimetry, molecular weight cut-off, hydraulic permeability, and equilibrium water content. The microscopy techniques are the impermeable kind of characterization techniques that tells about the membrane characteristics by analyzing only the membrane surface. On the contrary, permeation techniques are permeable techniques in which some material in a liquid or gas phase is permeated through the membrane so as to evaluate the membrane characteristics.

4.2.1 Microscopy Techniques

Microscopy is the field of science that deals with the use of microscopes to view objects that are not visible under normal light and sight. The microscopy techniques can be anywhere from millimeters to nanometers in range. The microscopy techniques are further divided into three categories: optical, electron, and scanning probe microscopy. The three are briefly described next.

Optical microscopy is the field of microscopy that uses visible light and one or more lens to produce a magnified image of an object. Optical microscopy, because of the principle use of light, is also known as light microscopy. An object is placed in the focal plane of the lens so as to get a magnified image of the object (Figure 4.4). The beam of light can either transmit through the sample or reflect off the sample surface (in the latter case it is known as reflected light microscopy). There are many applications of an optical microscope in the fields of physics, biology, and medical science for testing of different samples, but in the case of membrane science, optical microscopy is rarely used. This is because of their limited resolution, lower magnification, and poor sample surface view.

Electron microscopy is the field of microscopy that uses an electron beam instead of visible light, as is the case in optical microscopy, for the construction of a magnified image of an object under study. The advantage of electron microscopy over optical microscopy is its higher magnification and resolution. This is the outcome of the use of electron beams instead of visible light, and as a result much smaller specimens can be seen with greater detail. Electron microscopes generally classified under two categories: scanning and transmission electron microscopes (Figures 4.5 and 4.6). The difference

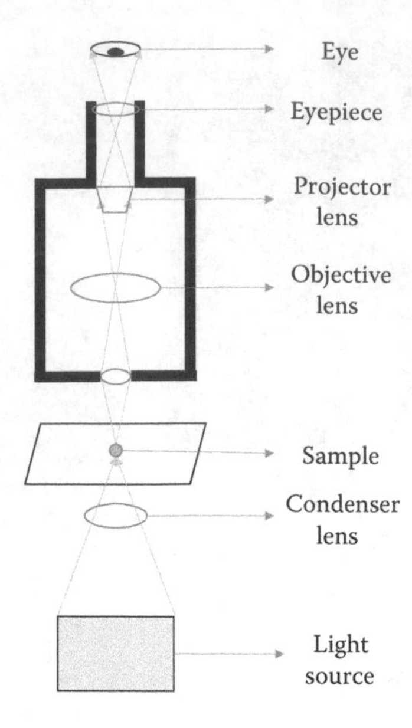

FIGURE 4.4 Working principle of an optical microscope.

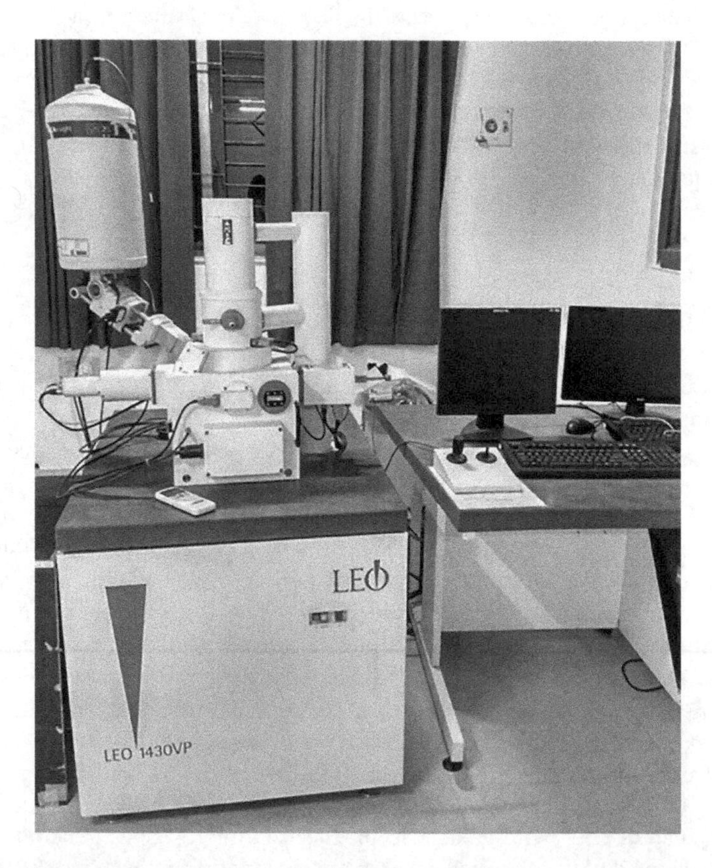

FIGURE 4.5 Scanning electron microscope. (Courtesy of CIF, IITG.)

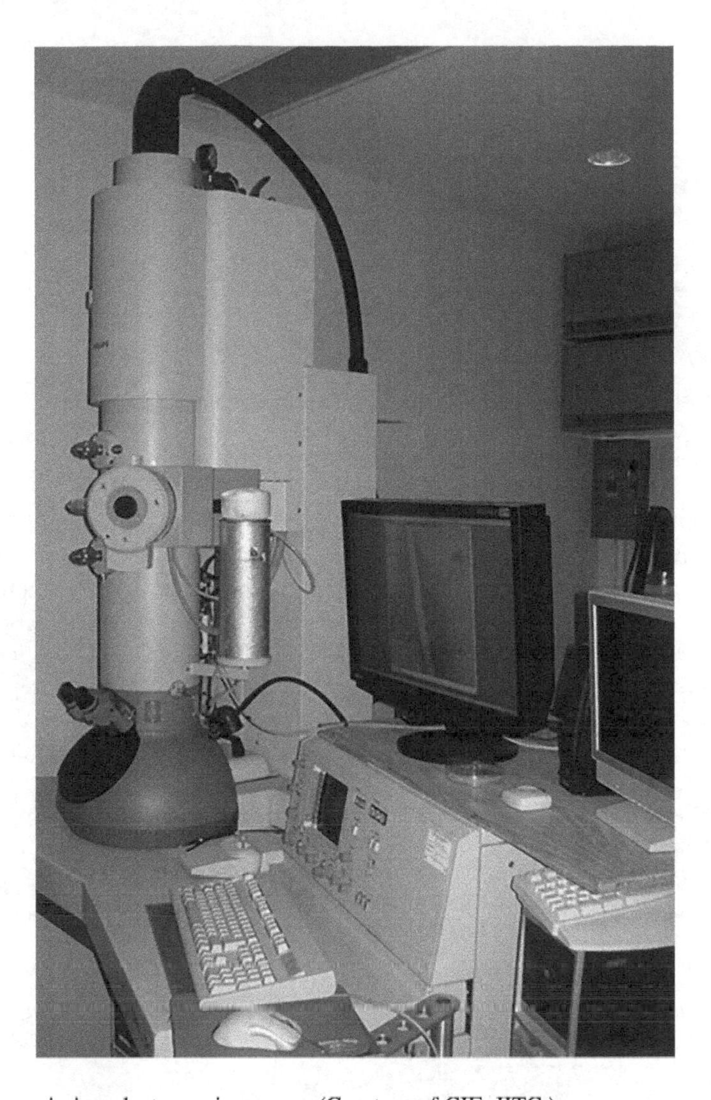

FIGURE 4.6 Transmission electron microscope. (Courtesy of CIF, IITG.)

between the two microscopes can also be seen from their names. Scanning electron microscope means it scans, and transmission electron microscope means it transmits. The actual difference lies in the method of image acquisition. The scanning electron microscope uses a very fine electron beam that is focused to a fine point on the sample surface and detects the reflected electrons from the sample surface to create an image. These reflected electrons are known as secondary or backscattered electrons. On the other hand, the transmission electron microscope uses a broad static electron beam over the sample and collects or detects the transmitted electrons from the sample. Thus, it can be said that the scanning electron microscope shows the sample surface and the transmission electron microscope can show the inside of the sample (providing the sample is very thin). In addition to advantages, electron microscopy also has disadvantages, such as laborious sample preparation, higher cost, need for a special operator, and special housing and maintenance requirements. In general, a transmission electron microscope is not used for membrane characterization due to its thin and trivial sample requirements. On the other hand, a scanning electron microscope has sample requirements that are convenient and easy.

Scanning probe microscopy is the field of microscopy that uses a probe to scan samples to get an image of the same. In this technique a sharp conducting probe is brought near to the sample surface

FIGURE 4.7 Atomic force microscope. (Courtesy of CIF, IITG.)

(in some cases contacted with the sample surface) and the surface of the sample is scanned. Some examples of this microscopy are atomic force microscopy (AFM) and scanning tunneling microscopy (STM). This microscopy provides higher magnification as well as a 3D view of the samples. This technique not only delivers high magnification images but also specimen properties, response, and reaction on touch or stimulation of the samples. Other advantages of the techniques are the observation of the sample in its native state, and easy, fast, and efficient sample preparation. The only drawback of this technique is that the image acquisition takes more time as compared to other microscopy techniques. Generally, AFM (Figure 4.7) is used in the field of membrane science.

4.2.1.1 Scanning Electron Microscopy

The first true scanning electron microscope was invented by Menfred von Ardenne in 1937. The principle of scanning electron microscopy (SEM) is shown in Figure 4.8. It can be seen that a narrow

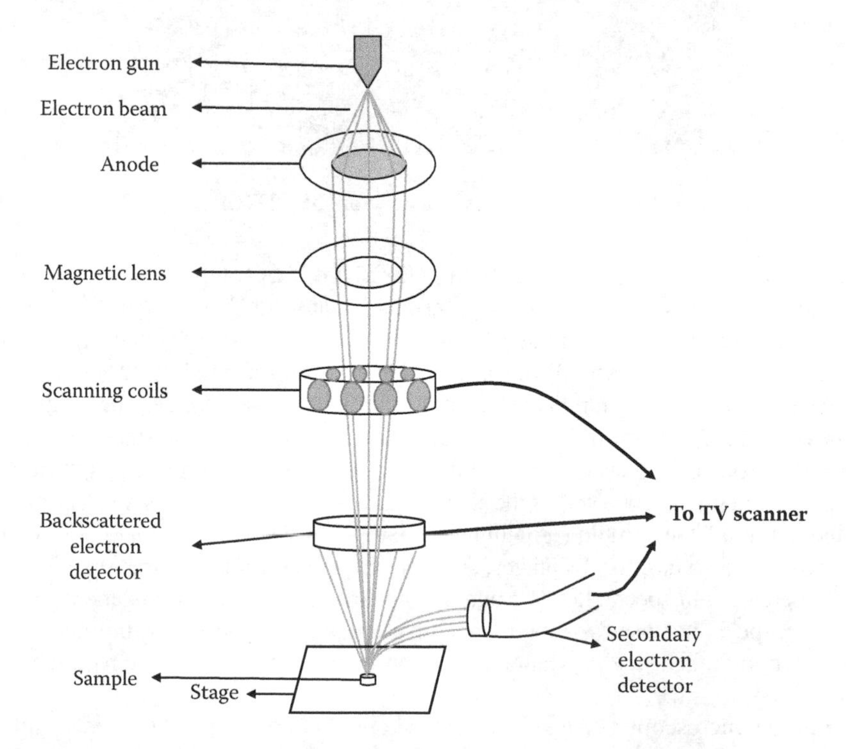

FIGURE 4.8 Working principle of scanning electron microscope.

beam of electrons from an electron gun (generally, a thermionic electron gun) is targeted on the sample surface with an acceleration voltage of 1 to 20 kV. The bombarded electrons, known as primary electrons (PEs), interact with atoms of the sample and provide energies to the electrons of the atoms. These excited electrons are emitted from the sample, and now known as secondary electrons (SEs) are detected by the secondary electron detectors. The electron beam scans the sample surface in a line-by-line rectangular pattern known as a raster scan pattern. The common material used for thermionic electron guns is tungsten since it has the highest melting temperature and lowest vapor pressure among all the metals and also because of its low cost. SEM has different modes of operation, and detection of secondary electrons is the most common. Other modes are detection of backscattered electrons (BSEs) and x-rays. To detect these different electrons or x-rays, different detectors are employed and the SE detector is the standard detector present in all SEM machines. BSEs are the electrons that are emitted from the sample by elastic scattering. Unlike SEs, BSEs are emitted from the depth of the sample, and thus carry less energy and result into poorer images as compared to SE images in terms of resolution. Due to their source of origin and relevance to the sample atomic number, BSEs are used for elemental analysis along with sample characteristic x-rays. Thus, BSE images can provide details of the elemental distribution in a sample. Characteristic x-rays are emitted from a sample when an inner shell electron is emitted by the PE, which results in filling of the vacancy by a higher energy electron and release of energy in the form of x-rays. The detection of these sample characteristic x-rays is used in elemental analysis of the sample, which provides information about the composition and number of elements present in a sample. This is known as the energy dispersive x-ray spectroscopy (EDS or EDX) technique of sample analysis and SEM-EDS when the cause of origin of x-rays is SEM or in simple words when SEM is combined with EDS. Currently, SEM has a resolution near to 1 nm. Generally, the samples are analyzed under vacuum conditions. Membrane samples are also analyzed under high vacuum conditions. In some cases they are observed under low vacuum conditions or wet conditions (in the case of environmental SEM [ESEM]) depending upon the sample type. The analysis of pollutants present in wastewater ESEM can be used. ESEM helps to study the samples in their native state and due to the absence of harsh conditions the sample remains safe. Other attachments are also available for SEM, like attachments for observing samples at various temperatures ranging from cryogenic conditions to higher temperatures, but are seldom used for membranes.

The sample size should be small enough to fit on the stub or sample holder. The sample should also be conductive in nature or at least at the surface and must be electrically grounded to avoid accumulation of electrostatic charge. The samples, mainly nonconductive, when analyzed under SEM face the problem of charging or burning due to high intensity of the electron beam focused at a small sample surface area. This damages the samples and makes it nearly impossible to take good images of the samples. Therefore, a conductive adhesive is used to place the sample on the stub. Carbon tape is the most commonly used conductive adhesive used for placing the sample on the stub. Further, the samples are coated with an electrically conductive material by the sputter coating method. Gold, gold–palladium alloy, platinum, chromium, iridium, tungsten, and graphite are some common conductive materials used for sample coating [1].

SEM permits vibrant and characteristic 3D images of the samples due to the use of a very narrow electron beam, which results in a very large depth of field. The extensive range of magnifications available from 100× to 100 k× allows the sample to be observed under much detail. Therefore, detailed analysis can be carried out for membrane samples, top surface, cross section, or the bottom side. This makes it easy to see any asymmetry in the structure and morphology of the membranes, and thus helps in the selection of membranes with the desired structure and morphology. The pore size, pore size distribution, and porosity of the membranes can also be obtained by using these images, since the pores and their geometry are very clearly visible.

SEM operates at a comparatively higher operating voltage (≥20 kV) and wider electron beam, which results in charging, burning, and lower resolution (at higher magnifications) of membrane samples. The top surface images require high magnification and that is why SEM is not preferred and

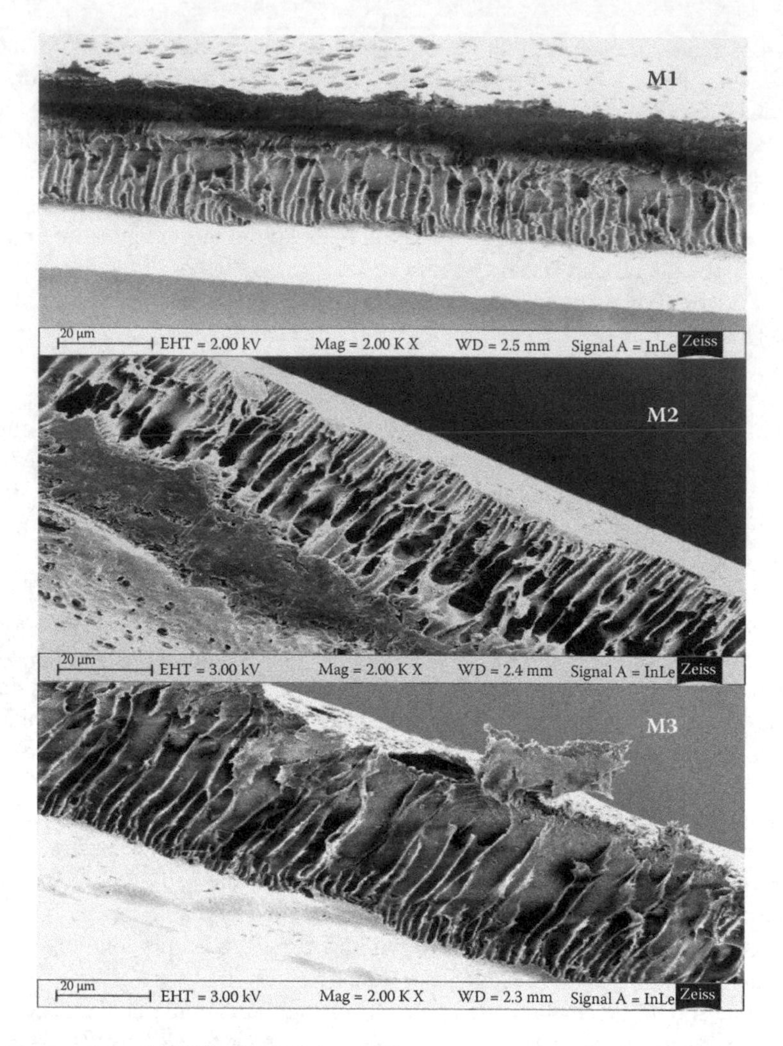

FIGURE 4.9 Cross-sectional SEM micrographs of PEG-modified polysulfone membranes.

is commonly used for acquiring cross-sectional images of membranes. Figure 4.9 shows cross-sectional images of polysulfone membranes. PEG of molecular weight 400 Da, 2000 Da, and 6000 Da were used to modify the polysulfone membranes M1, M2, and M3, respectively. The membranes show better pore and channel formations with higher molecular weight PEG. To obtain good quality top surface images of the membrane samples, field emission electron microscopy is commonly used, which has better resolution at higher magnifications.

4.2.1.2 Field Emission Scanning Electron Microscopy

A field emission scanning electron microscope is shown in Figure 4.10. Erwin Wilhelm Muller invented it in 1936, with a resolution of 2 nm. He was the first person to observe atoms experimentally because of this invention. A field emission scanning electron microscopy (FESEM) is a higher version of SEM in terms of magnification and resolution. The basic working principle and mode of working is the same. The main difference, which is also the main reason for other differences among SEM and FESEM, is the electron gun used. In FESEM, a field emission gun is used as compared to the thermionic gun for SEM. The field emission gun needs no thermal energy for

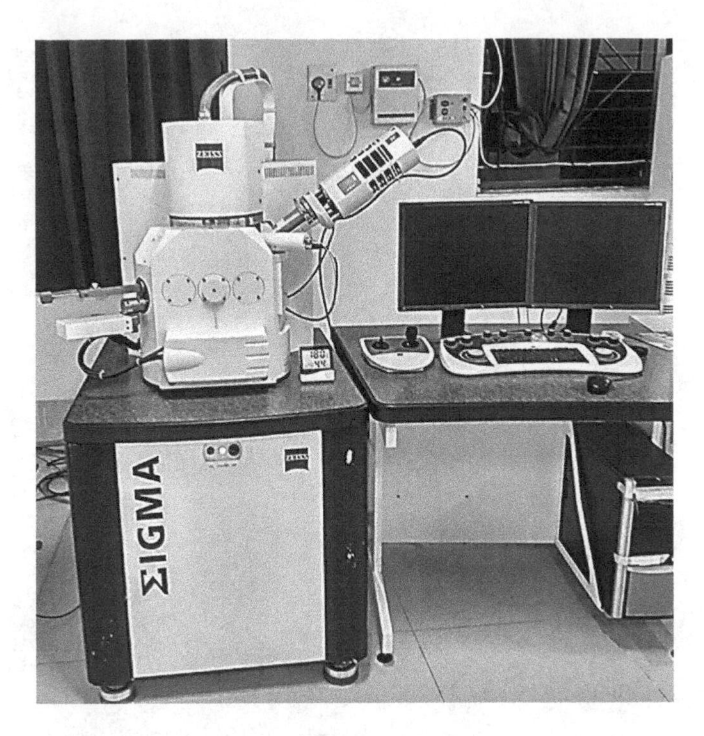

FIGURE 4.10 Field emission scanning electron microscope. (Courtesy of CIF, IITG.)

electrons to surpass the surface potential barrier. The electrons are emitted due to the application of a very high electric field to a metal surface under ultrahigh vacuum. The metal electrons surpass the surface barrier due to tunneling instead of a thermionic effect. The field emission gun electron beam is the most intense; it is also brighter than the thermionic gun. The electron beam is narrower as compared to SEM and thus results in a better image quality and higher magnification. The FESEM has near atomic resolution (0.2–0.6 nm). In FESEM the common mode of operation is the detection of SEs. BSEs and x-rays are also detected and used in elemental analysis of the samples.

There are fewer chances of burning and charging with FESEM as compared to SEM due to the use of a low operating voltage (0.5–10 kV). Still, the samples are sputter coated with conductive materials as in the case of SEM for extra protection and better image acquisition. The high resolution, magnification, and the resultant better image quality are the reasons the top surface images of the membrane samples are preferred to be taken by FESEM. The greater in depth and better image quality makes it easy to analyze the morphology and structure of the membrane samples. The pores and their geometry can be seen very easily and clearly, which further helps in the calculation of pore size, pore size distribution, and porosity of the membranes. Figure 4.11 shows top surface images of polysulfone membranes, where the pores can be seen very clearly. The polysulfone membranes M1, M2, and M3 are modified with PEG of different molecular weights 400 Da, 2000 Da, and 6000 Da, respectively. Among the three membranes, membrane M3 shows improved pore formation, pore size, and porosity as compared to the other two membranes. Thus, it can be stated that the higher molecular weight additives are better for the modification of membranes. The results of Figures 4.2, 4.9, and 4.11 are in agreement with the statement.

4.2.1.3 Confocal Laser Scanning Microscopy

Confocal laser scanning microscopy (CLSM) or confocal microscopy is a microscopy technique used to obtain 3D images of the samples. Marvin Lee Minsky, the cofounder of Artificial Intelligence Laboratory at the Massachusetts Institute of Technology, was the inventor of the confocal microscope.

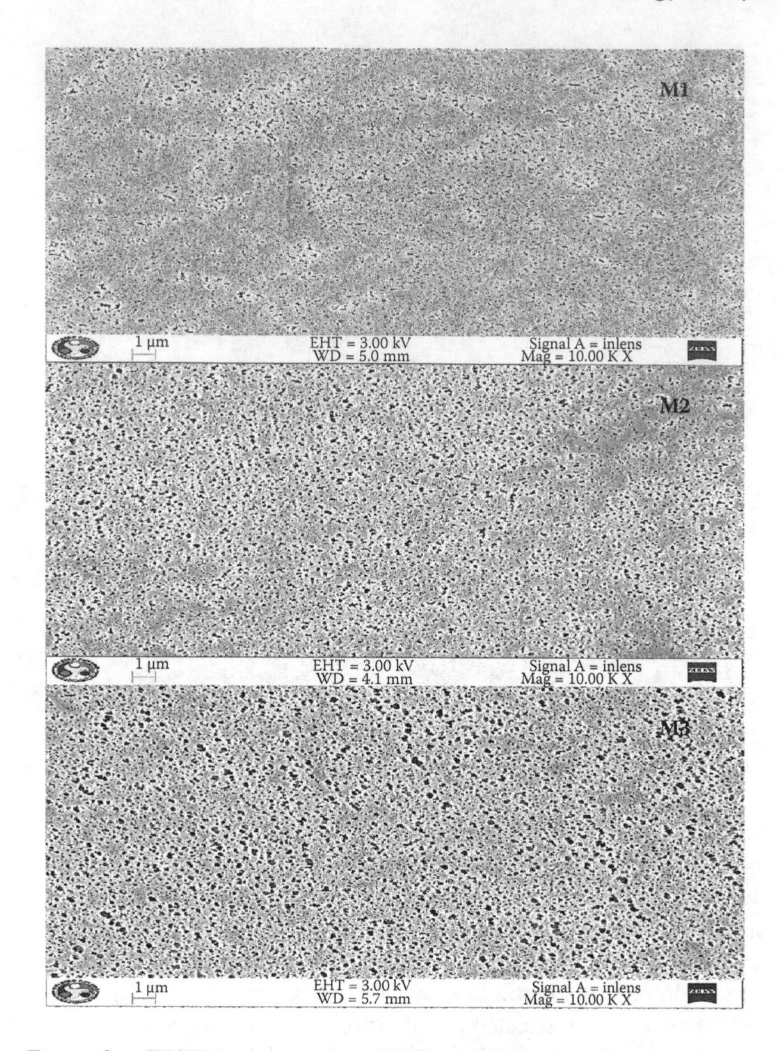

FIGURE 4.11 Top surface FESEM micrographs of PEG-modified polysulfone membranes.

The confocal microscope, a predecessor to CLSM, was patented by him in 1957. The principle CLSM is totally different from other microscopes or microscopy techniques (Figures 4.12 and 4.13). In CLSM, a laser light is used to scan the sample surface so as to produce a 3D image. The use of laser has given its name CLSM. The directed laser beam on the sample surface (known as the focal plane) is reflected and detected by the detector. A pinhole aperture is placed just in front of the detector so as to block the light reflected from out of the focal plane. Therefore, only the light reflected from the focal plane in the sample is detected. The pinhole aperture can block most of the reflected light. Hence, a strong laser beam is required for illumination. With CLSM, photomultiplier tubes are used as detectors. These tubes convert light signals to electrical signals for image production. Raster scanning is used for SEM and FESEM for the scanning of the sample. The resolution of CLSM depends upon the size of the laser beam focal spot. A resolution of 0.2 µm can be achieved. CLSM also helps to view not only the sample surface but also inside of the sample, depending upon the transparency of the sample. Therefore, not only the surface structure and morphology but also inside of the membranes can be analyzed by CLSM.

The best part of CLSM is that it can be used to detect fluorescence and is mostly used only in this configuration, because fluorescently labeled sample features can easily be shown compared to

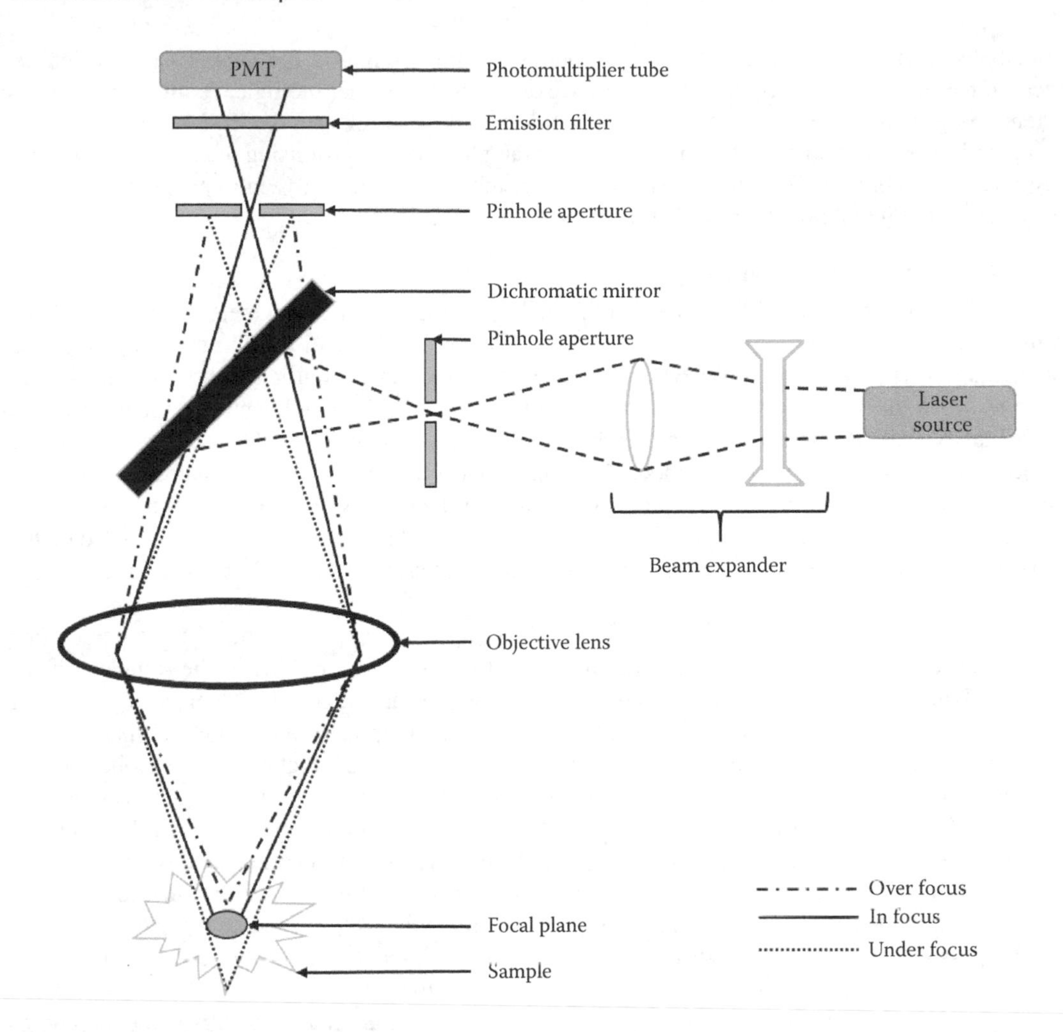

FIGURE 4.12 Working principle of confocal laser scanning microscope.

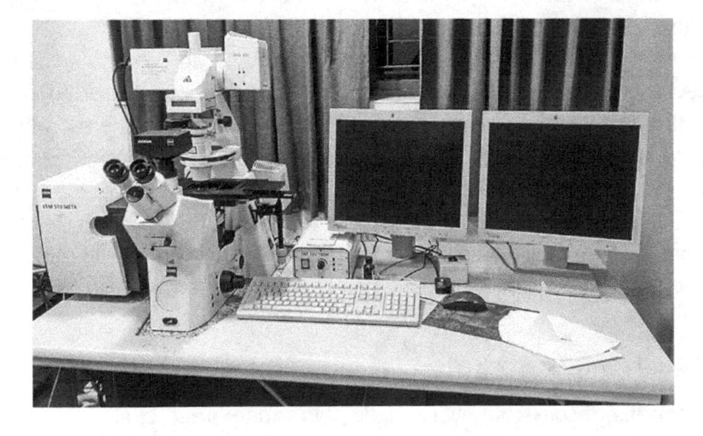

FIGURE 4.13 Confocal laser scanning microscope. (Courtesy of CIF, IITG.)

nonlabeled features. Although major applications of this technique are in biological sciences, membranes can also be analyzed for the presence of microbes or other biological entities. Membrane science is growing fast and vast in various fields and applications, thus CLSM is an important characterization technique to be considered. For example, this is a promising technique for stimuli-responsive membranes. The stimuli response of a membrane can be analyzed by tagging the stimuli response of the membrane with fluorescence.

4.2.1.4 Atomic Force Microscopy

Atomic force microscopy (AFM) is used for the analysis of the top surface of materials especially thin films. The atomic force microscope was invented in 1982 by IBM scientists. (The scanning tunneling microscope was also invented by IBM scientists Gerd Bining and Heinrich Rohrer for which they were awarded the 1986 Nobel Prize in Physics.) In the AFM technique a cantilever is used with a probe; this sharp probe scans the surface of the samples. The tip of the probe scans the sample surface in contact, noncontact, or tapping mode. The tip is deflected by the forces created according to Hooke's law [4], when the tip of the probe and sample surface is near or in contact with each other. This deflection of the cantilever in turn changes the reflection of a laser beam that was targeted on the top surface of the cantilever and detected by a detector. This variation in the laser beam is the measure of the applied forces.

In contact mode, the probe tip is in touch with the sample surface. The change in reflection of the laser beam due to deflection of the cantilever is used to measure the outline of the sample surface. The disadvantage of using the contact mode is that the sample surface could be damaged by the tip or the probe tip could be damaged by the high roughness of the sample. To avoid this, another mode of AFM is available, which is known as a noncontact mode. In a noncontact mode, the probe tip does not touch the sample, but the cantilever vibrates slightly above its resonance frequency above the sample surface and in turn the probe tip also vibrates on the sample surface. Forces like van der Waals decrease the resonant frequency of the cantilever. The oscillation amplitude is kept constant by using a feedback loop that helps in changing the distance of the tip from the surface. This difference between the probe tip and sample surface at each point helps AFM software to produce a topographic image of the sample surface. The disadvantage of the noncontact mode is that it will not be able to give accurate results with a moist sample, because the probe tip sticks to the moisture when making an approach for the detection of short range forces. An alternative to this is the tapping or dynamic contact mode. In this mode, the cantilever is mounted with a piezoelectric element that helps it to oscillate near to its resonance frequency. The forces decrease the amplitude of the cantilever as it approaches the sample surface. The amplitude is kept constant by regulating the height of the cantilever. The advantage of the tapping mode over contact mode is that it sustains the quality of the contact mode without damaging the sample.

In the AFM analysis, no pretreatment of samples is required and samples can be observed under their native state. Also, since AFM operates at very low forces it can be used to analyze soft samples as well. With membranes, AFM is used to analyze the top surface of the membrane samples. This reveals the extent of smoothness or roughness, pores, pore geometry, pore size, and pore size distribution in membrane samples [5]. Figure 4.14 shows a 3D AFM image of a polysulfone membrane modified with PEG 6000 Da. The high molecular weight additives give rise to high membrane roughness as compared to low molecular weight additives. In the figure, the dark spots are depths or pores present in the membranes and the raised bright spots are the walls of the pores.

4.2.2 Permeation Techniques

In the preceding sections, discussions were made about membrane characterizations with different microscopy methods. Those techniques revealed information about the membranes which is visible to the eye. The morphological and structural properties are pore size, pore geometry, pore size

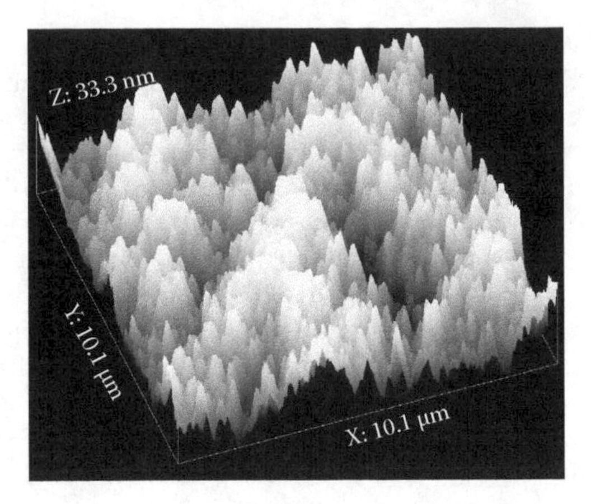

FIGURE 4.14 AFM 3D micrograph of a PEG-modified polysulfone membrane.

distribution, and pore number. The techniques are superficial in the sense they can only analyze the membrane surface and not give in-depth detail about the sample. Blind pores are also considered for calculations of pore size distribution and pore number. The instruments used for these analyses are very expensive and the methods of sample preparation are not easy. There is also a special requirement of operators for the analyses. These are some of the shortcomings of the characterization techniques previously discussed. Therefore, there is a need for techniques that are easy, cheap, and accurate. Permeation-based characterization techniques answer this call.

Permeation-based characterization techniques for membrane characterization are, first, nondestructive in nature, easy to use (and thus require no special operator), cheap, and need minimum special equipment. These technique uses either pure water or a gas for the analysis of a membrane and gives results like pore size, pore size distribution, pore number, and porosity of a membranc with maximum accuracy, as the blind pores are not taken into account.

Though permeation techniques are accurate to a maximum extent, they also have some shortcomings. The main shortcoming is the assumptions that are made to fit the data in the respective equations for the calculation of various membrane parameters. These assumptions are not always correct and thus the calculated values differ from the actual values. Still, the variations are not much and give a proper account of a membrane under consideration. Thus, these techniques are useful for comparison purposes and analysis of various parameters of membranes. The following sections cover the techniques of gas permeation, liquid–liquid displacement, pure water flux, hydraulic permeability, and equilibrium water content.

4.2.2.1 Gas–Liquid Permeation

Gas permeation is the technique in which a gas, usually an inert gas, is used for the characterization of membranes. The membranes are permeated with a gas at different pressures and times to evaluate different membrane parameters. The liquid–gas combination is also used to analyze the membrane parameters for better and accurate results. The main advantage, as discussed in the preceding section, is that this technique does not count dead-end pores in the analysis because the gas or liquid is not able to permeate through the blind pores. This technique analyzes the membranes under actual membrane process conditions, which is also good in itself. In the previously discussed techniques, this was not the case and the actual picture about the membrane performance could not be realized. Therefore, this technique is reliable and accurate for the membrane characterization. This section is divided into the gas-based gas permeation method, liquid-permeation based bubble-point method, and the combination of the bubble-point and gas permeation method. The following sections also explains the working principle, advantages, and limitations of the individual methods.

4.2.2.1.1 Gas Permeation Method

The gas permeation method or test is a standard procedure to determine the membrane pore size. In this method the mean pore radius is calculated by using gas permeability as a function of the mean transmembrane pressure. A gas is permeated through the membrane and the gas flow rate across the membrane is measured at different transmembrane pressures with the help of a rotameter. Figure 4.15 shows the scheme for this method. To calculate the mean pore radius, the following assumptions are made about the mathematics of gas transport through the membranes [6–10]:

1. The pores in the skin layer are cylindrical pores of radius r and length l.
2. The sum of the Fickian diffusion flow in the dense polymeric skin, the Knudsen, and Poiseuille flows in the pores.
3. The permeability coefficients are independent of pressure.
4. A skin of uniform thickness rests on a porous support with a negligible flow resistance.

The total gas permeation rate (m^3s^{-1}) is given by the following equation:

$$Q_g = L_g\Delta p(1-\varepsilon)\frac{A}{l} + \frac{2r_m\varepsilon}{3l}\sqrt{\frac{8RT}{\pi M}}\frac{\Delta pA}{P_0} + r_m{}^2\varepsilon P_m\frac{\Delta pA}{8\eta_g lP_0} \tag{4.1}$$

where the first term of represents the Fickian diffusion flow, the second term the Knudsen molecular flow, and the third term the Poiseuille compressible flow. Also, L_g represents the permeability coefficient of the gas used, r_m the mean pore radius (m), ε the porosity, l the pore length (m), R the universal gas constant (8.314 m³ Pa mol^{-1} K^{-1}), T the absolute temperature (K), M the molecular weight of the gas, ΔP the transmembrane pressure (Pa), A the membrane area (m^2), P_o the atmospheric pressure at which Q_g is measured, P_m the mean pressure (Pa) of the upstream and downstream of the membrane, η_g the viscosity of the gas (Pa.s), p_1 the applied pressure (Pa), and p_2 the downstream pressure (Pa). By neglecting the Fickian diffusion flow, the gas flow rate through a membrane is expressed in terms of Poiseuille flow and Knudsen flow, and the governing equation is as follows:

$$Q_g = \frac{2r_m\varepsilon}{3l_r}\sqrt{\frac{8RT}{\pi M}}\frac{\Delta pA}{P_0} + r_m{}^2\varepsilon P_m\frac{\Delta pA}{8\eta_g l_r P_0} \tag{4.2}$$

$$\Delta p = p_1 - p_2 \tag{4.3}$$

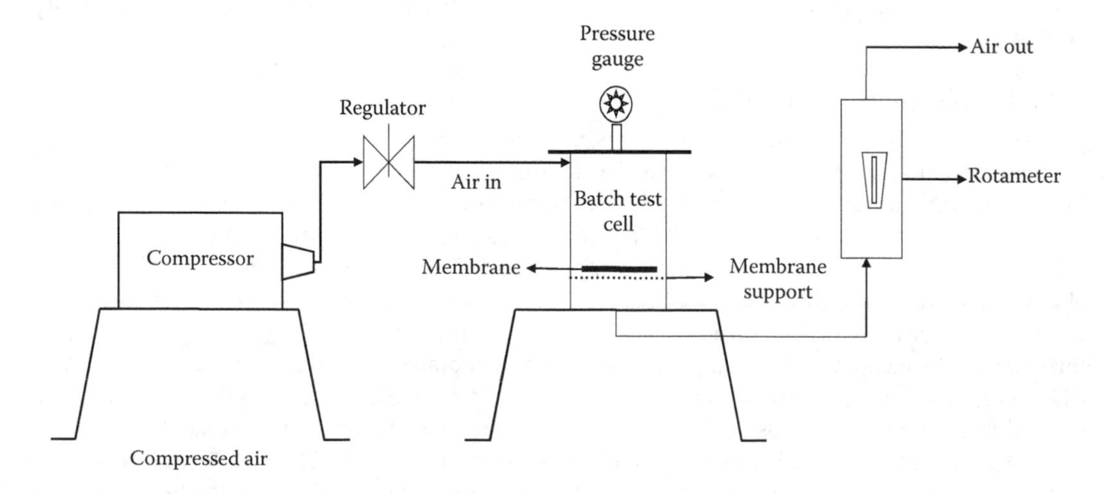

FIGURE 4.15 Schematic diagram of a gas permeation setup.

After rearranging, Equation 4.2 can be written as

$$\frac{Q_g P_0}{\Delta p A} = \frac{2 r_m \varepsilon}{3 l_r} \sqrt{\frac{8RT}{\pi M}} + \frac{r_m^2 \varepsilon P_m}{8 \eta_g l_r} \tag{4.4}$$

where

$$P_m = \frac{(p_1 + p_2)}{2} = \left(\frac{\Delta p}{2} + p_2 \right)$$

Since, the gas viscosity is negligible, its dependence on pressure is also negligible. Therefore, $\frac{Q_g P_0}{\Delta p A}$ is a linear function of the mean pressure, P_m. The term $\frac{Q_g P_0}{\Delta p A}$ is known as the "pressure normalized" gas flux. Thus, Equation 4.4 can be written as

$$y = a + b P_m \tag{4.5}$$

where

$$y = \frac{Q_g P_0}{\Delta p A} \tag{4.6}$$

$$a = \frac{2 r_m \varepsilon}{3 l_r} \sqrt{\frac{8RT}{\pi M}} \tag{4.7}$$

and

$$b = \frac{r_m^2 \varepsilon}{8 \eta_g l_r} \tag{4.8}$$

According to Equation 4.5, a straight line will be obtained for every experiment after plotting values of y against P_m. The intercept and slope of this plot gives the value of constants a and b of Equations 4.7 and 4.8. The mean pore radius can be obtained from Equations 4.7 and 4.8 as

$$r_m = \frac{32 b \eta_g}{3a} \sqrt{\frac{2RT}{\pi M}} \tag{4.9}$$

The effective porosity, ($\frac{\varepsilon}{l_r}$), can be calculated from Equation 4.8 as

$$\frac{\varepsilon}{l_r} = \frac{8 b \eta_g}{r_m^2} \tag{4.10}$$

Absolute porosity, ε, can be calculated from Equation 4.10 by considering the pore length (l_r) equivalent to the membrane thickness. The thickness of the prepared membranes can be calculated from the cross-sectional images of the membranes.

4.2.2.1.2 Bubble-Point Method

In 1908, Bechhold discovered a method, popularly known as the bubble-pressure or bubble-point method, to calculate the maximum pore size of a given membrane. The method measures the pressure point at which a gas bubble passes through the membrane. This method is nondestructive in nature for the sample, needs simple apparatus, is an easy procedure, and useful for the integrity test.

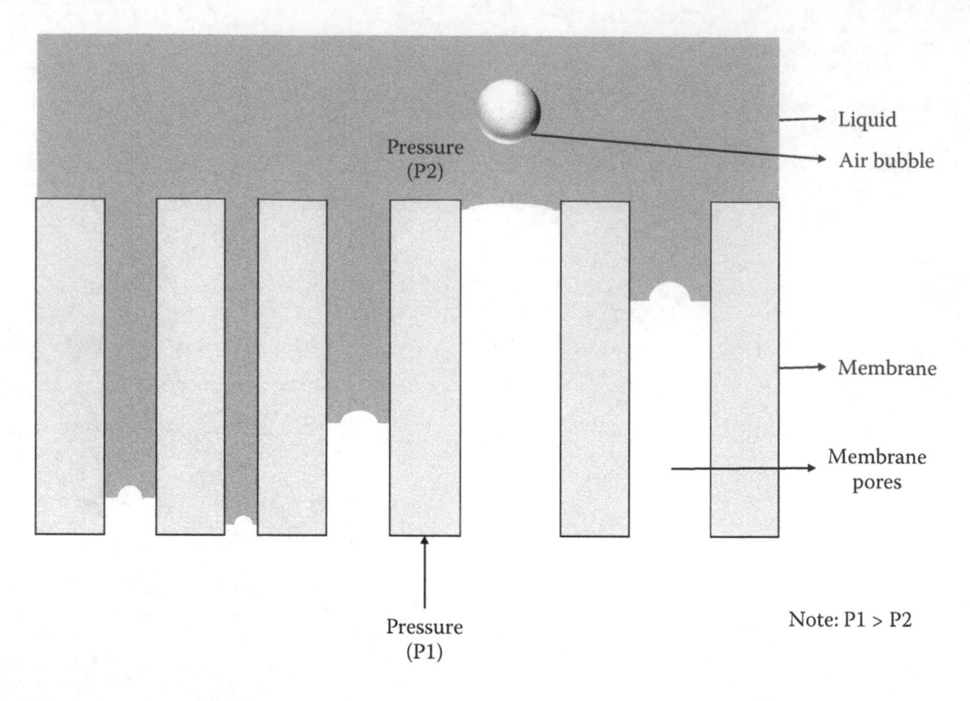

FIGURE 4.16 Principle of bubble-point method.

The principle of this method is shown in Figure 4.16, where it can be seen that the top surface of the membrane is wetted with a liquid and the bottom surface is in contact with a gas. Primarily, the pressure needed to pass the gas through the liquid-filled membrane is measured. A gas bubble passes through the membrane when its radius is equal to that of a membrane pore. The gas first passes through the largest pores and later through the small pores with an increase in pressure. It is well known from Cantor's equation (Equation 4.11) that the pressure required to pass a gas through a liquid filled membrane is inversely proportional to the pore radius. The pore radius can be calculated by using Equation 4.11 as the pressure value is known [11]:

$$P = \frac{2\,\gamma}{r} cos\theta \tag{4.11}$$

where P represents the pressure (Pa), r the pore radius (m), and γ the surface tension at the liquid–air interface (N/m).

The limitation of this method is that it can be used effectively for the measurement of large pores only. For small pores, ultrahigh pressures are required. This is because of the high surface tension at the liquid–air interface. This can be taken care of by using liquids of lower surface tension. Equation 4.11 shows that the type of liquid used does not affect the method. Though this is the case, different pore radius values were obtained with varying liquids. The reason is the wetting effect; various liquids wet a membrane to different extents. Generally, i-propanol is used as a standard liquid for this method. Factors like pore length and rate of pressure increase, and the affinity between the membrane material and liquid used also influence the method.

4.2.2.1.3 Bubble-Point with Gas Permeation Method

The bubble-point and gas permeation methods can be combined to mask the limitations of the methods used on their own. First, the gas permeation method is followed and gas flow is measured, as a function of applied pressure, across the dry membrane. Then, the bubble-point method is followed, the membrane is wetted with a liquid and gas flow is measured across the membrane, again as a function of applied pressure. The membrane pores at low pressures will be filled with liquid, and thus the gas flow through the membrane will be very low, since the diffusion of a gas through a liquid is low at low pressures. At the bubble point, the largest pores will be devoid of any liquid and the gas flow will increase due to the convective flow through the empty pores. According to the Laplace equation, a further increase in pressure will empty the smaller pores and so on [1,11]. So, theoretically, at the end (at the highest pressure) the gas flow should be equal to the gas flow of the dry membrane. The presence of any variation in the two gas flows is due to the small pores present in the membrane.

4.2.2.2 Liquid–Liquid Displacement Porosimetry

The liquid–liquid displacement (LLDP) method is mainly used with ultrafiltration and nanofiltration membranes. LLDP is a combination of the bubble-point and solvent permeability [12] methods. This method also helps in the calculation of membrane pore size, pore size distribution, and pore number [11,12]. In this method, first, the membrane is wetted with a wetting liquid. Assuming that all the pores are filled with the wetting liquid, the membrane is then brought in contact with another liquid that is nonwetting for the membrane and also immiscible with the wetting liquid. This nonwetting liquid is permeated through the membrane with an applied pressure, displacing the wetting liquid occupying the membrane pores. Because the liquids are immiscible, an interfacial tension arises between the two liquids at their subsequent boundaries. The nonwetting liquid is only able to displace the wetting liquid when the applied pressure value attains the value given in Equation 4.11. The mechanism of this method is based on the assumption that at low pressures the largest pores are invaded and at high pressures the smallest pores are invaded by the nonwetting liquid, as shown in Figure 4.17.

Here, an assumption has made that at pressure P_i the nonwetting liquid permeates only through the pores of radius $r \geq r_i$ ($\approx 2\gamma/P_i$). The smaller radius pores r_k become permeable at pressure P_k ($>P_i$). Therefore, it is feasible to get a flux versus pressure plot, the trend of which is depicted in Figure 4.18. Flux F_i represents the pores for which $r \geq r_i$, and flux F_k to the pores for which $r \geq r_k$. Thus, the flux $F_{i,k}$ at pressure $P_{i,k}$ is the result of pores with radii between r_i and r_k. The flux will keep increasing until the pressure P_n is not reached. This point is reached when all the pores are occupied by the nonwetting liquid. If the pressure P_n is crossed, than the membrane flux F becomes directly proportional to the applied pressure [13,14].

P_n depends on both the pore radius as well as interfacial tension between the wetting and nonwetting liquid. It is important to select liquids with low interfacial tensions, γ, so that the smallest pores can also be measured without applying very high pressures. Generally, a water–isobutanol–methanol (25:15:7 v/v) mixture is used as the wetting liquid and deionized water as the nonwetting liquid. The interfacial tension between these two liquids is low (0.35 mPa.m (dyne/cm)) at 20°C [15].

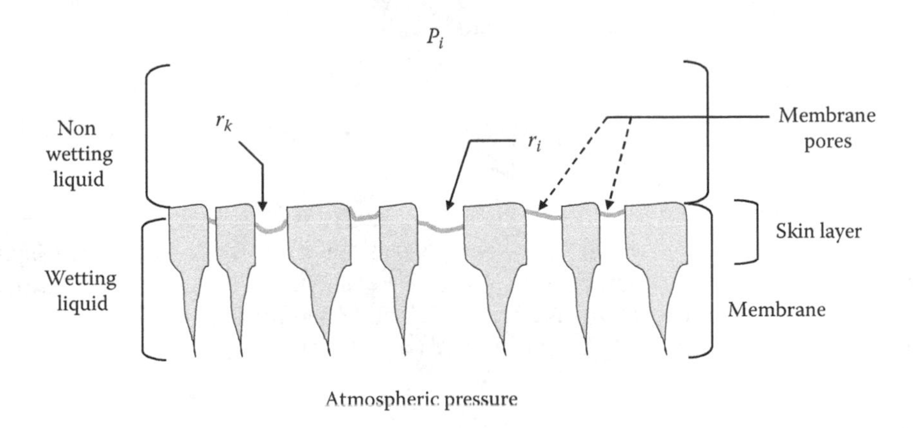

FIGURE 4.17 Principle of liquid–liquid displacement porosimetry method.

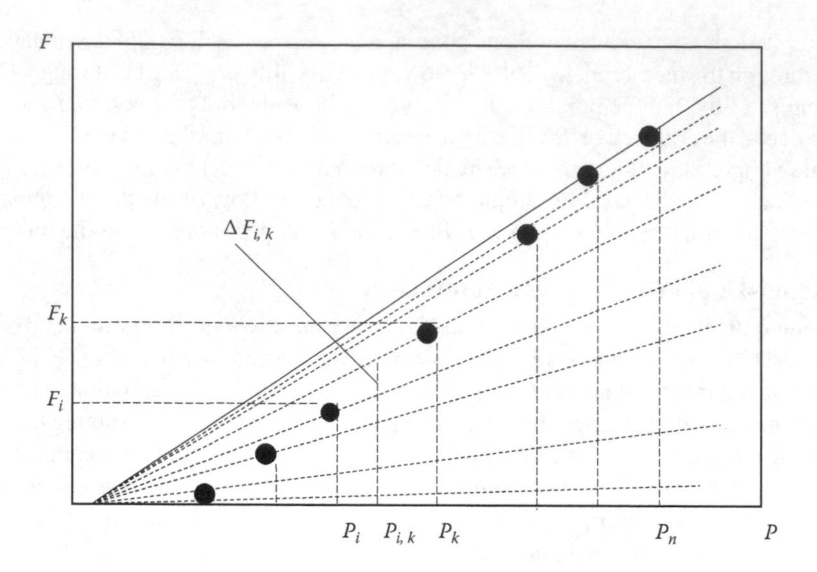

FIGURE 4.18 Flux versus pressure plot depicting the relation between pressures and pore radius.

The data of variation of the flux with respect to pressure yields the pore size distribution of the membrane by using the following equation:

$$L_n = \Sigma L_{i,k} = \Sigma \frac{F_{i,k}}{P_{i,k}} \tag{4.12}$$

where $F_{i,k}$ represents the flux at pressure $P_{i,k}$, and $L_{i,k}$ the partial permeability coefficient of the pores with a radius between r_r and r_k at pressure $P_{i,k}$. This corresponds to a mean radius:

$$r_{i,k} = \frac{r_i + r_k}{2} \tag{4.13}$$

Combining Equations 4.11 and 4.12, the permeability versus pore radius curve may be obtained from the flux versus pressure data. Also, from the following equations the pore number versus pore radius and pore area versus pore radius curves can be obtained:

$$N_{i,k} = \frac{d\eta}{2\pi\sigma^4} P_{i,k}{}^3 F_{i,k} \tag{4.14}$$

and

$$A_{i,k} = \pi r_{i,k}{}^2 N_{i,k} \tag{4.15}$$

where $N_{i,k}$ represents the pore density (i.e., the number of pores having radii between r_i and r_k per unit membrane surface area), d the length of the pore, and η the viscosity of the wetting liquid. $A_{i,k}$ represents the pore area having radii between r_i and r_k. Equations 4.14 and 4.15 are obtained from Cantor's equation (Equation 4.11) and the Hagen-Poiseuille permeation equation (Equation 4.16) assuming cylindrical pores and laminar flow as given next [1–3]:

$$F_{i,k} = \frac{\pi N_{i,k} r_{i,k}{}^4 P_{i,k}}{8d\eta} \tag{4.16}$$

The total pore area $A_{t,p}$ and the total number of pores per unit area of membrane N_t can be calculated as

$$A_{t,p} = \Sigma A_{i,k} \tag{4.17}$$

$$N_t = \Sigma N_{i,k} \tag{4.18}$$

The mean pore radius r_m can be calculated as [16]

$$r_m = \frac{\Sigma N_{i,k} r_{i,k}}{\Sigma N_{i,k}} \tag{4.19}$$

The limitation of the LLDP is that the calculated values of pore size and pore size distribution come out to be marginally different from the actual values. This might be due to the assumptions made during the analysis regarding pores, namely, the pores are cylindrical in nature and the membrane thickness is uniform throughout the membrane. However, the information comes in handy while comparing different membranes.

4.2.2.3 Molecular Weight Cut-Off

Molecular weight cut-off (MWCO) is a notation used to characterize the pore size of the membranes. The MWCO of a membrane is defined as the molecular weight that is 90% rejected by the membrane [1]. For example, a membrane with 68000 MWCO means solutes of molecular weight more than or equal to 68000 Da will be 90% rejected by the membrane. The different types of MWCOs of a membrane, that is, sharp or diffuse, are shown in Figure 4.19. In this method the membrane is characterized for its MWCO by using different molecular weight solute particles, such as poly (ethylene glycol), poly(vinyl pyrrolidone), and dextran. The solutions of these different molecular weight solutes are prepared and permeated through the membrane. The solute molecular weight for which the membrane gives at least 90% rejection is taken as the MWCO of the membrane and the solute rejection is given by

$$R = \frac{C_f - C_p}{C_f} \tag{4.20}$$

where C_f and C_p are the solute concentrations in feed and permeate, respectively.

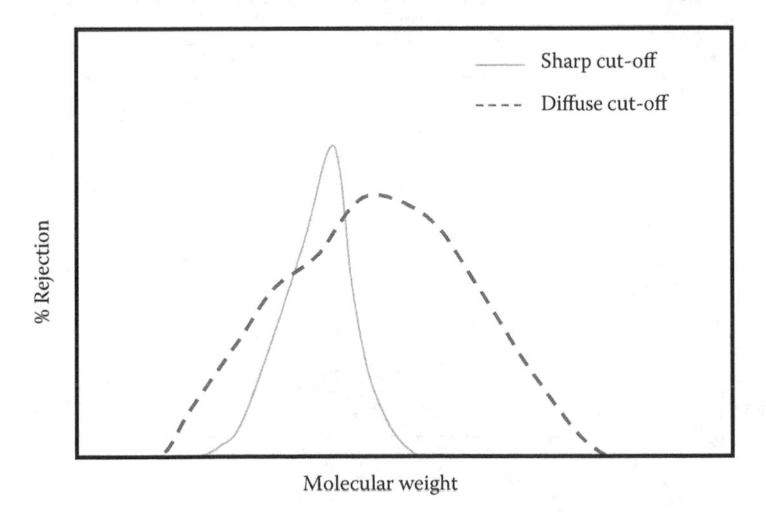

FIGURE 4.19 Sharp and diffuse molecular weight cut-off of membranes.

This method is not accurate and that is why the membrane cannot be defined by only the MWCO. There are other parameters that are important, including the shape and flexibility of the solute particles, inter- and intrainteractions in the medium, and concentration polarization or fouling. The latter have a great effect on the separation characteristics of a membrane. However, this method has its own limitations; for example, if two or more unlike solutes are used for calculating the MWCO of a membrane, they will give different results even though they are of same molecular weight. This is because of their difference in shape and inter- or intrainteractions. Therefore, to minimize the errors and increase the efficiency of the method, certain conditions have to be followed, including the use of a low feed concentration, low operating pressures, a solute having broad molecular weight distribution, and low adsorption tendency.

4.2.2.4 Water Permeation

Water permeation is the base of all the membrane applications and is measured in terms of a membrane's flux (L/m^2h). Therefore, a membrane with a better flux profile is considered as the best. To analyze a membrane's permeation profile, pure water is used as the feed. Since there are no impurities present, the membrane process is free from any kind of anomaly and will give defect-free results for the membrane. The membrane flux is further based on other membrane measures like membrane pure water flux (PWF), compaction factor (CF), permeability (P_m), and hydrophilicity.

The PWF can be calculated using the following equation:

$$F_W = \frac{Q}{A\Delta t} \tag{4.21}$$

where F_W represents the PWF (L/m^2h), Q the total volume of water permeated (L), A the total effective area (m^2), and Δt the permeation time (h).

The CF can be calculated as a ratio of the initial PWF and steady-state PWF:

$$CF = \frac{PWF_{initial\ state}}{PWF_{steady\ state}} \tag{4.22}$$

Figure 4.20 shows the PWF of two different polysulfone membranes (P1 and P2) modified with two different molecular weights (2000 Da and 5000 Da) of methoxy poly(ethylene glycol) (mPEG). It can be seen from the figure that the PWF declines abruptly at the start and almost attains steady state after 60 min. This is because of the membrane compaction and pore shrinkage, which take place

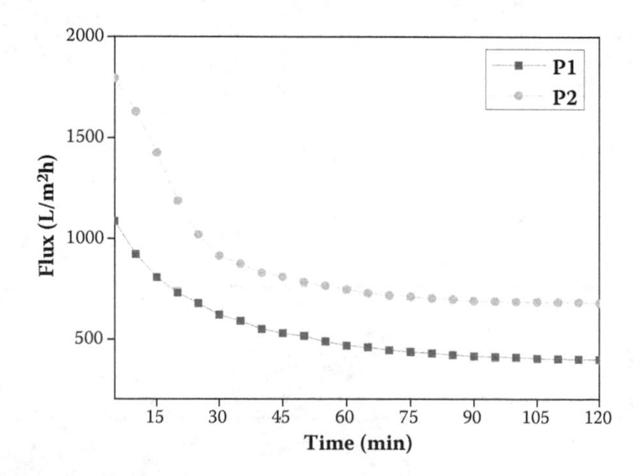

FIGURE 4.20 Pure water flux of mPEG-modified polysulfone membranes depicting compaction and flux loss.

at the start of the experiment. This is a common phenomenon seen in every membrane, especially the polymeric membranes. The figure also depicts that the PWF is higher if high molecular weight additives are used in the modification of the membrane. This result is in agreement with the fact that the use of high molecular weight additives result in a membrane with higher porosity, since a porous membrane usually has a better PWF profile.

4.2.2.5 Hydraulic Permeability

Hydraulic permeability (P_m), also known as permeability, defines a membrane's permeation property. P_m is directly proportional to the porosity of a membrane. Therefore, a membrane with high permeability results in higher flux, and vice versa. The PWF (F_W) (L/m^{2}h) of a membrane is calculated at different transmembrane pressures (ΔP) at a constant time interval. Figure 4.21 shows the resultant graph of the P_m study of two mPEG-modified polysulfone membranes (P1 and P2). The P_m (L/m^{2}hkPa) can be determined by the slope of the curve between the PWF and transmembrane pressure. It can also be calculated as

$$P_m = \frac{F_W}{\Delta P} \tag{4.23}$$

The hydraulic resistance (R_m) (m^{2}hkPa/L), which plays a vital role in the membrane permeation and fouling profile, is inversely proportional to P_m and can be obtained from the inverse of P_m. The higher the R_m of a membrane, the lower is the P_m as well as PWF of the membrane, and vice versa.

4.2.2.6 Equilibrium Water Content

Equilibrium water content (EWC) is important in membrane studies, as it is directly related to the porosity of a membrane and indirectly deciphers the extent of hydrophilicity or hydrophobicity of a membrane [7,17]. For calculation of EWC, a membrane token is taken and soaked in DI water overnight. After mopping with tissue paper they are weighed on a weighing balance and then dried in a vacuum oven for 24 h at 60°C. The membranes are again weighed in the dry state. The following relation is used to calculate the EWC at room temperature [7]:

$$EWC\,(\%) = \frac{W_w - W_d}{W_w} \tag{4.24}$$

where W_w represents the weight of a wet membrane (g) and W_d the weight of a dry membrane (g).

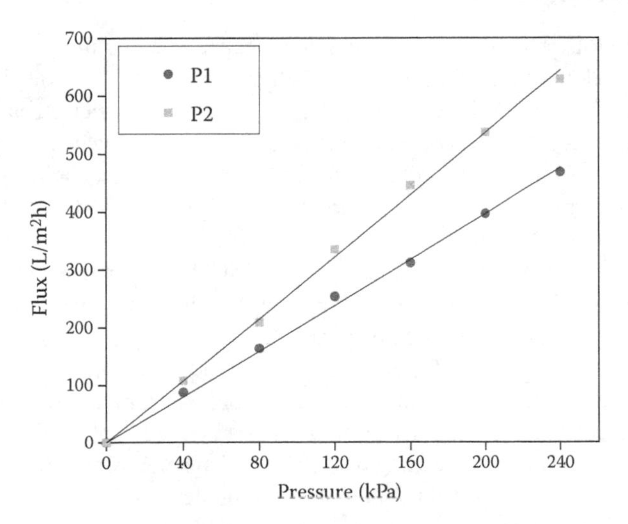

FIGURE 4.21 Effect of transmembrane pressure on mPEG-modified membrane flux.

4.2.2.6.1 Membrane Porosity

Porosity is the property of a membrane that tells about its permeation and separation quality. A good membrane always has a good porosity with well-defined pores. Porosity can be calculated with the help of following relation [7]:

$$Porosity = \frac{W_w - W_d}{\rho_w \times V} \tag{4.25}$$

where ρ_w represents the density (Kg/m^3) of pure water at room temperature and V the volume of the membrane (m^3). Cross-sectional images of the membranes are used to get the membrane thickness for membrane volume calculations.

4.2.2.6.2 Membrane Contact Angle

The extent of a membrane's hydrophilicity is measured in terms of membrane static contact angle; the lesser the membrane static contact angle, the higher the hydrophilicity of the membrane, and vice versa [7,18]. A drop of DI water is to be placed on the membrane surface with the help of a micropipette, at room temperature, then the images of the drop are to be taken with the help of a digital instrument, and then the membrane static contact angle can be deciphered.

4.3 OTHER TECHNIQUES

In the preceding sections, techniques for the morphological and structural aspects of membranes were discussed. The membrane parameters of pore size, pore geometry, pore size distribution, and pore number can be calculated by the techniques discussed. In this section techniques will be discussed that will infer information regarding the membrane material, stability, crystallinity, and functional groups. Further, these techniques will also help in the confirmation of modifications carried out in the membranes.

The techniques are diverse in nature and are also based on different principles. Accurate results regarding the membrane constituents are yielded by using these techniques. There are many techniques for confirming the membrane integrity; some are discussed in the succeeding sections.

4.3.1 X-Ray Photoelectron Spectroscopy

In 1887, Heinrich Rudolf Hertz discovered the photoelectric effect, which was explained by Albert Einstein in the year 1905. The explanation of the effect earned Einstein the 1921 Nobel Prize for Physics. In the year 1907, P. D. Innes recorded the first x-ray photoelectron spectroscopy (XPS) spectrum. It was Kai Siegbahn whose continuous efforts led to the invention of a proper XPS instrument. In 1969, in collaboration with Siegbahn, Hewlett-Packard engineers developed the first commercial monochromatic XPS instrument. Siegbahn received the 1981 Nobel Prize for Physics for his extensive efforts to develop an XPS instrument.

Electron spectroscopy is a technique that resembles SEM in principle but differs in the species bombarded on the material surface. It is also similar in working principle to x-ray spectroscopy but the two differ by the type of species bombarded and detected. For x-ray spectroscopy it is the electron bombarded and characteristic x-ray detected from the elements, and for electron spectroscopy it is the x-ray photon bombarded and characteristic electrons emitted from the element of a material are detected. The photoelectrons are similar to x-rays and carry chemical information of the elements present in a material, therefore they are known as characteristic electrons. Characteristic electrons are emitted only from the top surface of the material up to the depth of 10 nm due to their low escape energies (20–2000 eV), as compared to x-rays that are emitted from a much higher depth. Therefore,

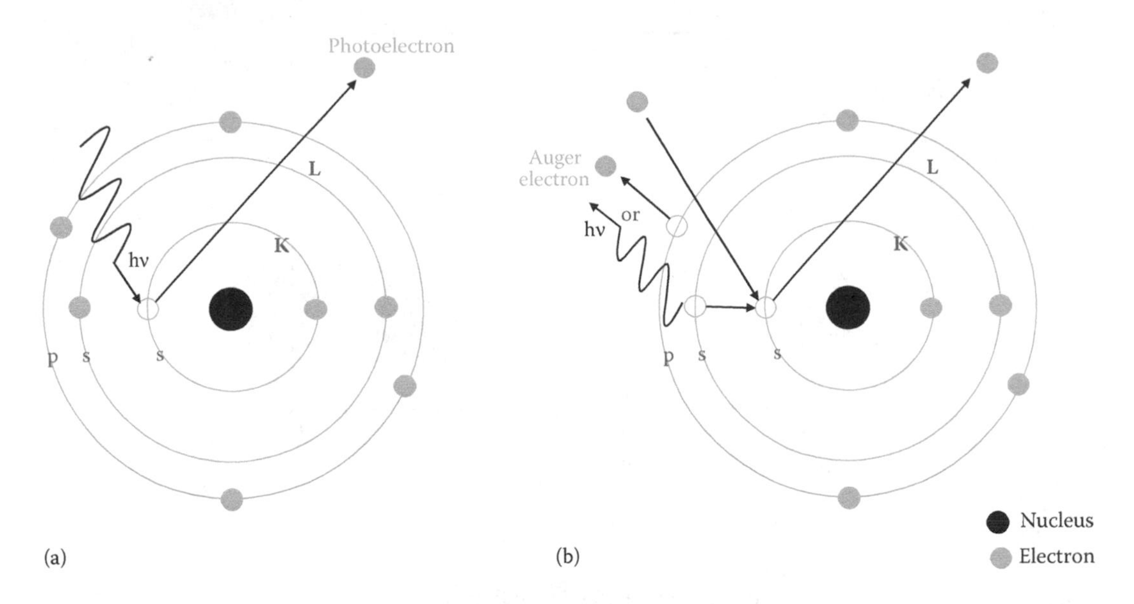

FIGURE 4.22 Mechanism of (a) photoelectron and (b) Auger electron generation.

electron spectroscopy can be called a surface chemical analysis technique. Further, there are two types of electron spectroscopies: auger electron spectroscopy (AES) and x-ray photoelectron spectroscopy (XPS). The origins of auger and photoelectrons are different, but both carry the same information about the chemical elements of a material. The generation of these two types of electrons is shown in Figure 4.22.

The photoelectrons are emitted when an atom absorbs the x-ray photon. The x-ray photon carries enough energy to emit an inner shell electron like from a K shell. If a K shell electron emits as a photoelectron with a known kinetic energy E_K, then the binding energy (E_B) of the photoelectron can be calculated by the following equation [19]:

$$E_B = h\nu - E_K - \varnothing \tag{4.26}$$

where Ø represents the energy required by the electron to escape the material surface, h the Planck's constant, and ν the frequency. These binding energies are characteristics of atomic electrons and thus used to identify the elements. XPS identifies elements by using the photoelectron spectra of the element binding energies. The XPS spectrum of pure TiO_2 is a plot between intensity versus binding energy, as shown in Figure 4.23. TiO_2 is a membrane additive widely used for the modification of membranes for various applications [20].

XPS is a powerful surface chemical analysis technique, and thus very beneficial in the field of membrane science. The membrane samples can be analyzed for their composition under XPS. It is very useful for researchers working for the development or modification of membranes with new materials. The technique is very accurate in its results and used widely in the fields of materials development and modification.

4.3.2 Thermogravimetric Analysis

Thermogravimetric analysis (TGA) is a thermal analysis technique that measures mass change in a material as a function of temperature. The thermal events in a material can bring change to its properties like structure, mass, phase, and stability. TGA is an easy and simple to use technique as

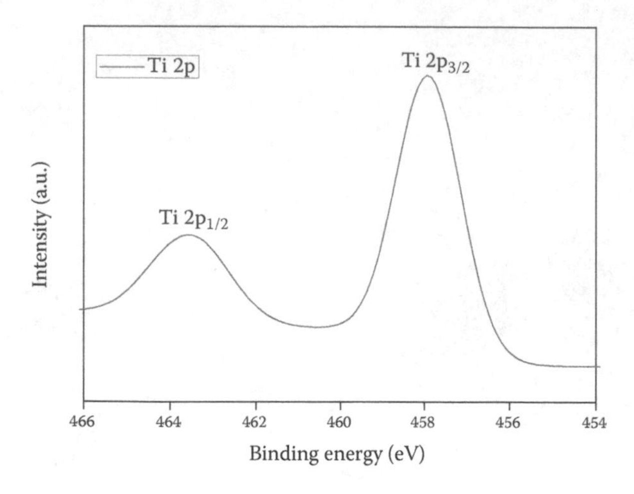

FIGURE 4.23 XPS spectrum of TiO_2.

compared to other techniques discussed in the preceding sections (i.e., SEM, EDS, and XPS). Samples of any kind and in any form can be used for analysis.

The TGA instrument is shown in Figure 4.24. A microbalance is the heart of the TGA instrument as it measures the mass change, which is the base of this technique. The null point type microbalance, the Cahn microbalance, is a common type of microbalance used in most TGA instruments based on its property to keep the sample in a vertical position when its mass change occurs. In the TGA instrument the sample is located in the center of the furnace over a microbalance. The furnace is cylindrical in shape with heating elements present in its walls. A thermocouple is present under the sample holder with a gap in between. This arrangement eliminates any possibility of a temperature gradient to occur between the thermocouple and heating elements and helps them to reach thermal equilibrium swiftly. Generally, nitrogen gas is used to fill the furnace to maintain an inert atmosphere.

In TGA, mass change is monitored with respect to increase in temperature. TGA is mainly used to analyze a material's decomposition and stability as a function of temperature in a scanning mode or as a function of time in isothermal mode by means of mass change. TGA curves can be plotted as percent mass change versus temperature or time. Figure 4.25 shows the TGA of a neat polysulfone membrane (P0) where mass change is plotted against increasing temperature. In the field of membrane science, TGA is considered important because of its ability to show compositional effects on the stability of a membrane. The neat and membranes modified with some type of additive can be

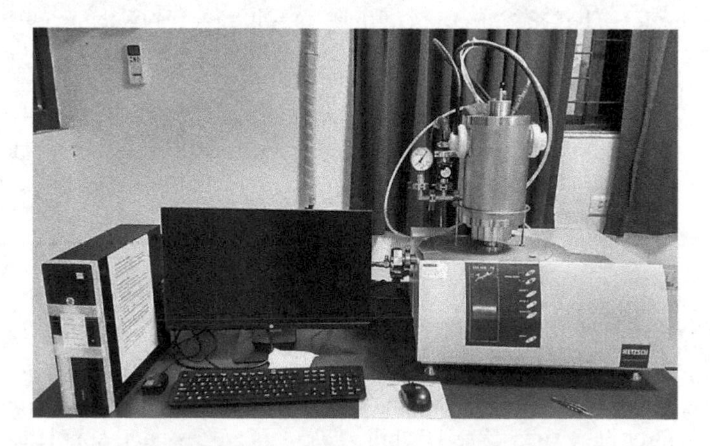

FIGURE 4.24 Thermogravimetric analysis instrument. (Courtesy of CIF, IITG.)

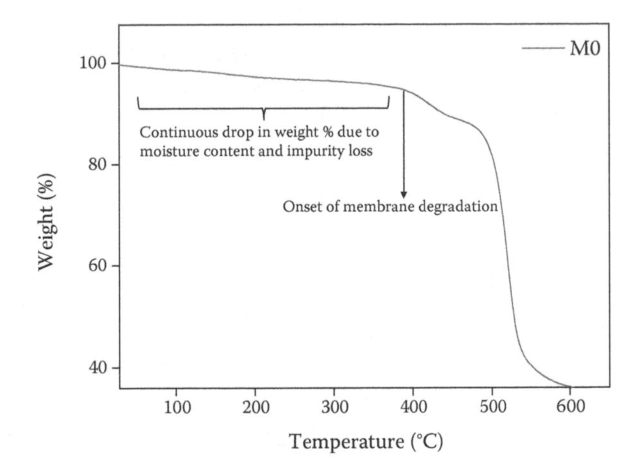

FIGURE 4.25 TGA spectrum of a polysulfone membrane.

compared for the change in stability. Additives like high molecular weight polymers, metallic nanoparticles, and zeolites are reported for increasing the stability of a membrane by using TGA as the technique of analysis [21,22].

Another thermal analysis technique that resembles TGA is differential scanning calorimetry (DSC). This quantitative technique uses the phase change of materials as the basis of analysis. The DSC is carried out over a temperature range by heating or cooling the sample at a constant rate. Some common thermal events or phase changes in DSC are solid phase transformation, glass transition, crystallization, and melting. The "differential" from the name of the technique states that the analysis is based on the different thermal events between the sample and reference. The DSC instrument measures the heat flow difference between the sample and reference. These techniques are useful specifically for polymeric membranes as polymers exhibit changes with respect to temperature change. With DSC in addition to temperature of glass transition, crystallization and melting of the membrane can also be determined. Membrane properties like crystallinity, polymer content, curing status, and stability can also be determined.

4.3.3 X-Ray Diffractometry

X-ray diffractometry (XRD) is a material characterization technique. It is mainly used for the analysis of crystal structures of powder samples from its inception; therefore, it is named as XRD. In XRD, a single wavelength x-ray beam is used for the analysis. The x-ray beam incident angle is continuously changed during the analysis so as to get a spectrum of diffraction intensity versus angle (angle between the incident and diffracted beam). The crystal structure and quality of the sample can be analyzed by comparing the obtained data with the XRD database available. The database contains over 60,000 diffraction spectrums of known crystalline materials. Figure 4.26 shows an XRD plot of pure TiO_2, in which relative intensity is plotted against 2θ (degree). The membranes or membrane materials can be analyzed for the extent of crystallinity, which is again a measure of strength and stability. The neat or modified membranes can be compared with the materials used, and the increase or decrease in crystallinity can be analyzed. Therefore, XRD analysis is crucial in the field of membrane science.

4.3.4 Fourier Transform Infrared Spectroscopy

Fourier transform infrared spectroscopy (FTIR) is a vibrational spectroscopy technique widely used to analyze samples for the presence of different functional groups. In vibrational spectroscopy, the interaction between electromagnetic radiation and nuclear vibrations of molecules is examined to

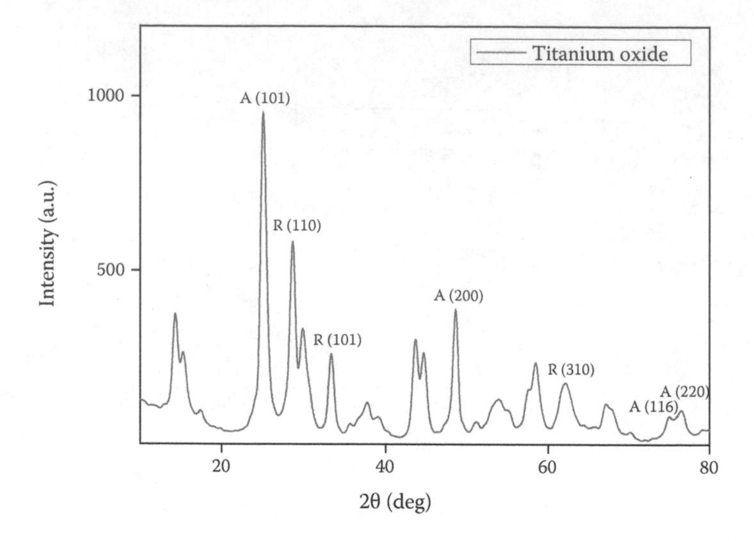

FIGURE 4.26 XRD spectrum of TiO_2, a commonly used membrane modifying agent.

analyze the structure of a molecule. Electromagnetic waves with long wavelengths (10^{-7} m), which are usually infrared, are used in vibrational spectroscopy. Vibrational spectroscopy can be used for the analysis of a liquid or solid sample. FTIR uses the Fourier transform method to obtain an infrared spectrum [19,23]. The heart of the FTIR instrument is the Michelson interferometer. The infrared radiations enter the Michelson interferometer and are later detected by a detector. The Fourier transform is used to convert the interferogram to an infrared spectrum, since the detector does not receive an infrared spectrum from the Michelson interferometer. The resultant infrared spectrum is a plot between light intensity and wave number. The final infrared spectrum includes both the sample (single beam spectra) and background (generally, KBr or atmosphere) spectra. Elimination of the background ratio of both the sample and background has to be made, which results in a transmittance spectrum as shown in Figure 4.27. The presented figure is the FTIR spectrum of two poly(2-acrylamido-2-methyl-1-propanesulfonic acid) (AMPS) modified polysulfone membranes (A1 and A2). The tagged peaks in the FTIR spectrum confirm the presence of polysulfone and AMPS in the prepared membranes.

In general, transmittance is the ratio of intensities given by

$$T = \frac{I}{I_o} \tag{4.27}$$

where I represents the intensity of a single beam sample spectrum and I_o the intensity of background spectrum. This transmittance is the y-axis of the plot as shown in Figure 4.27, represented as %T. The FTIR spectrum can also be presented as a plot between absorbance (A) and wave number. The absorbance can be calculated from transmittance as [19]

$$A = -\log T \tag{4.28}$$

In the transmittance spectrum the vibration band peaks point downward and for the absorbance spectrum they point upward. It is better to use the absorbance spectrum for quantitative analysis, since the transmittance spectrum is not linearly proportional to concentration. In membrane characterization, the FTIR technique is utilized to validate the chemical composition of the materials used for the membrane preparation based on the presence of vibrational band peaks for the functional group in the spectrum. The membrane prepared from different materials can be differentiated on the

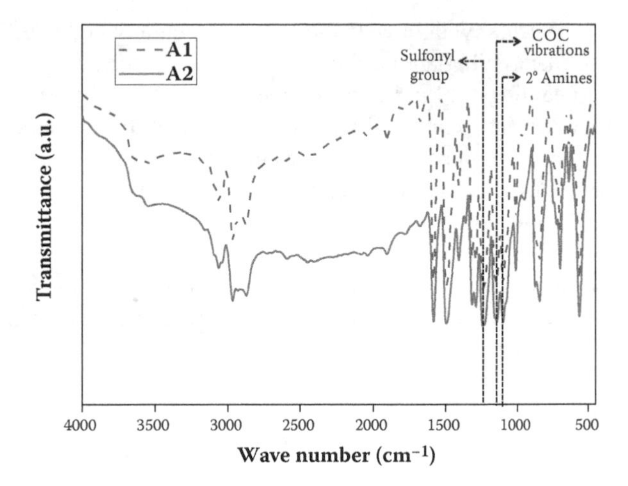

FIGURE 4.27 FTIR spectrum of AMPS-modified polysulfone membranes.

basis of the presence or absence of characteristic peaks of different materials used. On this basis the presence or absence of an additive in a membrane can be analyzed. The stability of the additive in a membrane can also be confirmed by FTIR analysis of membranes before and after membrane operation. The presence of vibration band peaks for the additive used, even after using the membrane for a longer period, shows the additive is perfectly embedded in a membrane [7]. Thus, FTIR analysis comes out to be an effective characterization technique for membrane science.

STUDY QUESTIONS

1. Are there different characterization techniques for the characterization of porous and non-porous membranes? If yes, then kindly explain why and how.
2. Amanda needs to know the membrane pore morphology. What characterization techniques do she need to use for this purpose?
3. Vinoth prepared polymeric as well as ceramic membranes, and wants to characterize them. What things should he consider for choosing the characterization techniques for the characterization of his membranes?
4. Deepak went to an international conference and was attending the poster presentation session. There he learned of a very novel membrane application and was interested in the materials used for that particular membrane. Deepak asked the author, who showed him characterization graphs of the membranes, which satisfied and answered Deepak's queries. What characterization technique could the author have used? What major inferences can be drawn?
5. Manish, a new research scholar, recently prepared mixed matrix membranes by using polysulfone and silver nanoparticles and was interested in the morphological study of the membranes as well as the raw materials. He is undecided among SEM, FESEM, and TEM techniques. Which techniques would you suggest to him and why?
6. Anita prepared a set of polymeric and ceramic membranes, and completed the morphological studies of the top surface of the prepared membranes. The morphological studies revealed that the membranes have uniform pore size distribution, but when she started permeation studies she got unsatisfactory results. What might be the reason and what she should do to confirm the problem?
7. Deepti calculated the molecular weight cut-off of a polysulfone ultrafiltration membrane with poly(ethylene glycol) of different molecular weights. But the membrane rejection of a novel

protein, having a molecular weight well above the molecular weight cut-off of the membrane, was not up to the mark. What are all the possible reasons for this outcome?

8. Smith has data regarding the hydraulic permeability and equilibrium water content of a membrane. With these data, can he predict the porosity of the membrane? If yes, then how?
9. Lokesh was adamant with one of his lab mates that the prepared membranes are hydrophilic, as he has blended a hydrophilic additive in the preparation of his membranes. What techniques should he use to prove his point? If the result comes out to be negative, then what might be the reason for it?
10. Why are there so many characterization techniques for the confirmation of the presence of a membrane material in a membrane? Is there any difference in their qualitative or quantitative results?

REFERENCES

1. M. Mulder, *Basic principles of membrane technology*, Dordrecht, Kluwer Academic Publishers, 1991.
2. S. P. Sutera and R. Skalak, The history of Poiseuille's law, *Annual Review of Fluid Mechanics*, 25, 1–19, 1993.
3. B. J. Kirby, *Micro- and nanoscale fluid mechanics: Transport in microfluidic devices*, Cambridge University Press, 2010.
4. A. T. Hubbard, *The handbook of surface imaging and visualization*, CRC Press, 1995.
5. N. N. Li, A. G. Fane, W. S. W. Ho, and T. Matsuura, *Advanced membrane technology and applications*, John Wiley & Sons, 2011.
6. B. Chakrabarty, A. K. Ghoshal and M. K. Purkait, Effect of molecular weight of PEG on membrane morphology and transport properties, *Journal of Membrane Science*, 309, 209–221, 2008.
7. R. Singh and M. K. Purkait, Evaluation of mPEG effect on the hydrophilicity and antifouling nature of the PVDF-co-HFP flat sheet polymeric membranes for humic acid removal, *Journal of Water Process Engineering*, 14, 9–18, 2016.
8. B. Chakrabarty, A. K. Ghoshal and M. K. Purkait, SEM analysis and gas permeability test to characterize polysulfone membrane prepared with polyethylene glycol as additive, *Journal of Colloid and Interface Science*, 320, 245–253, 2008.
9. H. Yasuda and J. T. Tsai, Pore size of microporous polymer membranes, *Journal of Applied Polymer Science*, 18, 805–819, 1974.
10. P. Uchytil, X. Q. Nguyen and Z. Broz, Characterization of membrane skin defects by gas permeation method, *Journal of Membrane Science*, 73, 47–53, 1992.
11. A. J. Burggraaf and L. Cot, *Fundamentals of inorganic membrane science and technology*, Elsevier, 1996.
12. G. Capannelli, I. Becchi, A. Bottino, P. Moretti and S. Munari, Computer driven porosimeter for ultrafiltration membranes, in *Characterization of porous solids*, edited by K. K. Unger, J. Rouquesol, and K. S. W. Sing, Amsterdam, Elsevier, 283–294, 1988.
13. G. Capannelli, F. Vigo and S. Munari, Ultrafiltration membranes—Characterization Methods, *Journal of Membrane Science*, 15, 289–313, 1983.
14. P. Abaticchio, A. Bottino, G. Capannelli, and S. Munari, Characterization of ultrafiltration polymeric membranes, *Desalination*, 78, 235–255, 1990.
15. M. K. Sinha and M. K. Purkait, Preparation and characterization of novel pegylated hydrophilic pH responsive polysulfone ultrafiltration membrane, *Journal of Membrane Science*, 464, 20–32, 2014.
16. B. Chakrabarty, A. K. Ghoshal and M. K. Purkait, Effect of molecular weight of PEG on membrane morphology and transport properties, *Journal of Membrane Science*, 309, 209–221, 2008.
17. B. Chakrabarty, A. K. Ghoshal, and M. K. Purkait, Preparation, characterization and performance studies of polysulfone membranes using PVP as an additive, *Journal of Membrane Science*, 315, 36–47, 2008.
18. F. L. Huang, Q. Q. Wang, Q. F. Wei, W. D. Gao, H. Y. Shou, and S. D. Jiang, Dynamic wettability and contact angles of poly(vinylidene fluoride) nanofiber membranes grafted with acrylic acid, *eXPRESS Polymer Letters*, 4(9), 551–558, 2010.
19. Yang Leng, *Materials characterization: Introduction to microscopic and spectroscopic methods*, John Wiley & Sons, 2008.

20. A. K. Pabby, S. S. H. Rizvi, and A. M. S. Requena (Eds.), *Handbook of membrane separations: Chemical, pharmaceutical, food, and biotechnology applications*, 2nd edition, CRC Press, 2015.
21. N. Hilal, M. Khayet, and C. J. Wright, *Membrane modification: Technology and applications*, CRC Press, 2016.
22. M. Duke, D. Zhao, and R. Semiat, *Functional nanostructured materials and membranes for water treatment*, John Wiley & Sons, 2013.
23. P. R. Griffiths and J. A. De Haseth, *Fourier transform infrared spectrometry*, 2nd ed., John Wiley & Sons, 2007.

5 Polymeric Membranes and Their Applications

5.1 INTRODUCTION

Advances in membrane science has led to the use of membrane processes in every field of science. They have been successful in replacing the conventional processes like distillation, crystallization, solvent extraction, and evaporation [1]. Polymeric membrane processes, due to their inherent properties like low cost, no use of additives or chemicals, low energy consumption, and easy to scale up made them famous and widely acclaimed in different sectors of various industries. Polymeric membranes are different in their structure, function, driving force, and mechanism of separation and this variation allows them to fit in different applications of various industries. They are used on a large scale for desalination; in industrial effluent treatments; recovery of valuable compounds; in purification, concentration, and fractionation of different products in the food and drug industry; and in the health care and pharmaceutical industries as an artificial kidney and controlled drug release. Polymeric membranes can also be tailor made so as to add specific separation properties or responsiveness to certain external stimuli. The polymeric membrane industry is interdisciplinary and thus has great potential for professionals from different fields, including chemists for the development of membranes and membrane materials, mathematicians to develop mathematical models explaining the separation and transportation mechanisms of the membranes, and chemical engineers for the upscaling or development of membrane processes for large-scale industrial use.

In this chapter, polymeric membranes, with their advantages and different applications, are discussed in detail.

5.2 ADVANTAGES OVER CERAMIC MEMBRANES

Polymeric membranes are the most widely used membranes and dominate the membrane market throughout the world for desalination, the food and health sectors, and gas separations. Polymeric membranes are thin films of 10 to 100 μm thickness. Polymeric membranes can be prepared from almost all the available polymers, but the practical process demands limit the number of polymers be used for the preparation of polymeric membranes. Some of the widely used polymers are cellulose acetate, polyacrylonitrile, polyamide, polyetherimide, polyethersulfone, polypropylene, polysulfone, polyvinylidene fluoride, and polytetrafluoroethylene [1,2]. The polymeric membranes are superior to ceramic membranes in terms of the following properties:

- Polymeric membranes are easy to cast and use. They can be easily used in different membrane modules and thus are able to provide greater surface areas per unit area of the module. This is not the case with ceramic membranes and they cannot be casted easily for their use in different membrane modules.
- Polymeric membranes are more economical than ceramic membranes. Also, the raw materials as well as the preparation costs are very low.
- Theoretically, any polymeric material can be used for the preparation of polymeric membranes, but this is not the case with ceramic membranes. They require special and specific materials for their preparation.

- The information related to polymeric membranes is ubiquitous and thus it is not hard to master the various facets of the polymeric membranes. For ceramic membranes, the information is not widely available, as they are quite young and not much data are available for them.
- The polymeric membranes can also be scaled up very easily as compared to ceramic membranes. The difficult fabrication process and the problems faced during their fabrication make it difficult for the ceramic membranes to be scaled up easily.

But along with their versatility and positive options polymeric membranes also carry some negatives or disadvantages. The disadvantages of polymeric membranes compared with ceramic membranes are

- Polymeric membranes carry low resistance against harsh chemicals and atmospheric conditions. This is not the case with ceramic membranes, as they are quite resistant to harsh chemicals and atmospheric conditions. This limits the use of polymeric membranes in various applications and give space to ceramic membranes to takeover.
- Polymeric membranes also do not have resistance against harsh thermal, mechanical, and operational conditions. The pH range of activity for polymeric membranes is also very limited and thus they are also not resistant to corrosion. The incapability of polymeric membranes to handle such conditions make them unfit for use under high temperatures, pressure, stress, and corrosive conditions. Therefore, ceramic membranes are nowadays quite famous for their resistance to such harsh conditions and are replacing polymeric membranes in various fields of membrane applications.
- The low capability of polymeric membranes to handle harsh conditions reduces their life span. The average life span of a polymeric membrane is 12 to 18 months, whereas ceramic membranes can last up to many years in continuous operation conditions. This is also a big drawback for the use of polymeric membranes for commercial as well as industrial applications.

In the recent times, polymeric membranes have been modified to increase their chemical, thermal, and mechanical resistance. However, these modifications are also not resulting in enough resistance toward the chemical and corrosion conditions. Therefore, their use under these conditions is still questionable.

5.3 POLYMERIC MEMBRANE APPLICATIONS IN VARIOUS FIELDS

Polymeric membranes, due to their advantages, such as performance and economics, are widely used in membrane separation processes. The efficiency and performance of the membranes depend upon the chosen polymer, surface chemistry, and pore morphology. The availability of plentiful polymers with diverse properties and preparation techniques to impart desired properties to the membranes make polymeric membranes suitable for most membrane separation processes. The availability of membranes ranging from microfiltration to reverse osmosis based upon their pore sizes also makes them versatile.

Based on their pore morphologies, membranes are classified as porous and nonporous (dense) [1,3]. These two types of membranes contain different separation properties. Porous membranes are generally used for microfiltration and ultrafiltration membrane processes for the separation of different kinds of liquid feeds. Nonporous membranes are mostly used in reverse osmosis and gas separation applications. Porous membranes follow the size exclusion principle for the separation process, which means the pore size of the membrane used is less than the feed component's size. Nonporous membranes follow the solution–diffusion approach for the permeation of the feed components. The driving forces of pressure, concentration, temperature, and electrical are used for both kinds of membranes.

In this section the applications of both porous as well as nonporous polymeric membranes in various fields and industries are discussed. They range from desalination to protein purification and wastewater treatment to osmotic power plants. This diversity of applications, low energy consumption, no need of extra chemicals, efficiency, and performance make membrane separation processes and in particular polymeric membranes so popular. This is the reason that they are used on such a large scale covering each and every industry.

5.3.1 Desalination

The 21st century has seen immense growth in terms of population, industrialization, and urbanization. This has led to the development of two critical issues: availability of clean water and energy. These two entities are the most basic requirements of modern humans. The unthoughtful growth and use of resources has resulted in this crisis and the world is still going toward this crisis with great speed. The use and depletion of conventional sources at such a speed has effects on both the human population and the environment. Therefore, there is a need to develop and use methods, technologies, and processes that provide the planet with sustainable clean water and energy.

Oceans are a great source of energy that can fulfill the present and future requirements of the different species on the planet. The only thing required is the sustainable methods to harness water and energy from the oceans. Membrane processes, as discussed in Chapter 1, provide the required solution. They are energy-efficient, cost-effective, and sustainable processes. Desalination of seawater or brackish water by using membranes gives the required combination of a sustainable method, technology, and process. Presently, reverse osmosis is the leading desalination process worldwide. It has surpassed conventional desalination technologies, including the multistage flash technique. The new reverse osmosis technologies—membrane distillation, capacitive deionization, electrodialysis, and forward osmosis—will continue to keep reverse osmosis as the most widely used technology for desalination. The continuous process improvements also add to the wide acceptance of reverse osmosis as a desalination technique. The improvements include better membrane material, module and process design, pretreatment methods, and energy recovery. The availability of membranes with better permeability and biological, chemical, and mechanical resistance makes it possible to perform reverse osmosis economically.

Presently, the reverse osmosis desalination process is dominated by thin film composite membranes. It consists of a polymeric web structure as support (120–150 μm), microporous film (~40 μm), and a thin selective film layer (0.2 μm) on the surface [3]. The microporous layer is used to give extra support to the thin-film composite membrane, since the polymeric web structure is porous as well as irregular. The selective film layer is ultrathin so as to decrease the resistance to permeation. It is generally made by polyamide via interfacial polymerization method from 1,3-benzenediamine and trimesoyl chloride. The membrane module widely used in reverse osmosis is the spiral-wound membrane module, because it offers a high membrane surface area, easy scale up, interchangeability, and low cost. The improvements in its design and dimensional changes in the spacers, feed channels, and vessels along with better materials have improved both the fouling as well as pressure losses. The spiral-wound polyamide membranes have more than 90% sale share and the hollow fiber asymmetric cellulose acetate membranes comes second in overall sales share. Currently, modular design research is concentrating on the hydrodynamic parameters so as to decrease the concentration polarization.

The first membrane to be used for desalination was the cellulose acetate asymmetric membrane developed by Loeb and Sourirajan [4], which also made it possible to commercialize desalination by reverse osmosis. Later came cellulose triacetate, which was stable at high temperatures in a wide range of pH, and had better biological and chemical resistance. Then came the first asymmetric noncellulosic membrane for desalination, the aromatic polyamide hollow fiber membrane developed by Richter and Hoehn. This membrane was later commercialized by DuPont with the name B-9 Permasep® and it was used for the desalination of brackish water. The membrane is resistant to biological and most chemical attacks, but it seems that over time the membrane becomes susceptible to chlorine and ozone poisoning. Therefore, there rose a need for the development of a membrane resistant to chlorine and ozone. Polypiperazine-amide asymmetric membranes answered the demand and they were developed as chlorine-resistant membranes. The problem with these membranes, as well as membranes made from sulfonated and carboxylated polysulfone, was low salt rejections, which deprived them of their commercialization as suitable desalination membranes. Later, Teijin developed polybenzimidazoline (PBIL) membranes with better permselectivity even in tough operating conditions, but they were also susceptible to pressure compaction and chlorine poisoning. The solution for these problems came in the form of anisotropic membranes, also known as composite membranes, where a porous sublayer was coated with a selective top layer by a two-step casting method. The porous sublayer provides mechanical strength to the top layer and the top layer provides selective salt rejection and better permeate flux. Francis [5] casted the first thin film composite membrane by float-casting the cellulose acetate thin membrane over water surface and later annealed and laminated it on a pre-formed cellulose acetate porous membrane. The thin film composite membranes got commercialized after extensive empirical studies and developments. Polysulfone was found to be the material of choice for these membranes due to its better resistance to biological, chemical, and mechanical stresses. Thin-film composite membrane were developed from polysulfone by using acid polycondensation and interfacial polymerization methods. Nowadays, better desalination membranes are produced from surface modifications, optimization of polymerization reactions, and by adding metal oxides or nanoparticles or nanotubes for better performance and efficiency. Ceramic, mixed matrix, and biomimetic membranes have been prepared and research is being carried out to develop a better desalination membrane. A better membrane can be developed by trial-and-error testing of better polymers, selection of better polymerization reactants, and better characterization of the membrane morphology with advanced tools and techniques. Apart from the development of a better desalination membrane, challenges for desalination are feed characterization, process and material development, sources of renewable energy, maintenance of stringent water standards, and brine management.

5.3.2 Wastewater Treatment

Water is the most treasured and sought-after resource in the world. It plays a vital role in the nurturing of life on planet Earth. Water is omnipresent in all industrial and commercial processes. Therefore, it is used on a very large scale and at a very high rate. The consumption of water would not be a problem if it was present in abundance. Just over 70% of the Earth's surface is covered with water, but only 2.5% is freshwater and the rest is present in the form of oceans. However, much of the freshwater is also not available, as most of it is in the form of snowfields and glaciers. Only 1% of freshwater is available for human use and consumption. This 1% is also not completely accessible to humans as it is present in the form of groundwater and only 0.3% can be accessed, which is present in the form of rivers, lakes, and swamps. That is why water is said to be precious. Therefore, it is very necessary to use water wisely and make 100% effective and efficient use of it. Further, problems like industrialization, urbanization, and population explosion worsen the current water scenario by polluting most of the available 0.3% freshwater. Plus, water pollution is not limited to surface waters. Current practices pollute groundwater as well. At the current pace of water pollution, it is not hard to imagine countries at war over water. If not controlled, then the problem of water pollution will worsen further and a global water scarcity problem will arise. Thus, it is important to devise new methods and practices to provide solutions to the problems associated with the water pollution.

Polymeric membranes, because of their high efficiency and low cost, come out to be a suitable solution to control water pollution. The problems associated with the polymeric membranes, including low life span and low chemical, thermal, and mechanical resistance, are trying to be eliminated by extensive research and studies. Membrane researchers have put a lot of effort into the betterment of both the flux, selectivity, and antifouling nature of the polymeric membranes. Many new methods and materials are developed for the modification of polymeric membranes for imparting improved properties to the polymeric membranes. Membranes like composite and mixed matrix are developed to mend the limitations of the polymeric membranes. Membranes are modified by taking ideas from nature for the fabrication of bioinspired polymeric membranes with improved properties of permeation, selectivity, and antifouling. This will help in the increased sustainable and economical reuse of the industrial wastewaters. Thus, industries can be made self-dependent to an extent for their water needs.

Industries such as textile, tannery, sugar, food, and health care, and municipalities are the major users of water. They consume a large quantity of water for different processes. The main source for their water need is the groundwater pumped by using a borewell or from a well. Municipal water or water from other sources like rivers, lakes, and ponds (transported by using tankers or pipelines) is also used. The increasing salinity of groundwater is making groundwater unfit for use in textile industries. In addition to this, municipal water charges are also increasing due to various reasons, and transportation charges are also increasing due to the hike in oil prices. Therefore, water scarcity is a concern, and alternative measures have to be taken to overcome this problem. There may be new cost-effective water resources or an increase in reuse of water. According to the present-day situation, the last option is the best and needs to be exploited by using advanced techniques, since conventional techniques will not be cost effective. Therefore, membrane separation processes are the best to be employed for the reuse of industrial wastewaters. They are efficient as well as economic.

Generally, industries mix or collect wastewaters from their different branches for treatment. This results into wider spectrum of the feed and its components, which makes the treatment by using a single process difficult. Therefore, it is better to segregate wastewaters from different branches of an industry and then any treatment process (e.g., membrane separation processes) can be employed. This will reduce the burden on the treatment process and altogether will increase the efficiency, performance, and economics of the process. This will also make the treatment process easy and fast, since specific membranes for a particular kind of feed can be employed. These membranes can easily treat these segregated wastewaters qualitatively and quantitatively according to the prescribed water quality. The other advantage of employing membrane separation processes for the treatment of wastewaters is that they not only efficiently remove the feed components but also help in recycling some of the precious components from the feed.

In this section, the use of polymeric membranes in different industries will be discussed. The importance is given to the different challenges and achievements of the polymeric membranes faced in different types of feeds from various industries.

5.3.2.1 Textile Industry

The textile industry is a major user of water for its different processes. The main sources or processes of a textile industry that generate wastewater are dyeing, bleaching, washing, and finishing. The feed in its chemical composition is different for each process. Table 5.1 includes the characterization information of the different textile wastewater streams originating from different processes of the textile industry. The dyeing process together with the bleaching process (generally, washing in both processes) generates the highest amount of wastewater in the textile industry. It is estimated that 60% of the total water is consumed in the washing step alone of these processes. The high consumption of water in this step is due to the need of efficient and swift removal of unfixed dyes and other components from the textile to get a high-quality product.

Wastewater from the dyeing process contains a high amount of dyes and salts, and has a high chemical oxygen demand. High temperature and high pH are also characteristic features of dyeing process wastewater. Mainly there are two methods of treatment for textile wastewaters or in general for any industrial wastewaters. First is the mixed wastewater treatment conventional method, where the wastewaters from all the processes of the industry are treated together at a common place.

TABLE 5.1
Composition of Textile Wastewater from Its Different Processes

Process	Components
Desizing	Enzymes, starch, waxes, and ammonia
Scouring	Fats, soaps, oils, waxes, enzymes, surfactants, disinfectants, and sodium hydroxide
Bleaching	Hydrogen peroxide, sodium silicate, acids, sodium phosphate, surfactants, organic stabilizers, high pH
Mercerizing	Sodium hydroxide and high pH
Dyeing	Color, surfactants, salts, metals, sulfide, formaldehyde, reducing and oxidizing agents, and urea
Printing	Color, waxes, metals, spent solvents, and urea
Finishing	Chlorinated compounds, waxes, resins, acetate, softeners, stearate, hydrocarbons, and spent solvents

This mixture leads to the increase in the complexity of the feed to be treated. Therefore, it is difficult to get good results with this method. The second wastewater treatment method, the segregated method, is better and somewhat newer compared to the mixed wastewater treatment method. The capital cost for this method is high, but overall process efficiency and performance are great. A feed can be treated efficiently and perfectly by this method due to its less complex nature. The perfect conditions and treatments can be employed as the characteristics of the feed are known and the feed does not consist of different or a very wide spectrum of feed components. This makes the method fast and efficient.

5.3.2.1.1 Mixed Wastewater Treatment Method

The mixed wastewater treatment method is the traditional method of wastewater treatment used by industries for treating their wastewaters. Prior to the application of membrane separation processes, the wastewater is given chemical and biological pretreatments. An aerobic and/or anaerobic biological pretreatment method is applied prior to the membrane separation processes.

Membrane separation processes are capable of complete color removal from textile wastewaters. This can be achieved mainly with the use of nanofiltration and reverse osmosis membranes. The efficiency of the color removal further depends on the molecular weight of the dye molecules. Generally, the molecular weight comes in the range of nanofiltration membranes' molecular weight cut-off, but the presence of other feed components in the mixed wastewaters makes it difficult for the nanofiltration membranes to perform at their best. Therefore, the mixed wastewaters are first treated with ultrafiltration membranes and their permeant is used as feed for the nanofiltration or reverse osmosis membrane separation processes. The ultrafiltration membrane reduces the overall load on the nanofiltration and reverse osmosis separation processes by removing feed components of molecular weight very much higher than the molecular weight cut-off of the nanofiltration and reverse osmosis membrane separation processes. The resultant permeant from the reverse osmosis membrane separation process is far better as compared to the nanofiltration membrane separation process. This is because the nanofiltration membrane separation process is not capable of separating the monovalent ions from the feed and thus it cannot be reused as a process water. The reverse osmosis membrane separation process considerably reduces the color, conductivity, and total organic carbon from the mixed wastewaters. Thus, the resultant permeant from a reverse osmosis membrane separation process can be reused as process water.

Researchers all over the world have tested nanofiltration and reverse osmosis membrane separation processes for the treatment of mixed wastewaters. They have concluded that the permeant from these processes is of high quality and can be reused without further treatment steps. In addition to membrane separation processes, researchers have used a combination of membrane separation processes and other physicochemical processes for the neutralization of chemicals and the results came out to be wonderful. The results also confirmed that the resultant treated mixed wastewaters can be used in the dyeing steps of the textile industry without any problem. This shows that the membrane separation processes are promising for the treatment and reuse of mixed wastewaters from textile wastewaters.

5.3.2.1.2 Segregated Wastewater Treatment Method

The segregated wastewater treatment method is a modern method of wastewater treatment. This method, unlike the mixed wastewater treatment method, receives feeds from individual steps or processes of an industry as the source of wastewater and these feeds are treated individually or separately. These individual textile processes include dyeing, rinsing, and bleaching. The wastewaters from these processes of the textile industry are treated in situ and reused in situ, which means the wastewaters are taken as feed from these individual processes and after treatment reused by that particular textile process itself. This strategy increases the treatment process's quality and efficiency. This will also reduce the overall consumption of water in the textile industry. The efficiency and performance of this method is far better than the mixed wastewater treatment method. Researchers have shown that the permeant from the membrane separation processes is directly used for wetting, dyeing, and rinsing purposes without any problem and the obtained results were fascinating. Therefore, it can be said that the membrane separation processes employed for textile wastewater treatment by the segregated wastewater treatment method is a promising combination for the reduction of overall water use in a textile industry.

5.3.2.2 Sugar Industry

The sugar industry is a very important industry due to its product, "sugar," which has a universal use. The sugar industry is huge and is highly energy intensive, which makes it a great user of natural resources to fulfill its energy demands. The main user of the total energy consumed is the evaporation of water from the sugar juice. The membrane processes have the ability to reduce the energy consumption of the sugar industry due to their energy, cost, and performance effectiveness. Nowadays, membrane processes like microfiltration, ultrafiltration, nanofiltration, and reverse osmosis are employed on a large scale in the sugar industry. The number of applications where membrane processes can be implemented in the sugar industry is reduced because of some of the problems, including the high volumes pumped, high viscosity, and high osmotic pressure of sugar juices. Therefore, membrane processes are limited only to clarification, purification, and recovery. Some of the important and major applications of membrane processes in the sugar industry are discussed in the subsequent sections.

5.3.2.2.1 Clarification and Concentration of Raw Sugar Juice

The raw sugar juice is clarified by the use of a microfiltration and ultrafiltration membrane. These membrane processes are successful here, because of the low viscosity of the juice, dissolved solid concentration, and temperature. This also reduces the need of any prefiltration technique. The membrane processes help in the removal of proteins, colors, starch, gums, and colloids. Research has shown that membrane processes in the cross-flow mode with optimum transmembrane pressure and velocity is best in terms of flux and reduced fouling.

The reverse osmosis membrane process is used for the concentration of the sugar juice. The conventional method of sugar juice concentration—evaporation—is a very energy intensive process. Therefore, the reverse osmosis membrane process helps in reducing the energy consumption to a great extent. The problem associated with the reverse osmosis membrane process is that the energy demands increase for the concentrated sugar juice because of its higher osmotic pressure. Therefore, there is a need in the development of better membranes and membrane technology, though to reduce the pressure on the membrane process, a combination of both the reverse osmosis membrane process and evaporation are used for the concentration of sugar juice.

5.3.2.2.2 *Clarification of Press Water*

Press water stream is the stream originated from the pressing unit, where the pulp is pressed after the extraction process. The press water consists of sugars (60%–80%), salts (20%–40%), colloids, and other impurities with a large amount of water. Nanofiltration and reverse osmosis membrane processes are used for the purification of press water. The permeate of these processes can be sent to the extraction unit and the concentrate to the crystallization unit.

5.3.2.2.3 *Brine Recovery*

The sugar industry uses resins for the removal of colorants like melanin, melanoidines, caramels, and polyphenols. These colorants are adsorbed on the resins and the resins are regenerated by using a brine solution. This results in the development of a big pollution problem, since the stream originated from this process is high in salinity, colored organic matter, and chemical oxygen demand. Ultrafiltration membranes were used for the clarification of this stream to remove the solids from it, but nowadays nanofiltration membranes are widely used for the brine recovery. These membrane processes are capable of removing the solid content as well as other impurities and allows NaCl to pass through it. This permeate is again used for the removal of the colorants and thus reduces the total brine consumption.

Therefore, membrane processes reduce the consumption of water, brine, energy, space, time, and cost. Still there is a need of development of better membranes so as to increase their efficiency and performance.

5.3.2.3 Food Industry

The membrane processes with their recent developments turned out to be a critical technology for the food industry. Nowadays, many food and beverage processes are not possible without the use of membrane processes. The main food-based industries that use membrane processes are dairy, beverage, and food additives. Membrane properties like selectivity, low temperature operation, no need for chemical additives, easy to scale up, low on energy consumption, and low cost make them perfect for use in food industries. In addition to being an integral part of the food industries, membrane processes are also used for the treatment of wastewater generated in these industries. This makes the industries capable of reusing the treated water and helps in the recovery of important resources. This further increases the importance of membrane processes in the food industry. Mostly pressure-driven microfiltration, ultrafiltration, nanofiltration, and reverse osmosis membrane processes are used in the food industry. Other than these, in some of the food industries membrane contactors, pervaporation, electrodialysis, and membrane bioreactors are also used.

Understanding membrane processes is a must for any industry to utilize them efficiently; therefore, it is important to know the advantages as well as the disadvantages of different membrane processes. The major disadvantage of any membrane process is fouling. It results in the decline of membrane performance and efficiency, and therefore it is important to be taken care of at right time. Membranes in the food industry are highly prone to fouling due to the nature of the feeds. Thus, cleaning of membranes should be done at a regular interval. The performance and efficiency of a membrane can be reestablished by cleaning it by permeating a cleaning liquid through it by means of flushing and recirculation. The common cleaning liquids are diluted acidic, alkaline, or enzymatic solutions. There is also a need to take extra precautions against the problem of fouling. This includes the selection of module type, membrane, and plant operation. For example, the selection of a membrane module with correct feed channel height could prevent the channel blockage. It is assumed that the channel height should be 10 times the size of the largest feed particle for plate-and-frame or tubular membrane modules to avoid channel blockage. Similarly, in the case of spiral wound membrane modules, the channel height should be 25 times of the size of the largest feed particle to avoid channel blockage. On the other hand, it should be taken care that the more open channels require larger pumps to pump the feed. It is also assumed that it is better to operate the modules under a turbulence flow profile, since it will reduce fouling of the modules due to the effect of turbulence. But this approach also has the side effect that higher pressure drops are generally seen with turbulence flow profiles, which leads to increased energy consumption and costs.

The most effective way of tackling the problem of fouling is by the use of low fouling membranes. The hydrophilic membranes are best for the food industry, as they are of less fouling nature. In addition to the antifouling nature, they should also be capable of handling a wide range of pH (which is common in the food industry), have chemical and mechanical resistance, and be capable of operating at high pressures and in some cases at higher temperatures. Membranes with these profiles are best for the food industry, but they are very sparse in number. Therefore, there is a need for the development of membranes with such properties. It is important for researchers to bring all the desired properties in a single membrane. Future membrane enthusiasts should also think and work on this track.

In the subsequent sections, the importance of membrane processes, functions, applications, and properties is discussed in detail.

5.3.2.3.1 Dairy Industry

The dairy industry utilizes membrane processes in its various processes for purification, concentration, and fractionation of different products. The most talked about use of membrane processes in the dairy industry is the processing of whey, a waste product from cheese industry, into a valuable product. The increased use of membrane processes in the dairy industry is due to their most important property: selectivity. Because of selectivity, membranes are capable of separating and purifying different valuable milk products without the use of heat or additives. In addition to pressure-driven membrane processes, electrodialysis is also employed for the demineralization of whey. Some of the important uses of membrane processes are discussed next:

- Bacteria and spore free milk production—It is important to remove any bacteria or spore present in the milk before its consumption or use in the production of different dairy products. Pasteurization is the conventional technique where milk is heated at a specific temperature for a specific time period to remove the bacteria or spores. This process alters the organoleptic and chemical composition of the milk. On the other hand, the microfiltration membrane process provides a solution for the removal of bacteria and spores without using heat. This maintains the organoleptic and chemical properties of the milk and thus it can be used in its natural flavor and composition for the production of various milk-based products or direct consumption.
- Skim milk production—Membrane processes are used for the standardization, concentration, and fractionation of skim milk. The standardization, as can be inferred from the term, is the standardization of the milk to maintain the natural protein content of the milk. This is important for customer acceptance and satisfaction. It can be carried out by using an ultrafiltration process without the need of adding milk powder, casein, or whey protein externally to the milk. On the other hand, membrane processes like microfiltration and ultrafiltration are also used for the concentration of the milk so as to produce milk protein concentrate. This milk protein concentrate is further used in the production of various other milk products with variable amounts/concentrations of milk proteins. The reverse osmosis membrane process is also used for the concentration of milk, as an alternative to evaporation for the removal of water from milk without disturbing the solid content profile of the milk. The product of this process is generally used in ice cream production. As a matter of fact, the microfiltration membrane process is also used for the fractionation of milk protein from skimmed milk. The membrane process is operated at a constant transmembrane pressure and results in the separation of micellar casein from whey proteins, which is used in cheese production. Later, the whey protein permeate can be concentrated by using the ultrafiltration membrane process to get whey protein concentrate. By using ion-exchange membrane process products like lactoferrin, β-lactoglobulin,

and α-lactoalbumin can be separated from this whey protein concentrate. Therefore, membrane processes can be said to be a technology that helps in processing milk without wasting a single part of it and making use of every part or constituent of it by converting into a product.

- Whey protein processing—Whey is a by-product of the cheese production process. It mainly consists of water and is high in mineral and sugar content, therefore it has a high biological oxygen demand that makes it difficult to dispose in the environment. The membrane processes came out to be a boon, since they have the capability to convert this waste product into a valuable product and solved the problem of its disposal. The microfiltration membrane process is used to remove fat and bacteria from whey, after which its concentration, demineralization, and fractionation can be done. The whey can be concentrated directly by using the reverse osmosis or nanofiltration membrane process. This reduces the water content and total volume, which makes it easier to transport or perform evaporation and spray drying. Whereas the ultrafiltration membrane process, if utilized for the concentration of whey, allows the total nutritional and functional proteins to be extracted from whey. Therefore, the ultrafiltration membrane process is the most widely used whey concentration process for the production of whey protein concentrate. Furthermore, the nanofiltration and reverse osmosis membrane processes can be used for ultrafiltration whey permeate demineralization and preconcentration, respectively, before the production of lactose from its crystallization. This produced lactose that can be used in beverages as an ingredient.
- Cheese production—Membrane processes are also extensively used in the production of cheeses such as cream cheese, feta cheese, and quark. The membrane processes are basically used for the concentration of milk for cheese production. The permeate form of this concentration of milk by the ultrafiltration membrane process can be used by the reverse osmosis membrane process as a feed for the recovery of lactose, and the permeate of the process can be used for flushing the setup after pasteurization or UV treatment.

5.3.2.3.2 Beer Industry

Membrane processes are useful in the beer industry for their role in the separation, sterilization, and purification of beer. The main constituents for beer production are water, malt, hops, and yeast. The beer process starts with the crushing of malted barley seeds. The malt is then mixed with fermenting water and is heated. This results in the formation of a thick, sweet liquid known as wort. This wort is then boiled in a wort kettle. Before the inception of the fermentation process, the wort had to be clarified so as to remove the hot trub (precipitated proteins, hops, and malt). Here comes the role of membrane separation processes, since it is important in a beverage industry to maintain the flavor and quality of the beverage and membrane processes are best for this purpose. The microfiltration membrane process is used for the separation of hot trub and wort for better a fermentation process for the production of beer. Usually, tubular membranes are used for this purpose. The fermentation process takes place in two stages. First is primary fermentation where the yeast grows by multiplication and carries out the fermentation of the fermentable sugars by converting them into alcohol and carbon dioxide. This results in the formation of green beer. The second fermentation stage is the secondary fermentation, maturation, in which all the available fermentable sugars are converted into alcohol and the produced beer is saturated with carbon dioxide. After this stage of fermentation, the produced beer is to be separated from the biomass and other solids. The process of separation is carried out with the help of a separator and diatomaceous earth, or cross-flow microfiltration. Nowadays, the continuous microfiltration membrane process using hollow fiber membranes is used in the beer industry for the separation process. Therefore, the membrane process helps in the removal of yeast, other microorganisms, and haze from the produced beer without affecting its quality and flavor. It is better than the conventional kieselguhr filtration and also reduces the associated health risk and waste disposal problems present in case of conventional kieselguhr filtration. Finally, the clarified beer is sent to the bright beer cellar or, as per need, the alcohol concentration is reduced by using the reverse osmosis membrane process. After adjusting its alcohol content, the beer is pasteurized for the removal of any harmful microorganisms and to increase the life span of the beer, but nowadays it is replaced with membrane-based sterilization processes. Membrane processes are capable of sterilizing beer without affecting its natural flavor, freshness, and quality. Beer sterilized by membrane processes is also known as "cold beer," as there is no use of heat in the beer treatment. It is highly appreciated by consumers and costs on the same level as flash pasteurized beer.

The microfiltration membrane process can also be used to separate and utilize both the beer and yeast present at the bottom of the fermenters. If this is successful, then the amount of beer equals 1% of the total yearly production of a beer factory. This amount varies a lot depending upon the capacity of the factory. Also, membrane processes like pervaporation and membrane contactors are finding new opportunities in the beer industry. Membrane contactors are used for carbon dioxide removal followed by nitrogenation, so as to get a dense foam head, removal of oxygen for beer preservation, and to increase the life expectancy of the product and production of deoxygenated water for the dilution of high gravity brewed beer. Similarly, pervaporation is used for the recovery of flavors from the original beer so as to add them to a nonalcohol beer.

5.3.2.3.3 Wine Industry

The three initial steps in wine making are crushing, pressing, and centrifugation of the harvested grapes to obtain the grape must. The important thing here is the sugar content of the must that has to be controlled or adjusted for better wine production. One of the conditions of alcohol production is that it is directly proportional to the amount of sugars present. This can be achieved by using the reverse osmosis membrane process. For higher sugar content, ultrafiltration and nanofiltration membrane processes are used in combination. The ultrafiltration permeate high in sugars and acids is concentrated by using nanofiltration. The reduced sugars and acid-rich permeate of the nanofiltration membrane process is fed to the fermentation tank.

Unlike beer, after primary fermentation the wine is clarified by using microfiltration or diafiltration. The secondary fermentation process starts in the oak barrels or stainless steel tanks, where the clarified wine is stored. Later, after the completion of the fermentation process the wine is again clarified either by using conventional kieselguhr filtration or microfiltration. Usually, in the microfiltration membrane process, polymeric hollow fibers or tubular ceramic membranes are used. Depending upon the wine type, membranes with pore sizes in the range of 0.2 to 0.65 μm are used for the separation, purification, and sterilization of the produced wines. Electrodialysis is also used in some cases before bottling so as to remove the tartrate salts (potassium and calcium), which may affect the wine quality by precipitating during the wine storage. The conventional method for this purpose is cooling of the wines, which hastens the precipitation of tartrate salts (potassium), but the limitation of this method is that all the tartrate salts (e.g., calcium tartrate) are not removed from the wine. Therefore, electrodialysis is used, which removes all the tartrate salts efficiently and effectively.

The nanofiltration and reverse osmosis membrane processes can be used for alcohol content regulation in the produced wines. It is important because the sugar content in the grapes chosen for wine production is high, resulting in the production of wine with high alcohol content. This high alcohol content masks the taste of other flavors present in the wine. Therefore, it is important to regulate the alcohol content to a proper limit. The water from the nanofiltration and reverse osmosis membrane processes can be dealcoholized by using membrane contactors, and the dealcoholized water can be reused in the wine production or other processes. Membrane contactors are also considered as an option for the direct removal of alcohol from wines instead of using nanofiltration or reverse osmosis. The pervaporation membrane process is also capable of dealcoholization of wine but is not considered because it results in the loss of the characteristic flavors of a wine. The microfiltration membrane process can also be applied for the recovery of wine from the wine lees. Wine lees are the sediments present at the bottom of the wine storage tanks during wine production in the form of yeast, seeds, or fining agents along with some amount of wine. These sediments are developed while transferring wine from one storage tank to another. The advantage of the microfiltration membrane process for this application is its simplicity, high efficiency, and performance.

Some of the other advancing applications, which are unlike the standard membrane applications in the wine industry, are used either for rejuvenating the produced wines or for adding further enriching flavors, colors, and antioxidants to the produced wines. These applications help in enhancing the produced wine qualities, which increase their demand in the market and thus increase the net profit of the industry. This positive output raises the morale of the people working in the field and thus raises chances of further development in the industry for better results and achievements.

5.3.2.3.4 Fruit Juice Industry

Membrane processes are successfully used in the fruit juice industry for the separation, purification, concentration, and sterilization of fruit juices like apple, orange, mosambi, and pineapple. The fruit juices are extracted from the fruits by crushing the fruit followed by pressing. Conventionally, time-consuming and costly processes are used for the clarification of these extracted fruit juices by using enzymes, gelatin, bentonite, or other filtration methods. On the other hand, membrane processes, in particular ultrafiltration, provide better, fast, cost effective, and efficient solution for the clarification of extracted fruit juices. The end product is completely natural with natural taste, flavor, and nutritional values. The membrane processes are very efficient and fast. Thus the total recovery reported is also very high (almost 98% to 99%) and cost effective as compared to the conventional methods. After the clarification process, the extracted fruit juices are concentrated and sterilized before packing and transportation. Conventionally, the evaporation process is used for concentrating the fruit juices and pasteurization to sterilize the concentrated fruit juices. These processes are time consuming and they deteriorate the juice, thus affecting its quality by decreasing its flavor and nutritional values. Therefore, membrane processes are a better alternative, particularly reverse osmosis. The reverse osmosis membrane process is used for the concentration of the extracted fruit juices followed by an optional pasteurization process. Reverse osmosis in combination with evaporation is also used for better results in terms of time, efficiency, and performance. Other evolving uses of membrane processes in fruit juice industries include enzyme-based membrane reactors for the reduction of enzyme costs in the pretreatment of the extracted juices; membrane distillation and osmotic evaporation integration for quality, cost, and energy-effective results; aroma recovery from the fruit juices by using pervaporation; and electrodialysis for the deacidification of fruit juices.

5.3.2.3.5 Food Additives

Food additives are important commodities in food industries due to their use for enhancing the food product properties including regulation of taste, flavor, and product appearance. Because membrane processes work under ambient conditions of temperature and pressure, they are perfect for the production, separation, purification, and concentration of food additives. Some of the food additives are animal blood plasma, gelatin, and carrageenan. These are the by-products of animal or plant-based food industries. Membrane processes make it possible to recover these products from their sources without affecting their natural properties. Usually, the ultrafiltration process is utilized followed by reverse osmosis or a combination of reverse osmosis and nanofiltration for the separation, purification, and concentration of these products.

5.3.2.4 Tannery Industry

The tannery industry is one of the major water-using industries. Tanneries convert raw materials (mostly animal skins) into products like shoes, clothes, upholstery, and leather goods. This conversion from a raw material to a full-furnished product needs a series of chemical and mechanical treatments. The water use mainly occurs in the chemical steps of treatments, where the raw material is treated with different chemicals like acids, bases, tannins, chromium salts, sulfides, dyes, and solvents in an aqueous phase or solution. Therefore, a huge amount of wastewater is generated in this industry, which, if treated, can be the source of water for the same. The wastewater coming from different processes of a tannery consists of high amount of organic and inorganic content, which results in severe water, air, and land pollution problems, and thus the tannery industry is considered as one of the major polluting industries. This is made worse by the use of a very low level of technology in the processing of the raw material. Proteic and lipidic components from the raw material mainly comprise the organic pollutants, and some of the organic pollutants are added during the processing of the raw material like tannins. The different chemicals used in the tannery industry comprise inorganic pollutants that are not completely fixed or washed off during the processing due to the low level of technology and efficiency of the processes in the tannery industry.

The wastewater from all of the tannery industry processes combined and submitted to physical, chemical, and biological treatments results in the formation of sludge that is difficult to reuse and thus dumped into the environment causing environment pollution. Therefore, to make the treatment process more efficient, nowadays the wastewater from different tannery industry processes is treated separately. In addition to this, membrane processes are used extensively for the tannery wastewater treatment. This results not only in the treatment of the wastewater but also helps in the recovery and recycling of primary resources. The membrane processes are also employed to remove the salt and biomass content from the global water of the tannery industry at the end treatment process. This will result in the sustainable growth of the tannery industry. The use of water and other primary resources will also be decreased to a great extent. In the following sections, the different processes of a tannery will be discussed along with the use of membrane processes.

5.3.2.4.1 Soaking

Soaking includes the treatment and soaking of skins in water along with some imbibing agents so as to increase the hydration of proteins, for the solubilization of denatured proteins; for the elimination of the salts used for the preservation of the skins; to hydrate and open dried skin contract fibers; and for the removal of blood, excrement, and earth attached to the skin.

The wastewater from this process consists of chemical additives, salts, and excrement. Generally, the wastewater is sent directly to the treatment plant. Membrane processes can be used for better treatment of the wastewater instead of sending the wastewater directly to the treatment plant. Ultrafiltration membranes are the best solution for this wastewater. Prior to the submission of feed to the ultrafiltration membrane process, the wastewater should be pretreated with the physical treatment methods of sedimentation and filtration so as to decrease the fouling and concentration polarization problems of ultrafiltration membrane. The ultrafiltration membrane process will help in concentrating the organic waste, and the clear permeate along with the salt content can be reused in the pickling process. This will reduce the water consumption as well as the cost of the process.

5.3.2.4.2 Unhairing–Liming

The unhairing and liming process includes the removal of hair as well as other nonessential components from the skin that are not going to be transformed into leather like subcutaneous adipose tissue. Also, this process (liming) results in saponification of natural fats and elimination of interfibrillar proteins, mucoids, and swelling of the derm. The liming process generally uses lime and sodium sulfide or sulfhydrate. The waste from this process consists of compounds released after degradation of hairs and epidermis, that is, sulfide, alkalis, and amines. The ultrafiltration membrane process results in the separation of the sulfide and lime in the permeate along with low molecular weight proteic compounds. The permeate can be used in the liming process. The other high molecular weight compounds are concentrated as retentate. The maximum amount of sulfide that can be recycled by using an ultrafiltration membrane process is in the range of 55% to 60%. The use of membrane processes in the treatment of the process wastewater again resulted in the reduction of water consumption, pollution, and cost of operation with good recovery of resources.

5.3.2.4.3 Deliming and Bating

The deliming process includes the removal of excess lime from the skins. Acids or acidic salts are used to remove the excess lime from the skins. The bating process involves the use of enzymes so as to make the skin soft. Proteolytic enzymes are used in this process that act upon the dermis fibrous structure and make the skin soft.

In this process, to increase the wastewater treatment efficiency by reducing the overall polluting load of the wastewater from the process, ultrafiltration membranes are employed. The membrane process results in the removal of fatty substances and a considerable reduction in the overall chemical oxygen demand of the process wastewater. The permeate from the ultrafiltration membrane process can be used as fresh bating bath or as washing water.

5.3.2.4.4 Degreasing

Degreasing includes the removal of fats from the skin. The presence of fat in a skin results in a leather with undesirable traits, such as hardness to touch, decreased physical strength, and dyeing flaws. Organic solvents are used in the removal of fats from the skins. These organic solvents are harmful to the environment, since they emit volatile compounds. Also, they present problems to the biological treatment processes. Therefore, it is important to get rid of these solvents and there is a need for safe and efficient solvents for this process.

The ultrafiltration membrane processes provide a solution. They are able to efficiently separate the fat from the wastewaters, and the resultant permeate, containing the organic solvents, can be reused in the process. Thus, the problems of wastewater treatment as well as the disposal of organic solvents are both successfully managed by using the ultrafiltration membrane process.

5.3.2.4.5 Pickling

Pickling involves the removal of residual lime from the skins by acidification and dehydration of the fibers. Acid solutions (sulfuric, formic, chloridic, lactic) were used for the treatment in the presence of suitable salt concentrations (sodium chloride, sodium sulfate, and salts of the acids used). The treatment of the wastewater from this process is not possible with conventional cleaning processes, since these treatment processes are not able to bring the salt concentrations below the required regulatory concentrations. For instance, the chloride concentration of these solutions is equivalent to 9 g/l, which is a problem for biological treatment plants. Therefore, membrane processes are a good option for the treatment of this process wastewater. The reverse osmosis membrane process is better because the osmotic pressure of the wastewater is equivalent to 1.2 to 1.5 MPa. Therefore, after pretreatment the pickling process wastewater can be submitted to a reverse osmosis membrane process for treatment. The salts are recovered in the form of retentate and the permeate can be reused for the same process after salt concentration adjustments or can be used in soaking baths or as washing water.

5.3.2.4.6 Chromium Recovery

The process of tanning uses various types of substances including chromium salts, oils, and aldehydes to avoid putrefaction of the skins. Chromium sulfate is the most widely used salt in the tanning industry for this purpose. The wastewater from chromium tannage contains about 30% of the initial salt used.

Chromium is a highly toxic and carcinogenic pollutant. Therefore, it is necessary to remove or convert it into its nontoxic state. Generally, wastewater from a chromium tannage is sent directly to the treatment plant where it ends up as sludge, which is difficult to treat and dispose of. The traditional method of chromium recovery is the precipitation of chromium salt with sodium hydroxide followed by the dissolution of chromium hydroxide into sulfuric acid. The recovery of the chromium by this method is not satisfactory and therefore there is a need for other efficient methods. The nanofiltration membrane process is a better choice for the removal of chromium from chromium tannage wastewaters. The membrane process is quite efficient for the removal of chromium and gives better results compared to other techniques. Chromium-free permeate can be used as washing water or in other processes of tanning, especially pickling due to the high concentrations of chloride. The retentate with high chromium concentrations can be utilized in the process again. This helps in the recovery and reduction of costs and resources of the process.

5.3.2.4.7 Vegetable Tannage

This process involves the stabilization of the skins by using vegetable tannins such as chestnut, memosa, and quebracho. Membrane processes like ultrafiltration and nanofiltration can be employed for the treatment of wastewater generated from this process. The nanofiltration membrane process can yield a retentate with increased concentration of vegetable tannin that can be reused in the process and the permeate water can be further sent for treatment to the wastewater treatment plant.

5.3.2.4.8 Dyeing

Dyeing of skins enhances their appearance and commercial value. The skins are treated with different dyes in aqueous solutions. Because no process is 100% efficient, a lot of dye used is sent off to the wastewaters. The amount of dye in the wastewater is also high due to the use of inferior techniques.

The dyes are toxic and harmful for the flora and fauna and some of them are even carcinogenic in nature; therefore, they should be properly removed from the wastewaters. The reverse osmosis membrane process is suitable for this purpose because of the small size of the dye molecules. The dyes and dyeing substances can be recovered in the retentate and the permeate water can be reused for different applications in the tanning industry, most preferably for the preparation of new dye baths.

5.3.3 Biotechnology and Health Care

Membrane processes over time have developed and matured. The use of membrane processes in the field of biotechnology and health care has also seen a great upward curve. Membrane processes are used in these fields for applications like product production and separation and analysis, immunoassays, sterilization of products or raw materials, sample preparations, drug delivery, hemodialysis, and blood oxygenation. Mostly microfiltration and ultrafiltration membrane processes are used in biotechnology and health care.

The microfiltration membrane process is primarily used for applications where the removal of an undesired entity is important, for example, sterilization of injectable biotechnology or pharmaceutical products involving the removal of viable particles like bacteria, virus, and fungus.

Developments in membrane technology have broadened the application of the microfiltration membrane process. Nowadays, it is used for the cell harvesting from bioreactors and also as a component of bioreactors, removal of blood cells from plasma, sample preparations, and protein purification. In these applications, microfiltration membranes are used as small discs in case of small applications and as stacks of sheets housed within cartridges for large-scale applications. These are generally used in multilayer fashion so as to reduce the risk of contamination of a product by increasing the particle retention capability of the membrane process. This is highly required in the biotechnology and health care fields, since the slipping of a bacteria or virus through a membrane could be a life-taking cause.

Microfiltration is boon for applications like sterilization of products that are heat sensitive and denatures at high temperatures like proteins. Microfiltration with pore diameters around 0.2 to 0.22 μm is used for the separation of bacteria from a solution. The bacteria removal efficiency of microfiltration is checked with a marker bacteria like *Brevibacterium diminuta* or *Serratia marcescens*. As compared to bacteria removal, virus removal is quite challenging; therefore, specialized membrane filters are produced that have properties similar to the nanofiltration membrane process. The efficiency for virus removal is checked by using solutions spiked with animal viruses (e.g., porcine parvovirus) or bacteriophages (e.g., coliphage $\phi\chi$-174). Test solutions containing these particles are permeated through the filters and then detected and identified by plating. Live, attenuated, and killed bacterial as well as viral vaccines are also commonly purified by the microfiltration membrane process.

Nowadays, bacteria, yeast, and mammalian and insect cells are efficiently harvested from a bioreactor by using the microfiltration membrane process. In addition to this, separation of blood cells from plasma by using microfiltration is also gaining importance. Microfiltration membranes are also used as the base for the immobilization of biomass by means of affinity entities grafted on their surface. This will increase the efficiency and performance of the bioreactors. For gases, they are used to submit sterilized gases to the fermenters and bioreactors.

Generally, the microfiltration membranes to be used for the sterilization of liquids are hydrophilic in nature, and the membranes dealing with gases are hydrophobic in nature. Also, the membrane modules should themselves be sterilizable.

Ultrafiltration membranes have pore sizes in the range of 5 to 100 nm. The main application of ultrafiltration membrane processes is in the purification or separation and analysis of proteins. They are also used for the purification of nucleic acids like plasmid DNA. Other important applications of the ultrafiltration membrane process are hemodialysis and blood oxygenation.

Thus, it can be said that membrane processes play an important role in the fields of biotechnology and health care. Two important applications—protein purification and hemodialysis—involving membrane processes will be discussed in the subsequent sections.

5.3.3.1 Protein Purification

The main usage of the ultrafiltration membrane process in protein separation is said to be for these following four forms:

- Protein concentration—Protein concentration is basically the removal of water from a protein solution. It can be a pretreatment step before applying the conventional protein purification techniques of precipitation, crystallization, and chromatography. It is also used for the adjustment of protein content in the pharmaceutical formulations.
- Desalting—Desalting is the removal of salts and other low molecular weight compounds from a protein solution. For example, with salt-induced precipitation, it is the removal of precipitating salts such as ammonium sulfate and sodium sulfate from protein solutions.
- Clarification—This is the removal of bacteria, viruses, or other undesired entities from a protein solution. Prion removal by using ultrafiltration is also gaining ground.
- Fractionation—This involves the separation of one protein from other proteins by using the ultrafiltration membrane process. This highly depends upon the feed pH, ionic strength, protein–protein and protein–membrane interactions, and permeate flux and mass transfer coefficient.

5.3.3.2 Hemodialysis

Hemodialysis is the technique to purify the blood of a person whose kidneys are not working properly. Membranes with better hemocompatibility and physiological functions imitations are needed in this process. The membranes should have better flux and selectivity. They should not allow any toxic material to bypass them. This requires membranes with stringent pores that can at the same time maintain the flux. Membrane materials with better hemocompatibility and antifouling properties are required, since these two are critical for the better performance of the hemodialysis process. Innovative membranes with self-cleaning, good hemocompatibility, and flux profiles were developed by using advanced or modified materials such as hydrophilic nanocomposites, carbon nanotubes, and block copolymers. There is still need for further developments in the membrane science for the development of compact, fast, and efficient hemodialysis membrane process.

5.3.4 Separation of Racemic and Azeotropic Mixtures

Liquid separation, whether it is for recovery of a valuable compound, removal of a solid, or purification, is an important task in every industry. Liquid separation becomes challenging when the compounds present in it are chiral in nature, for example, amino acids and sugars. In 1848, Pasteur separated two enantiomers of tartaric acid. These two enantiomers had similar chemical qualities and crystalline forms but showed different properties toward a polarized light in a solution. Compounds like amino acids and sugars are important with huge demands and thus need to be separated effectively and efficiently. Enantiomers are required in pharmaceutical, biotechnology, and food industries. The other important fact about them is only one of the enantiomers shows activity towards a function and the other shows no or a much weaker effect as well as may have side effects or even toxicity. For example, thalidomide is a racemic drug used to treat nausea during pregnancy. The D (dexter; on the right) isomer of thalidomide is a safe sedative, but the L (laevus; on the left) form causes severe birth defects and deformities. Conventionally, distillation is used for the separation of enantiomers but cannot be used for the separation of azeotropic mixtures, where the vapor phase has the same composition as a liquid phase. Also, this technique has a major drawback in that it is a highly energy intensive process and have limited entrainers for use. Therefore, there was the need for techniques that could carry out the separation of enantiomers and azeotropic mixtures. Membrane processes are both efficient and capable of effectively separating enantiomers and azeotropic mixtures.

Pervaporation is the most important membrane process used in the chemical and petrochemical industries. Pervaporation is effective and efficient in enantiomer and azeotropic mixture separations. The basics of the pervaporation technique are discussed in Chapters 1 and 2. In short, the difference in the partial pressures of the feed side and permeate side components is the driving force for the separation. An important fact is that the driving force is not affected by the relative volatility of the feed components. This makes pervaporation suitable for the separation of enantiomers and azeotropic mixtures. Pervaporation membranes need both high affinity as well as selectivity for a component. If the affinity is high, then the membrane will swell and its selectivity will decrease, and vice versa. Therefore, a membrane with optimum affinity and selectivity is required. The membrane material plays a crucial role in this and thus a better membrane material is required for the development of a better membrane.

Other membrane processes used for the separation of enantiomers and azeotropic mixtures are liquid membranes and ultrafiltration. These membrane processes are carried out by using enantioselective membranes (in which enantioselector components, which have affinity for a particular enantiomer, are added) and nonenantioselective membranes (in which the membranes are integrated with other chiral separation processes). The enantioselectors are either grafted or immobilized in the membrane for better separation properties. For example, in a supported liquid membrane, an enantioselector is immobilized and carries out the enantiomeric separation. The advantages of a supported liquid membrane over an emulsion liquid membrane and immobilized liquid membrane is that it is very thin and thus requires fewer chiral carriers or enantioselectors and is also fast. There are two types of solid enantioselective membranes. They are inherent chiral membranes and functionalized chiral membranes. The inherent chiral membranes are prepared from chiral material and the functionalized chiral membranes are functionalized with chiral materials. The solid enantioselective membranes can further be categorized into two categories based on their separation mechanism: diffusion-enantioselective membranes and adsorption-enantioselective membranes. Generally, the inherent chiral membranes act as diffusion-enantioselective membranes because they are not able to bind the enantiomers firmly, and functionalized chiral membranes are adsorption-enantioselective membranes because they can easily adsorb the enantiomers on their surface due to the presence of chiral groups, with an affinity for enantiomers, present on their surface. Some examples of chiral materials are norbornadiene, disubstituted acetylene, and diphenylacetylenes. They have optically active pinanyl groups and are used for the preparation of chiral membranes for the separation of enantiomers and azeotropic mixtures. Generally, antigens, antibodies, and enzymes are used as enantioselectors in the preparation of adsorption-enantioselective membranes.

5.3.5 Gas Separation

Gas separation based on membranes on the industrial scale is a developing field. The membranes are famous for this application because of their efficiency, low energy consumption, and low costs as compared to conventional technologies. Membranes were analyzed for gas separations for the first time by Mitchell in 1830. Mitchell observed that hydrogen-filled balloons deflate over time. He concluded that the deflation of balloons is happening because of the diffusion of hydrogen across the balloon wall. This was later confirmed by Graham in 1866 by reporting the gas permeation rates through films made up of natural rubber. Though the discovery of gas permeation through polymer films was accomplished in the early 19th century, the actual use of polymer films for gas separations came to light in the late 20th century after more than 100 years. The first polymeric membranes prepared to be used for gas permeation were very thick with high flux rates. The discovery of asymmetric membranes by Loeb and Sourirajan in 1963 brought to light the field of membrane science and then started the era of membranes. In 1980, the first commercial polymer membrane gas separation system came to market.

In recent times, industries are employing membrane gas separation processes for separation and recovery of hydrogen and ammonia; in applications like carbon dioxide–enhanced oil recovery, oxygen enrichment, nitrogen production, natural gas processing, air dehydration, landfill gas upgrading, and helium recovery; and in refineries and petrochemical industries. The polymer membrane gas separation market is expanding annually with great speed. The main reasons for this growth are the demand for high-quality products, regulatory pressures, depleting natural resources, and environmental and economic causes. Some membrane gas separation applications are explained in the following subsections.

5.3.5.1 Air Separation

Membrane gas separations are used for the separation of different individual gases from air. The sequence of individual gases in descending order of permeation is water vapor > carbon dioxide > oxygen > nitrogen. Therefore, nitrogen or any other gas can be easily separated from air. It is better to extract oxygen from compressed air, as it is easy, efficient, and economical. The first membranes used for gas separations were spiral wound, but nowadays hollow fiber membranes are trending. Membrane gas separations are best if the residual oxygen percentage of 1% is acceptable.

The membranes for air separation can be operated in either one of two modes: pressure or vacuum. The pressure mode is the standard mode of operation and thus widely used. In this method, the air is pressurized and given to the membrane module as feed. On the other hand, the permeate side is kept at atmospheric pressure. Thus, pressure difference plays the role of driving force for the separation of air in the desired individual gas. The pressure difference in this method is higher as compared to the vacuum mode. This results in the requirement of less membrane area for the purpose. In the vacuum mode, the feed is pressurized a bit higher than the atmospheric pressure and the permeate side is kept at vacuum. In this mode the retentate vents out at atmospheric pressure from the membrane module. This mode is more energy efficient than the pressure mode. In the case of oxygen enrichment, this mode is suitable due to its energy efficiency and the pressure mode is better for the separation of nitrogen. The two modes can also be used together for their synergistic effects. In this mode the feed will be pressurized, as is the case of pressure mode, and a vacuum will be created on the permeate side, as in the vacuum mode. This will result in increased feed to permeate and feed to retentate pressure ratios. The main applications of air separation are in the production of oxygen, water removed from air to be used in oil and gas drilling, regulated and controlled atmosphere, maritime transportation, gases for laboratory use, inert atmosphere, beverage dispensing, and inflation of tires.

5.3.5.2 Hydrogen Production

The membrane gas separation process was first employed commercially for the production of hydrogen. The increased demand of hydrogen resulted in the use of hollow fiber membranes made from polysulfone in the 1980s. Hydrogen gas is highly permeable and thus selectivity for other gases is also high. Membrane separation processes for the production of hydrogen are employed in ammonia purged gas streams, syngas, and refineries. Refineries have increased demands for hydrogen due to the formation of stringent environment and regulatory laws and heavier crude. The product purity can be maintained in membrane gas separation processes by maintaining the feed-to-permeate pressure ratio.

The resources of hydrogen are the air, hydropower, solar, and biomass, which come under the category of renewable sources. Fossil fuels, nuclear, and other sources of hydrogen come under the nonrenewable category. Hydrogen can also be produced from electrolysis of water by using solar energy, as it is both an economic and environment-friendly method. Other options for hydrogen production are different gas streams like carbon dioxide and hydrogen gas mixtures, and syngas. The steam reforming and water gas shift are used for the production of hydrogen from these gas streams. The best use of hydrogen is in fuel cells for energy production, as it is environment friendly and economic. The water gas shift, an old age technology, can be combined with membrane separation processes for the development of new applications of the processes. A membrane reactor with a high temperature-resistant and highly permselective membrane (for hydrogen or carbon dioxide) is required for the combined water gas shift and membrane gas separation processes. The said membrane reactor will continuously work for conversion of steam into products at high temperatures. This will reduce the overall requirement of steam.

5.3.5.3 Carbon Dioxide Separation

Increased industrialization, population, and urbanization have resulted in an increased amount of carbon dioxide emissions into the environment. Carbon dioxide threatens planet Earth and its inhabitants. Therefore, it is necessary and important to separate carbon dioxide from the processes before emitting it into the atmosphere. Membrane separation processes are ideal for this job due to their incredible properties. They are low on energy consumption, cost, chemical use, and complexity, and high on performance, efficiency, and selectivity. Therefore, it is good to employ membrane gas separation processes for the separation of carbon dioxide from the feeds as compared to the old, energy intensive, costly, inefficient, and complex conventional processes.

It is important to remove carbon dioxide from feeds like natural gas, landfill gas, casing head gas in case of oil recovery, and from the environment. The separation of carbon dioxide from these streams is explained next.

5.3.5.3.1 Natural Gas

In case of natural gas, methane (75%–90%) is the main constituent. Nitrogen, carbon dioxide, hydrogen sulfide, and water are present as impurities. These impurities should be removed for better purity of the natural gas. Therefore, membrane gas separation processes are employed for the removal of these impurities. The amine absorption technique is quite complex and costly, and is thus scrapped. Membrane processes are required with better flux and selectivity profiles. Therefore, first the carbon dioxide is separated from the high-pressure natural gas stream by using membrane gas separation processes; this is known as the "natural gas sweetening process." Then the amine absorption process is employed. This results in the production of pipeline quality gas. Thus, the membrane separation processes employed alone need improvements and there is a need for better membranes with good flux, stability, and selectivity profiles.

5.3.5.3.2 *Landfill Gas*

Landfill gas, also known as biogas, is produced under anaerobic conditions from the decomposition of organic matter at atmospheric pressure. Landfill gas consists of methane (54%–59%), carbon dioxide (40%–45%), nitrogen (4%), oxygen (1%), water vapor (1%), and hydrogen sulfide and halogenated hydrocarbons in trace amounts. Methane is the main component of the landfill gas and is recovered from a covered landfill. The pure methane from a landfill can be obtained by using membrane gas separation processes. First, the toxic gases like hydrogen sulfide and halogenated hydrocarbons present in trace amounts are removed by using the adsorption process and then the membrane gas separation processes can be employed for the separation of the remaining gases. In this later step, the gases are compressed and fed to the membrane system. This will increase the efficiency and overall flux of the membrane system.

5.3.5.3.3 *Head Gas*

In the case of enhanced oil recovery, carbon dioxide is used as the injection medium to increase oil production in the existing oil fields. The carbon dioxide used decreases the viscosity of the oil and make it easy for the oil to come up to the surface. The removal of oil from oil fields results in the formation of a casing head gas in addition to the oil stream. The head gas consists of mainly methane with a variety of hydrocarbons. The addition of carbon dioxide for the dilution of oil adds carbon dioxide to the head gas too. The carbon dioxide concentration in the head gas ranges from 40% to 90%. The natural gas and hydrocarbons in addition to the carbon dioxide, present in the head gas, are important and have to be retrieved from the gas stream. The retrieved natural gas is an additional fuel source and carbon dioxide is reused for the dilution of the oil by reinjecting the retrieved carbon dioxide to the oil fields. Therefore, it is necessary to remove the carbon dioxide from the head gas for proper utilization of the products. Carbon dioxide with purity of 95 mol% is sufficient to maintain its dilution powers and this can be easily achieved by using available membrane systems. The membrane systems discussed in the case of landfill gas can also be used in the further purification of retrieved natural gas from the head gas. These include removal of carbon dioxide, hydrocarbons, and dehydration. All of these process steps are efficiently feasible with the help of available membrane systems.

5.3.5.4 Gas–Vapor Separation

Membrane systems are also employed for the separation of vapors and gases, especially from air or nitrogen streams. Some famous examples where membrane systems are employed successfully are in the recovery of ethylene or propylene from polyolefin production and separation of gasoline vapors from air. The membrane systems employed for these processes use rubbery polymers, for example, cross-linked polydimethylsiloxane or reverse selective glassy polymers, like poly[1-(trimethylsilyl)-1-propyne] or polymethylpentene. In the case of gasoline vapors recovery silicone membranes are used, but the silicone rubbers exhibit weakness in the presence of gasoline vapors. Therefore, fluoropolymer membranes represent a better choice over silicone rubber membranes in terms of strength and safety.

5.3.6 Fuel Cells

Fuel cells are electrochemical devices that convert chemical energy directly into electrical energy. The main components of a fuel cell are a cathode, where reductive reactions occurs; an anode, where oxidative reactions occurs; and an ionically conductive electrolyte that separates these two electrodes from each other. The fuel cell operates when a fuel (e.g., hydrogen or methanol) is fed to the anode and an oxidant (commonly oxygen from air) is fed to the cathode. The processes of fuel oxidation and oxygen reduction occur spontaneously with continuous generation (at anode) and consumption (at cathode) of electrons. Heat is also generated during the fuel cell operation since the process is not

TABLE 5.2
Fuel Cells Classification Based on Electrolyte Used

Fuel Cell	Electrolyte
Alkaline	Potassium hydroxide
Direct methanol	Polymer membrane
Polymer electrolyte	Polymer membrane
Phosphoric acid	Phosphoric acid
Solid oxide	Yttria-stabilized zirconia
Molten carbonate	Potassium carbonate/lithium

100% efficient. An external circuit is used to extract the electricity produced from the fuel cell. A fuel cell seems to be similar in function and structure to a battery, but a distinct difference between the two is that the battery is a storage device, the power generation of which directly depends upon the stored (limited quantity) chemical reactant in it, and on the other hand, a fuel cell is an energy conversion device that has the capability of power generation as long as the reactants are provided. The amount of power generated by a fuel cell depends upon various factors, including fuel cell type, size, temperature of operation, and pressure of gases. Also, a single fuel cell only produces enough electricity for a small application. Therefore, it is necessary to club or combine many fuel cells, as per the energy requirements, in a stack form. The basic component that divides the fuel cells in various types is the type of electrolyte used in a fuel cell. These are listed in Table 5.2. From the table it can be seen that there are two fuel cells that use polymeric (cation exchange) membranes as electrolytes: the proton exchange membrane fuel cell and the direct methanol fuel cell. The cation exchange membrane plays an important multirole in these two fuel cell types. As it separates the two electrodes, it separates fuel and oxidant to prevent a chemical short circuit and it also helps in the transport of protons from the anode to the cathode.

5.3.6.1 Proton Exchange Membrane Fuel Cell

A proton exchange membrane fuel cell or solid polymer or polymer electrolyte fuel cell is a highly efficient fuel cell of low weight and volume in comparison to other types of fuel cells. This fuel cell uses a thin cation exchange polymer membrane as an electrolyte and platinum-coated electrodes. This fuel cell is clean and green in its operation and efficiency. It requires hydrogen, oxygen, and water for its functioning in the range of 80°C. The membrane electrode assembly is the key component of the proton exchange membrane fuel cell. This assembly is composed of a thin cation exchange polymer membrane (50–200 μm) and electrodes pressed directly onto the opposite sides of the membrane. The carbon paper or carbon cloth gas diffusion layers are placed adjacent to the electrodes for the proper distribution of the gases to the catalyst coated on the electrodes and removal of the products from the electrode sites. The fuel cell is made by sandwiching the membrane electrode assembly between two metal plates. The plates have channels engraved into them for the supply of fuel and air to the electrodes and for the removal of water. Generally, the thermodynamic potential of a proton exchange membrane fuel cell is equivalent to 1.23 V. Therefore, in case of higher needs two or more cells are stacked together and the required voltage can be generated.

The main advantages of proton exchange membrane fuel cells are their potential to be used in the automobile industry, power plants, and in small-scale power generation applications like in homes and small industries (stationary as well as portable). This will replace batteries that are widely used and have limited power-generation capacities. Also, proton exchange membrane fuel cells are compact and operate at low temperatures and high efficiencies. In addition to advantages proton exchange membrane fuel cells also have some limitations, which have to be rectified for their proper use, like cost of the membrane electrode assembly. The membranes and catalysts used are expensive. Also, it needs pure hydrogen as a feed for power generation, and it usually gets mixed with carbon monoxide and other gases due to their low temperature profile. Therefore, there is a need for membranes to be developed with better performance at higher temperatures that are cost effective.

5.3.6.2 Direct Methanol Fuel Cell

Direct methanol fuel cells are similar to proton exchange membrane fuel cells in structure and function. The only difference is the type of fuel used and therefore the electrode reactions. In the case of proton exchange membrane fuel cell, hydrogen is used as fuel, and for a direct methanol fuel cell, liquid methanol is used as fuel. The membrane electrode assembly and other prominent features are similar to the proton exchange membrane fuel cell. The direct methanol fuel cell is quite new as compared to the proton exchange membrane fuel cell.

The advantages of a direct methanol fuel cell are operation at high temperatures (60°C–140°C) as compared to a proton exchange membrane fuel cell (80°C), and no fuel storage problems as is the case with proton exchange membrane fuel cell. The main problem linked with a direct methanol fuel cell is its low efficiency and performance as compared to a proton exchange membrane fuel cell. This limitation can be eliminated by having a far better performing membrane that will give a platform for better transport and reaction of the reactants onto the electrodes and eventually help in increasing the efficiency and performance of a direct methanol fuel cell. A direct methanol fuel cell has similar applications as a proton exchange membrane fuel cell including in the automobile industry, auxiliary power for instruments and devices, portable electronics, and as a good candidate for the replacement of batteries.

5.3.6.3 Challenges for Fuel Cell Membranes

The main challenges to be overcome in the fuel cell membranes are the following:

- Increased ionic conductivity
- Enhanced chemical and mechanical stability at high temperatures in oxidizing and reducing environments
- Low swelling tendency
- Low crossover of fuel (hydrogen or methanol) and oxygen
- High compatibility with the coated catalyst on the electrodes
- Easy to process in membrane electrode assembly
- Low cost

The aforementioned challenges can be summarized in two points separately for the proton exchange membrane fuel cell and direct methanol fuel cell:

- Proton exchange membrane fuel cells have membranes with high proton conduction at high temperatures and low humidity condition.
- Direct methanol fuel cell membranes, which are sufficiently thin and highly conductive and at the same time can act as a barrier for the methanol crossover.

Presently the Nafion® membrane is the choice of fuel cell manufacturers and users. This membrane was developed by DuPont USA in the 1960s. Currently, there is no pure polymeric membrane available that can give high proton conductivity in anhydrous conditions. Therefore, there are membranes modified with other compounds so as to get membranes with the desired properties. Therefore, all the membranes are modified with compounds, which are high proton conductors or have better water retention capacities. For example, phosphoric-acid-doped polybenzimidazole, Nafion mixed with silica, sulfonated poly(ether ether ketone) combined with mixed-zirconium phosphate-phosphonate, inorganic solid acid membranes like cesium hydrogen sulfate or cesium dihydrogen phosphate, sulfonated poly(arylene ethers), sulfonated polyimides, and sulfonated fluoropolymers. These materials are helping in developing the required fuel cell membranes and are the base of further research for better fuel cell membrane materials. Therefore, it can be said that current research in membrane-based fuel cells is based on the discovery and development of better fuel cell membrane materials.

5.3.7 Osmotic Power Plants

Osmotic power, blue energy, or salinity gradient power is the energy obtainable from the difference between the salt concentration of saline (sea water) and fresh (river water) waters. It is a new form of energy conversion concept, which was recently converted into reality by the Norwegian power company Statkraft. In the year 2009, Statkraft installed and tested an osmotic power plant prototype in Oslofjord, Norway. Two membrane processes—reverse electrodialysis and pressure-retarded osmosis—can be used in an osmotic power plant for power generation.

The basic principle of osmotic power generation is osmosis. In this method, water from lower salt concentrations permeates through a membrane to the other side containing water with higher salt concentrations, which is a simple osmosis process. This process becomes interesting when a fixed volume compartment is present on the higher salt concentrations. This will result in the increase of pressure and this generated pressure can be used to rotate a turbine for power generation. When seawater is used the pressure may increase up to 2.6 MPa, which is equivalent to a water reservoir with height of up to 270 m. Therefore, it can be said that this process is a storehouse of clean and green energy.

Sidney Loeb, the architect of the modern-day membrane revolution, is the person behind the development of this idea into reality. Loeb, being a professor at the time at the Ben-Gurion University of the Negev, Be'er Sheva, Israel, thought of this idea while thinking about the mixing of freshwater from the Jordan River into the Dead Sea as wastage of freshwater as well as a source of potential energy. Loeb developed two techniques for the production of power from this energy: reverse electrodialysis and pressure-retarded osmosis. In the case of reverse electrodialysis, freshwater and salt solutions are permeated through ion exchange membranes. The power is generated by virtue of the chemical potential difference present between the two types of solutions. On the other hand, in the case of pressure-retarded osmosis, spontaneous permeation of dilute solution to the concentrated solution side generates pressure difference, which is utilized to generate power.

Recently, it is the pressure-retarded osmosis that has been utilized for power generation on a large scale by Statkraft, a hydropower company. In the pressure-retarded osmosis process, freshwater is given as feed on one side of the membrane or membrane modules (on a larger scale). The membrane modules are generally spiral or hollow fiber so as to increase the performance and efficiency of the process. On the other side of the membrane or membrane modules, pressurized concentrated salt solution (seawater) is fed. Due to the process of osmosis ~90% of the fresh water permeates to the pressurized concentrated salt solution side of the membrane. This results in the increase of volumetric flow of the pressurized concentrated salt solution into the membrane modules, resulting into the development of high pressures, and this pressure is used to generate power. It is to be noted here that the feed of the concentrated salt solution is double as compared to the freshwater feed. Therefore, membranes with high flux and selectivities are required, which can successfully permeate freshwater at a high rate and efficiently retain salts present in the salty water. The permeate of this process is brackish water, which is generated on the mixing or dilution of salty water with freshwater. One-third (approximately) of this generated permeate is used to generate power by feeding it to the turbine and two-thirds (approximately) of the permeate is recycled to the pressure exchanger to add pressure to the salty water feed. The salty water feed pressure is in the range of 1.1 to 1.5 MPa, which is similar to a water head with height equivalent to 100 to 145 m in a hydropower plant. On the other hand, freshwater feed is fed to the membranes at ambient pressures. The pretreatments of the feeds depend upon the water quality and generally applied for the seawater to filter the unnecessary feed components.

The important factor for this technique to be used with its full potential is the membrane and membrane theory. There is the need to test and validate the membrane theory and hypothesis related to a pressure-retarded osmosis membrane. According to Statkraft, there is need to develop membranes that can generate energy equivalent to 4 W/m^2 so that power can be produced at an optimal cost [6,7]. The commercial reverse osmosis membranes are not able to give more than 0.1 W/m^2, therefore, special pressure-retarded osmosis membranes have to be developed so as to meet the requirements of the process. Thus, there is a large scope for the membrane enthusiasts in this field and they can work on this problem for bringing the green energy initiative to reality.

The planet Earth, being populated and polluted, desperately needs such clean and green sources of energy. Pressure-retarded osmosis has huge potential for meeting the power requirements of the world, since in almost every country of the world a river meets an ocean. Thus, it shows great promise for the development and implementation of this process worldwide. The technologies developed in the field of hydropower and desalination can be utilized for pressure-retarded osmosis power generation with little modifications. Therefore, it is easy for membrane giants to invest in this membrane process too. Pressure-retarded osmosis is a prosperous opportunity for the membrane business fraternity and should be exploited to its fullest due to its green and renewable properties. It is also better than wind, wave, tidal, and hydropower in terms of efficiency and performance. Therefore, at this moment commercialization of osmotic power is very important and necessary.

5.4 STIMULI-RESPONSIVE POLYMERIC MEMBRANES

Polymeric membranes that are responsive to environmental stimuli due to the presence of porous substrates and functional gates carry interesting potential for various applications related to various fields. Membranes depending on the type—porous or nonporous—can be made responsive by grafting or blending responsive materials, respectively. The responsive materials have a tendency to respond to a particular type of external stimuli like temperature, pH, light, magnetic, or chemical. This response to the external stimuli brings change to the degree of swelling or shrinking of the membrane and membrane pores in particular, thereby bringing change in the selectivity and permeability of the membrane. This responsiveness is also used to selectively bind or release a target component. This will be fruitful in the development of membranes with selective adsorption properties, like those required for enantiomer and azeotropic mixture separations. The responsiveness property of the membranes also imparts properties like reduced or self-cleaning, high flux, and high selectivity. Therefore, responsive membranes consist of a plethora of opportunities to develop effective and efficient membranes for various applications. In the subsequent sections, membranes with different stimuli responsiveness are discussed.

5.4.1 TEMPERATURE-RESPONSIVE MEMBRANES

Temperature-responsive or thermoresponsive membranes are one of the famous stimuli-responsive membranes. These membranes are famous because of their easy regulation and operation, since the temperature (ex situ or in situ) can be artificially manipulated and easily designed. There are two types of temperature-responsive membranes. First are positive temperature-responsive membranes, which have the characteristic of increasing permeability with an increase in environmental temperature. The most widely used polymer for this type of functional membrane is the poly(N-isopropylacrylamide) (PNIPAM). If a membrane contains PNIPAM or other positive temperature-response-inducing polymers, then the membrane pores "open" when the environmental temperature is above the lower critical solution temperature (LCST) and "closed" when the environmental temperature is below the LCST. In the case of PNIPAM, the LCST temperature is 32°C. Second, are the negative temperature-responsive membranes. These membranes work opposite to the positive temperature-responsive membranes, because here the membrane pores open when the environment temperature is below the LCST of a thermoresponsive polymer used in the fabrication of the membrane and close when the environment temperature is above the LCST temperature of the used thermoresponsive polymer. Chu et al. [8] used an interpenetrating polymer network for the preparation of a negative temperature response membrane (negatively thermoresponsive). In their study, Chu et al. used upper critical solution temperature (UCST) instead of LCST, though the terms are interchangeable, but the latter is preferred by the researchers. They have used poly(acrylamide) and poly(acrylic acid) for the formation of an interpenetrating polymer network gel, which behaves as a negative temperature-responsive polymer. The two polymer chains behave like a zipper, which is known as the chain–chain zipper effect. The two polymer chains in a shrunken state forms intermolecular hydrogen bonds that are disrupted when the temperature is higher than the UCST of the interpenetrating polymer network gel. Therefore, the membrane pores or gates close below the interpenetrating polymer network gel and open above the UCST. The UCST of the said interpenetrating polymer network is somewhere between 20°C and 25°C.

Temperature-responsive membranes have immense scope in fields like controlled drug delivery and chemical separation, and in bioreactors. Therefore, this field of membrane science is also competitive for further research in the development of medical science, pharmaceuticals, and biotechnology.

5.4.2 pH-Responsive Membranes

pH-responsive membranes are a widely studied and developing category of stimuli-responsive membranes. These membranes are the choice of researchers and membrane enthusiasts because of the options provided by these membranes in terms of both materials and applications. Researchers all over the world are studying the structural and functional features of these membranes. These membranes are exploited for their better use in different membrane-based separation applications. They are modified in various ways to optimize and achieve the best response. The various preparation, modification, and stimulation methods are evaluated for the improvement or addition of some novel features to pH-responsive membranes.

pH responsiveness has been researched for drug and gene delivery, and has shown much future potential as a successful technique. Nowadays, pH responsive membranes are used or researched for their applications in various fields due to their unique applications. The basic way of pH-responsive membranes functioning is by the opening and closing of membrane pores due to the change in the environmental pH. The pH-sensitive groups will swell or shrink based on the environmental pH. The swelling or shrinkage also depend upon the type of material. Some of the materials, like materials containing a carboxyl group, swell at high pH values and materials containing a pyridine group swells at low pH values. The alkali-swellable carboxyl-containing materials swell at high pH values because of the dissociation of the carboxyl group into carboxylate ions, which results in a high charge density in the material and shrinks at low pH values because of the protonation of the carboxyl groups and domination of hydrophobic interactions. On the other hand, the acid-swellable pyridine-containing materials swell at low pH values due to the protonation of the pyridine groups, which leads to the increased internal charge repulsions among the protonated pyridine groups and shrink at higher pH values because of the decrease in ionization of the pyridine groups, which results in decreased charge repulsion and increased material internal interactions. Poly(acrylic acid) and poly(methacrylic acid) are the most widely used carboxyl groups containing pH-responsive polymeric materials, and poly (vinyl pyridine) is the most widely used pyridine group containing pH-responsive polymeric material. Some of the other widely used pH responsive functional groups are dibuthylamine, imidazole, and tertiary amine methacrylates.

The performance and efficiency of pH-responsive membranes can be analyzed by analyzing various membrane parameters, including membrane ion-exchange capacity, membrane water flux as a function of pH, membrane pH reversibility, and pH-based permeability control of solutions and rejection of solutes. The ion-exchange capacity of a membrane can be calculated by taking into account the mass of the anionic or cationic groups. This is known as theoretical ion-exchange capacity. Another method to measure the ion-exchange capacity is by means of titration. It is measured by titrating the membranes with a standard NaOH or HCl solution. The theoretical ion-exchange capacity of a membrane is always higher than the titrated ion-exchange capacity, owing to the fact that the pH-responsive groups are either embedded in, or grafted on, a membrane, and thus the degree of dissociation of the pH-responsive groups is not 100%. Generally, the larger the ion-exchange capacity of a membrane larger will be the permeability of the pH-responsive membranes at optimum pH. The membranes with very high ion-exchange capacities can also be used as ion-exchange membranes. The membrane water flux measurements as a function of pH reveal the behavior of membrane pores' opening and closing. In the case of alkali-swellable groups (for example, carboxyl group), the membrane water flux decreases due to the closing of the membrane pores at increasing pH values, and vice versa. On the other hand, it is opposite in the case of acid-swellable groups (for example, pyridine group), where the membrane water flux increases with increasing pH value, and vice versa. The membrane pH reversibility analyzes the swelling and shrinking cycle of a pH-responsive membrane by checking the reversible change in the membrane pores from closing to opening or vice versa with respect to change in pH values. Similarly, the pH-based permeability control and solute rejection are analyzed by changing the solution pH from lower to higher, or vice versa. The solution permeability as well as solute rejection changes with a change in pH values and confirms the pH responsive nature of the membrane.

pH-responsive membranes contain immense potential for various applications in diverse fields, including the chemical, food, and health care industries. These membranes are highly promising in applications like controlled release of drugs and chemicals, sensors, flow regulation, selective filtration and fractionation based on size and charge, and in self-cleaning surfaces of membranes. There is a lot to explore and design regarding these membranes and thus are promising membrane types for future innovations.

5.4.3 Light-Responsive Membranes

The light (photo)-responsiveness property, on the impression of light, brings morphological and functional changes to a material. A material can be light responsive in its nascent form or can be tailor-made by introducing light responsive materials as side chains, entrapment, or cross-linking. Some of the widely used light-responsive materials are azobenzene, spiropyran groups, and triphenylmethane groups. Azobenzene is famous because of its easy trans to cis isomerization. UV irradiation is the common source of this trans to cis transformation. The stable trans form of azobenzene changes to cis on UV irradiation and reverses back on visible light irradiation or by heating. Similarly, the triphenylmethane groups undergo dissociation and form ions (triphenylmethyl cations) on UV irradiation, and this reaction is thermally reversible. Spiropyran is also a famous photochromic group that forms a planar and conjugated chromophore upon ring opening due to the break of the C–O spiro bond. The open ring can be closed thermally or by a photochemical reaction.

Light is a clean and green stimulus that brings morphological and functional changes to light-responsive materials. Light-responsive materials have sparked membrane enthusiasts to use these materials for the modification of membranes for controlled morphological and functional changes. This will enable them to control the membrane's permeability, hydrophilicity (wettability), charge, shape, and antifouling nature. The light-responsive material goes through various types of conformational changes on light irradiation depending upon their constituents. They may be like α-helix to random coil or, vice versa, β-form to α-helix or random coil, left-handed α-helix to right-handed α-helix, and so on. The changes in the light-responsive moieties will help in controling as well as increasing the permeability of the membranes, such as by the ring opening and closing mechanism of light-responsive azobenzene. The conformational changes due to the effect of light also make the light-responsive moieties come to the surface of the membrane so as to increase the hydrophilicity of the membranes. For example, in the case of azobenzene, the trans to cis transformation and in the mero form of spiropyran groups bring a lot of change in their dipole moments. This leads to a dramatic change in their surface-free tension, which further results in increased hydrophilicity of the product. The azo-based materials can give a variation of 10° in the water contact angle of the smooth and flat membranes by the effect of light irradiations. Similarly, the light-responsive materials are efficient enough in the smart regulation of membrane pores or gates for selective removal of liquids, gases, metals, and ions. For example, a light-responsive membrane was prepared by using porous alumina coated with azobenzene containing poly(styrenesulfonate) and poly(acrylamide) copolymer. This light-responsive membrane has shown a 1.6 times increase in the permeability of SO_4^{2-} ions under UV irradiation. This shows that the light-responsive materials are fruitful when regulating membranes pore size, shape, and structure, thus regulating the performance and efficiency of a membrane. Similarly, these light-responsive membranes can be used for effective gas and liquid separations. The membranes can also be molecularly imprinted with light-responsive materials having specific structure and affinity for a particular metal like heavy metals or other target molecules, which can be used for successful separations of that particular substance from a feed. This can be done by forming a complex between a template molecule and functional monomer involving noncovalent interactions, for example, hydrogen bonding, metal-ion chelating interactions, ionic interactions, hydrophobic interactions, or by reversible covalent bonds, Schiff base, ketal and acetal, or ester bonds. The copolymerization of the monomer takes place in the presence of a porogenic solvent with a cross-linking agent. The template is removed after successful polymerization of the monomer, and this results in the formation of a molecularly imprinted polymer that contains specific binding sites similar to the template. These molecularly imprinted polymers are similar in function to the antibodies, enzymes, and receptors. Thus, when used in membranes they help in the removal of specific target molecules in the presence of a light stimulus.

The light responsive membranes have a great future due to their clean and green stimuli and properties. Therefore, further developments in the light responsive membranes are critical and much needed.

5.4.4 Magnetic Field Responsive Membranes

Magnetic field responsive membranes are membranes that respond to an external applied magnetic field. The membranes are prepared with a material, usually nanoparticles, that consists of electromagnetic or magnetic properties, for example, Fe_3O_4 nanoparticles. These magnetic nanoparticles are either attached to a polymer chain and grafted on a membrane surface or directly blended into the membrane matrix. These grafted or blended magnetic nanoparticles impart magnetic properties to the membranes. On the application of an external magnetic field, the membranes show structural as well as functional changes. The pores of the membrane may close or open due to the effect of the external magnetic field on the grafted or blended magnetic particles. Similarly, if the magnetic particles are modified with a thermoresponsive polymer, like PNIPAM, then the magnetic particles may induce stimulus into the thermoresponsive polymer upon application of an external magnetic field, since the magnetic particles heat up upon application of the external magnetic field. This will add dual response to the membrane and can be used for achieving better performance and response. The magnetic field responsiveness also improves the antifouling nature of the membranes, because the moment of the magnetic nanoparticles upon application of an external magnetic field cleans and does not allow the feed particles to be adsorbed on the membrane surface.

The main purpose of introducing the magnetic response into a membrane is the improvement of the selective separation and antifouling nature of the membrane. The magnetic materials are modified with either other responsive polymers or polymers with membrane improvement properties like hydrophilic polymers. This combination or combined modification of a membrane greatly improves the performance as well as the efficiency of a membrane. Many researchers used magnetic materials modified with block copolymers for the improvement of membrane performance, longevity, efficiency and easy cleaning. The magnetic materials used also impart strength to the prepared membranes in the cases where it is blended in the membrane matrix. Generally, such membranes are known as mixed matrix membranes. There is a lot of scope for further improvement and development of magnetic field-responsive membranes. There is also a need for the development of new magnetic materials and preparation methodologies for the improvement of membranes.

5.4.5 Chemical-Responsive Membranes

Chemical-responsive materials are the ones that respond to a chemical stimulus. They sense and respond to a specific chemical. When chemicals are used to modify a membrane, then the resultant membrane is known as a chemical-responsive membrane, since it inherits the property of chemical responsiveness from the chemical responsive materials. Mostly, chemical-responsiveness-based research is dedicated to drug delivery, but there are other applications where this property can be utilized, including sensors, and product recovery and separation. Chemical-responsive membranes, in addition to drug delivery, will be very useful in membrane bioreactors. In membrane bioreactors, they will increase the production and separation of a product by many times. The membranes will be responsive to the product and upon its production will open their pores, and the product will be separated the moment it is produced, which is efficient and selective. This will lessen the burden on the downstream processes and will cut the product cost by a huge margin. This will help to get a previously expensive product at a lower cost. The pharma, food, and biotechnology sectors will be the most influenced sectors by the development of chemical responsive membranes.

A classic example of chemical–responsive membranes is the self-regulated or controlled delivery of insulin to a diabetic patient. In this case, a membrane is made responsive to glucose and on the presence of glucose opens its pores for the delivery of insulin. This is achieved by making the membrane pH responsive with chemicals like poly(acrylic acid). These poly(acrylic acid) polymer chains are grafted on the membrane surface and pores, and an enzyme glucose oxidase is covalently immobilized on the poly(acrylic acid) chains. In neutral pH conditions, the membrane pores are closed due to the dissociation of the negatively charged carboxyl groups of the poly(acrylic acid) polymer. The repulsive forces between the negatively charged carboxyl groups extend the polymer chains and thus the membrane pores are closed. But the increased presence of glucose in the vicinity of the membrane or membrane surface activates the enzyme glucose oxidase, and this enzyme oxidizes the glucose to gluconic acid, which results in the decrease of pH in the local region. This further results in the protonation of the poly(acrylic acid) carboxyl groups and results in the opening of membrane pores due to the decreased intrarepulsions. This helps in the regulated delivery of insulin to the environment by a self-regulatory mechanism. Therefore, this is an important and vital development in case of medical treatments, where the patient does not have to take regular insulin shots and thus is a great relief for the patients. This is all possible because of the development of chemical-responsive membranes. Similarly, membranes can be made responsive for the delivery of other drugs, production and separation of products, and other industries. Therefore, there is a large scope in the development and use of chemical-responsive membranes in the present and future.

STUDY QUESTIONS

1. Is it possible to prepare a membrane with all of the available polymers? What are the disadvantages of this, if any?
2. What is the membrane process that turned out to be a boon for Saudi Arabia? Explain its principle and mechanism.
3. Rita, a young municipal officer, thought of an overall membrane process-based water treatment plant for her city. What membrane processes should she use for effective municipal wastewater treatment and in what sequence? Explain the role played by each membrane process in the treatment plant.

4. Sahil is an industrialist who wants to reduce the overall cost of the treatment of wastewater from his textile and tannery industries. What processes and practices should he use to achieve the goal?
5. Name the industry where membrane processes have converted a previously waste product into a useful product. Explain the process and use of the product.
6. How can membranes be used to reduce the overall biological oxygen demand of a feed?
7. Explain the possibility of membrane bioreactors for wine production.
8. Why is there a need for juice clarification and how is it achieved by using membrane processes?
9. Randeep developed excellent ceramic membranes for the removal of chromium. In which industry can these membranes be most useful and how?
10. Jonathan is a doctor and uses membrane processes on his patients. Name the membrane processes Jonathan is using and what is their purpose?
11. Christina had a limited supply of ethanol and in a hurry to attend a phone call, she by mistake mixed ethanol with water. Which membrane process can she use to separate the mixture? What are the properties of the membranes required for this membrane process?
12. How membranes are useful in gas separations?
13. What are the role and challenges for membranes in fuel cells?
14. Is it possible to generate energy by using membrane processes? If yes, then how?
15. What are the role of a stimulus in a stimuli-responsive membrane?

REFERENCES

1. M. Mulder, *Basic principles of membrane technology*, Springer, 2007.
2. B. K. Dutta, *Principles of mass transfer and separation processes*, Prentice Hall of India, 2007.
3. R. W. Baker, *Membrane technology and applications*, 2nd ed., John Wiley & Sons, 2004.
4. S. Loeb and S. Sourirajan, Sea water demineralization by means of an osmotic membrane, in *Saline Water Conversion—II* (Advances in Chemistry Series vol. 38), 117–132, 1962.
5. P. S. Francis, Fabrication and evaluation of new ultrathin reverse osmosis membranes, NTIS Reports No. PB-177083, National Technical Information Service, US Department of Commerce, Springfield, VA 22161, 1966.
6. S. E. Skilhagen, J. E. Dugstad, and R. J. Aaberg, Osmotic power—Power production based on the osmotic pressure difference between waters with varying salt gradients, *Desalination*, 220, 476–482, 2008.
7. S. E. Skilhagen, Osmotic power—A new, renewable energy source, *Desalination and Water Treatment*, 15, 271–278, 2010.
8. L. Y. Chu, Y. Li, J. H. Zhu, and W. M. Chen, Negatively thermoresponsive membranes with functional gates driven by zipper-type hydrogen-bonding interactions, *Angew. Chem. Int. Ed.*, 44, 2124–2127, 2005.

6 Applications of Ceramic Membranes

6.1 INTRODUCTION

In general, ceramic membranes are composite membranes made up of various inorganic materials such as α-alumina, γ-alumina, zirconia, silica, titania, and kaolin. They are composed of a macroporous support with one or two mesoporous layers and a thin selective microporous top layer [1]. Ceramic membranes are used in various industrial applications. Nowadays, efforts are made to replace polymeric membranes from the industrial applications with ceramic membranes. Compared to polymeric membranes, ceramic membranes possess superior chemical, thermal, and mechanical stability. The only drawback of ceramic membranes is their initial cost. There is a need to develop techniques for the production of ceramic membranes on a large scale and also to increase their packing density, which are the two areas where ceramic membranes remain behind the polymeric membranes. Advancements in ceramic membranes are making them the membranes of choice for various industrial applications. The cost of operation and maintenance is far lower as compared to polymeric membranes. The costs of pre- and posttreatments, membrane replacement and lifetime, energy consumption, and cleaning operations all show that the ceramic membranes are on par with the polymeric membranes.

This chapter stresses on the advantages and applications of ceramic membranes. The advantages over polymeric membranes are discussed in detail. In the following sections, various industrial applications of ceramic membranes are discussed in detail. This chapter will give insight to the readers about ceramic membrane applications and their role in changing the shape of various industrial products and in particular industries.

6.2 ADVANTAGES OVER POLYMERIC MEMBRANES

As discussed in the "Introduction" section of this chapter, ceramic membranes have an edge over polymeric membranes in terms of resistance to harsh environments, maintenance, and life span. These traits make them applicable to various industrial applications. Some of the major advantages of ceramic membranes over polymeric membranes are

- Resistant to chemical, mechanical, and environmental stresses—These properties make them highly corrosion resistive, high temperature and pressure resistive, and stable under a wide range of pH. This allows the use of ceramic membranes in various harsh conditions of pH, temperature, pressure, chemicals, and physical stresses. This also allows the use of strong cleaning agents for cleaning of membranes. Ceramic membranes are tolerant to very high doses of chlorine (up to 2000 mg/L in certain cases), wide pH range (0.5–14), and wide temperature range (350°C–500°C). Therefore, they can also be used in industries without pretreatment of the feed.
- Longer life span—Ceramic membranes due to their resistance to various harsh conditions of physical and chemical stresses have a very long life span, which goes up to 5 to 10 years. There are many examples of ceramic membranes where the installed ceramic system is operational even after 10 to 14 years of service.
- Antifouling nature—Ceramic membranes are less prone to fouling as compared to polymeric membranes. If fouling happens, then the best cleaning treatments can be given to the ceramic membranes without worrying about its deterioration or negative effect.

These properties make ceramic membranes beneficial for many industrial processes because they will reduce their energy consumption and cost. Also, they will increase the product quality and quantity, since these are all the positive traits of using membrane processes. However, ceramic membranes also have some major drawbacks due to which they are not approached for industrial-scale operations even though they have been in the market for so long. Some of the major ceramic membrane drawbacks are

- Not applicable for nanofiltration and reverse osmosis—Ceramic membranes are mostly available in the microfiltration and ultrafiltration range with pore diameters in the range of 0.010 to 10 μm. Therefore, ceramic membranes cannot be used for nanofiltration and reverse osmosis industrial applications, and these applications cover the major part of the industrial membrane processes.
- Initial high cost—The initial costs of ceramic membranes are very high. These costs include the cost of membranes, controls, pumps, and fittings. It is estimated that ceramic membrane costs are 10 times more than that of polymeric membranes. Though the running cost of a ceramic membrane system is lower than the polymeric membrane system, it is still important to reduce the initial ceramic membrane costs so that it can be readily used in the industries.
- Brittleness—Ceramic membranes are very brittle in nature. If dropped or subjected to undue vibrations, they get damaged very easily. This is also an area where work has to be done to improve the strength of ceramic membranes.

The advantages and disadvantages of ceramic and polymeric membranes show that polymeric membranes will be good for laboratory-scale use, since in laboratory-scale use it is the application of membrane separation that matters and not the life span and cost. However, for industrial-scale applications, cost and life span are the most important factors in addition to the separation effectiveness and efficiency. In most of the applications, the life span of polymeric membranes is equivalent to 12 to 18 months (increased up to 36 months by using optimal and timely cleanings) and 10 years for ceramic membranes [1–4]. As previously said, these traits of ceramic membranes give an edge to them over polymeric membranes and favor them to be the choice membranes for industrial applications. The only hurdle in the path of ceramic membranes is high initial cost due to which ceramic membranes are not applied on a large scale in industrial applications. Therefore, there is a need to develop low-cost ceramic membranes with long life spans. This will drive the use of ceramic membranes in the industry and will definitely be fruitful.

The advantages of both the membrane types can be utilized by preparing polymer–ceramic asymmetric composite membranes. These membrane types would have resistance to harsh conditions, and the long life span of ceramic membranes and separation properties of polymeric membranes. This membrane type will also partly overcome the disadvantages of ceramic and polymeric membranes. These membranes would have wider pore size ranges obtained by the deposition of a polymeric film layer over the top of a ceramic support, which has high chemical, mechanical, and thermal stability. The cost of the membrane can also be lowered by the optimum use of polymeric and ceramic membrane layers. Definitely, the performance of these membranes is high as compared to individual polymeric and ceramic membranes. These membranes can be used in applications where polymeric membranes are favored, but due to their drawbacks cannot be used effectively and efficiently. Similarly, in the case of ceramic membranes these membranes can be successfully employed. The main applications for which these membranes are required include nanofiltration, reverse osmosis, pervaporation, and gas separation.

6.3 CERAMIC MEMBRANES AND THEIR APPLICATIONS

In this section, the various applications of ceramic membranes in different fields are discussed, such as in fruit clarification, oily wastewater treatment, heavy metal removal from wastewater, protein fractionation, and dairy wastewater treatment. The capability of ceramic membranes to withstand harsh chemical and thermal conditions make them suitable for these applications. They are very much effective and efficient in their roles in these different applications. The different parameters, problems, and solutions of ceramic membranes associated with these applications are discussed in detail.

6.3.1 Juice Clarification

Juice clarification is important for fruit and vegetable juices as these beverages are of high nutritional values. The constituents of these beverages—vitamins, minerals, and antioxidants—are very essential for the apposite growth of humans. Conventional methods used for juice clarification are low-pressure evaporation and filtration by means of fining agents, such as diatomaceous earth or gelatin to get rid of suspended and colloidal particles. The drawback of these processes is the fact that many of the compounds that are essential factors for the regulation of the quality of the beverage, for example, flavor, aroma, sugar content, and acidity, are also removed or deteriorated in the process. The chemicals and temperatures used in these conventional processes are responsible for this loss of important compounds, since these compounds cannot withstand the harsh conditions provided by the chemicals and high temperatures.

Membrane technology is one of the best options for juice clarification from different fruits and vegetables, since membrane processes are chemical free, energy efficient, fast, easy to scale up, and need little or no downstream processes, and can be operated at low temperatures. Therefore, the quality of the fruit juice can be preserved as compared to other conventional processes. Membrane processes such as microfiltration, ultrafiltration, and reverse osmosis are explored for the application of juice clarification from fruits like mosambi, orange, pineapple, apple, grape, watermelon, lemon, and carrot. Citrus fruits (e.g., mosambi, orange, pineapple, and lemon) are the most sought-after fruits in the beverage industry due to their high nutrition, low cost, and widespread availability. The juice of citrus fruits is composed of both low molecular weight compounds (sugar, salt, acid, flavor, and aroma compounds) as well as high molecular weight compounds (cellulose, hemicellulose, and pectic polysaccharides) together with haze-producing proteins and microorganisms. The presence of pectic polysaccharides in the juices give rise to the cloudiness and postbottling haze formation. Also, pectic polysaccharides result in fermentation during storage of the juices for longer periods. Therefore, by using membrane processes, it is also important to remove these high molecular weight pectic polysaccharides and their derivatives from the juice, while retaining the low molecular weight compounds, which are essential for the nutrient value, flavor, and aroma of the juices.

Membrane scientists have explored various methods for the production of suitable fruit juices in convergence with membrane processes, such as enzymatic treatment. These methods further enhance the quantity as well as maintain the quality of the fruit juices. Different membrane processes were also explored for their permeability and selectivity to get better results. The output of these works is two important points: average membrane pore size and the fruit juice (feed) particle size. Therefore, it is important to select the perfect pore size membrane for a particular particle size containing fruit juice. It is important to consider the membrane pore size as compared to the molecular weight cut-off of the membrane, because the molecular weight cut-off of a membrane is ambiguous for the feed as well as the membrane. In most cases, due to the selection of a membrane for separation of feed containing a particular size particle based on its molecular weight cut-off gives irrelevant results. This is because of the presence of other factors, like shape and flexibility of the particles, inter- and intrainteractions of the particles with the medium (especially membrane material), extent of concentration polarization, and fouling. Therefore, it is better to select a membrane based on its membrane pore size. A tenfold increase has been seen in the membrane mosambi juice flux when the membrane is changed from a 10 kDa molecular weight cut-off to 0.2 μm and no change in the quality of the permeant. Thus, it confirms that it is best to choose a membrane based on its pore size as compared to the molecular weight cut-off.

Processes other than membrane processes are also used and explored with membrane processes to obtain better results in terms of juice clarification. Enzyme treatment and effects of temperature, electric, and magnetic fields are used for the betterment of the overall juice clarification process in terms of quantity and quality [5]. The enzymatic treatment is used to minimize the presence of pectic polysaccharides, which are the major compounds for the problems of haziness, cloudiness, and fermentation in the juices. These pectic polysaccharides downgrade the quality of a processed fruit juice; therefore, it is important to remove them from the juice prior to storage or bottling. Other processes such as the use of electric and magnetic fields are done so as to decrease the concentration polarization and membrane fouling to a great extent in case of fruit juice clarification. The enzymatic treatment came out to be the best treatment for the removal of pectic polysaccharides as compared to other conventional processes. It is also seen that the membrane flux increases with increasing the temperature to a few degrees (for example, from 21°C to 25°C); this will enhance the overall productivity of the membrane process without affecting the juice quality. At present, membrane researchers have studied the conditions of pectinase concentration, heating temperature, and extent of enzymatic treatment and tried to optimize them, so as to have a perfect juice clarification process.

In the literature it is observed that polymeric membranes were initially explored for juice clarification, but due to the problems associated with polymeric membranes, such as low capability to deal with harsh conditions presented by the acidic nature of the citric juices, which results in lower life span, and for any process to be commercially successful it is important that it should be long running. Therefore, these days research in fruit juice clarification is inclined toward the use of more corrosion-resistant ceramic membranes. In the coming sections, the use of ceramic membranes for the clarification of mosambi, orange, and pineapple fruit juice is discussed.

6.3.1.1 Mosambi

Mosambi (*Citrus sinensis* (L.) Osbeck) juice is considered as a health drink due to its nutrient content. Even doctors prescribe mosambi juice to their patients because of its wide availability, low cost, and, most important, as mentioned earlier, its nutrition. The fruit is not available throughout the year, therefore, it is important to clarify and store mosambi juice for year-round availability and to save the postharvest losses due to the fruit spoilage. Nowadays, customers demand fresh and safe fruits and vegetables. There is also need of pretreatment of mosambi juice before its clarification due to the presence of high concentrations of pectin, carbohydrates, and some proteins. The presence of these constituents increases the viscosity of the mosambi juice, which is not good for the consequent clarification processes. There are various ways to pretreat mosambi juice, including enzymatic pretreatment. Pretreatment processes make the juice less viscous by hydrolyzing and degrading pectic polysaccharides. This makes it easy for a membrane process to separate the pectic polysaccharides from the juice, which results in the reduction of membrane fouling as well as preservation of the natural freshness, aroma, and flavor of the juice. In addition to enzymatic pretreatment, other pretreatments are centrifugation and use of fining agents, such as gelatin and bentonite. Also, sometimes a combination of one or more pretreatment processes is preferred for better juice clarification.

In mosambi juice clarification, the mosambi juice is treated with enzymes for a particular time period and temperature. The enzymes are inactivated after the completion of the enzyme treatment process by heating the mosambi juice at 90°C for 5 min. Later, the enzyme-treated mosambi juice is centrifuged at desired g values (~ 2500) or rpm (~ 4000) to remove the suspended solids. Fining agents are also used in combination with the centrifugation process to enhance the removal of suspended solids from the juice. This will further enhance the membrane performance and helps in decreasing the overall membrane fouling. The enzyme-treated and centrifuged mosambi juice is fed to a membrane process for final clarification. Few things to note in these pretreatment processes are time and temperature of enzymatic treatment, concentration of the fining agents, and centrifugation time and rpm so as to make the process effective and efficient. Usually, for 5 L mosambi juice, enzymatic treatment is done for 100 min at a temperature of 40°C ± 2°C; gelatin up to 3000 ppm and bentonites in the range of 1000 to 4000 ppm are the fining agents used. Finally, centrifugation is performed at relative centrifugal force of 2000 to 3000 xg.

The mosambi juice clarified using a membrane process is assessed for its clarity, color, pH, acidity, viscosity, density, and presence of suspended solids and pectin concentration. The color and clarity measurements can be done by measuring absorbance at 420 nm and 660 nm, respectively, by using a spectrophotometer. The pH and presence of suspended (and dissolved) solids in the clarified mosambi juice can be measured by using a pH meter and refractometer, respectively. The presence of acids can be analyzed by titrating the clarified mosambi juice with 0.1 N sodium hydroxide (NaOH) to pH 8.2 and is represented as percent citric acid. Capillary viscometer can be used for the measurement of the viscosity of the clarified mosambi juice. The specific gravity bottle can be used to measure the density of the clarified mosambi juice. The presence of pectic polysaccharides is measured in terms of alcohol insoluble solids (AIS). First, 20 g of mosambi juice along with 300 ml of 80% alcohol is boiled and simmered for 30 min; thereafter, it is filtered and washed with 80% alcohol. The filtrate is dried at 100°C for 2 h and AIS is expressed as percentage by weight.

The effects of pretreatment processes on the quality of the mosambi juice, the membrane permeate flux of pretreated mosambi juice with different pretreatment processes, and the quality of the permeated mosambi juice from the membrane are important aspects of mosambi juice clarification. These aspects are necessary to be assessed for obtaining better quality clarified mosambi juice. Therefore, it is good to optimize the pretreatment and posttreatment parameters beforehand to make the overall mosambi juice clarification process successful.

6.3.1.2 Orange

Orange (*Citrus sinensis*) juice is a very popular and widely consumed healthy beverage. In recent times, the consumption of other fruit juices has remained almost stable except orange and apple. The consumption of orange juice has tripled as compared to other fruit juices, such as mango, plum, pineapple, tomato, and grape. Germany, the Netherlands, Norway, Finland, and Spain consume the most orange juice [5]. The increase in the consumption of orange juice is due to its associated properties, such as nutrient values, innovations in cultivation and harvesting processes, and worldwide advertisement by the companies.

The technology used for orange harvesting, whether done by large firms for industrial or done by small companies on a smaller scale, is almost same for everyone. The harvested oranges from the fields are delivered to the processing plants, washed, sorted (depending upon their size and present state), and processed. The processing of oranges results in three products, namely, pulpy juice (45–55 vol.%), peel (45–55 vol.%), and essential oils (0.2–0.5 vol.%). Pulpy orange juice consists of 11°Brix (dissolved solids) and 20% to 25% pulp. The raw suspension of orange juice consists of pulp, pectic polysaccharides, proteins, fibers, and cells. These constituents make the juice opaque.

In the first step of orange juice clarification, the pectic polysaccharides, responsible for the spoilage of the orange juice as well as binding of proteins and pulp, are hydrolyzed and degraded by using pretreatment processes, such as enzymatic treatment. The degradation of pectic polysaccharides also helps in reducing the overall viscosity of orange juice, which is good for the membrane processes as the rate and amount of fouling will be less. Later, the orange juice constituents, such as pulp, proteins, fibers, and cells, are easily removed by using mechanical processes, such as centrifuges, decanters, and finishers. Last, the pretreated orange juice is filtered through an ultrafiltration membrane process.

Microfiltration and ultrafiltration membrane processes are used for orange juice clarification by using either polymeric or ceramic membranes. Nowadays, due to the better operational properties of ceramic membranes, they are preferred for the orange juice clarification. Measures are taken to improve the membrane processes for the better orange juice clarification by using membranes with high membrane flux, which reduces the total membrane surface requirements. The fouling problem is dealt by using cross-flow filtration and membranes with better antifouling nature. The ceramic membranes used for orange juice clarification have pore sizes in the range of 20 and 200 nm, membrane flux of values ~500 L/m^2h at 150 kPa, and cross-flow velocity of ~4.5 m/s.

Membrane processes provide superior quality orange juice with its natural freshness, aroma, and flavor intact. Membrane-process-clarified orange juice is free from particles and microorganisms due to the separation or filtration properties of the membrane processes. Therefore, orange juice can be stored for a longer time and will not be degraded nor its quality deteriorated. The membrane processes in orange juice clarification or any fruit juice clarification reduce the use of enzymes or other chemicals/additives, operating costs, and operation time to a great extent. The yield, quality, and life of membrane-clarified juice is also very high.

6.3.1.3 Pineapple

Pineapple (*Ananas comosus*) is a tropical fruit of the Bromeliaceae family. Pineapple was once the subject of rivalry or matter of pride among European aristocrats due to its unavailability in the region, and thus it was grown in hothouses (these hothouses maintain the temperate climate), which increases their overall cost. In recent times, pineapple is still a famous fruit for its pleasant taste and aroma, and is still not available to a wide population. Therefore, there is always a high demand for fresh pineapples, pineapple juices, and pineapple concentrates for the production of different products and recipes. Pineapple juice is sweet, delicious, and healthy. It is a source of calcium, magnesium, copper, zinc, folate, vitamin C, proteins, and bromelain (a proteolytic enzyme). These constituents of pineapple juice are responsible for the cure of anti-inflammatory, anticancer, heart, fertility, asthma, osteoarthritis, manganese deficiency, and reduction of bloating and constipation-related problems. These properties show that the pineapple fruit juice market is worth millions of dollars. Therefore, techniques and processes for effective and efficient processing are required to maintain the freshness, aroma, and flavor of pineapple juice.

Membrane processes are nonthermal processes that are capable of processing the pineapple juice while maintaining the original flavors and aroma of the juice. The same process used in mosambi and orange juice processing is used for the clarification of pineapple juice, that is, the use of pretreatments, such as enzyme and centrifugation before the application of membrane processes. The reason is also the same, that is, to reduce the overall membrane fouling and to increase the overall pineapple juice yield. Also, other measures are taken to reduce fouling, such as gas sparging and stirring so that the suspended solids present in the pineapple juice will not accumulate over the membrane surface or plug the membrane pores. Similar advancements are required in case of technology, materials, and processes for the development of better fruit juice clarification processes.

6.3.2 Oily Wastewater Treatment

The petrochemical, chemical, leather, textile, and metal processing and finishing industries generate a large amount of oily wastewater. This oily wastewater is a cause of pollution. Recent strict legislation has made it mandatory to treat this oily wastewater before its discharge in the environment. The other major issue with the generation of oily wastewater in huge amounts is that it worsens the already-severe condition of freshwater availability. Therefore, an efficient treatment process will make it possible to reuse the treated water in the industries.

To date, there are various technologies, including gravity separation, absorption, air flotation, coagulation, and flocculation, available for the treatment of oily wastewater [5]. The problem with these technologies is that they are either inefficient in treating oily wastewaters with oil droplet sizes less than 20 μm, or they further generate secondary pollution. Therefore, efficient and effective techniques are required for the separation of emulsions and treatment of oily wastewaters. Membrane processes are recognized as an advance technology for the treatment of oily wastewaters. Membrane processes are simple and cost effective; generate discharge with acceptable quality; require no additives or chemicals; generate no secondary pollution; and are fast and efficient. The major issue with membranes is fouling, which restricts their wide use on the industrial scale for the treatment of oily wastewaters. Ceramic membranes, which are better than polymeric membranes in terms of resistance to fouling and various chemical, temperature, pressure, and environmental stresses, are a better choice for the treatment of industrial oily wastewaters. The resistance to harsh conditions makes them tolerant of cleaning processes with strong cleaning agents. This results in the longevity of the ceramic membranes with better performance and efficiency. Particularly, zirconia ceramic membranes have shown the best performance as compared to other ceramic membranes. Recent advances in ceramic membranes include their modification with better antifouling and stable materials. However, ceramic membrane performance in terms of oily wastewater treatment and antifouling nature is still not up to par. Therefore, further research is stressed for the development of ceramic membranes with better oily wastewater treatment potential and antifouling nature.

Recent advances in the development of better ceramic membranes are based on the use of advanced materials and in optimization of membrane structure and function. The permeation theory states that the permeation rate is directly proportional to the membrane pore size and inversely proportional to the membrane thickness. Therefore, optimization of these two factors is very important for the better treatment of oily wastewaters. Ceramic membranes are prepared with ultrathin configurations for better performance by using unique structures consisting of nanomaterials. Good use of advanced nanotechnology techniques is done for the development of such structures. These structures, such as carbon nanotubes, nanofibers, and nanowires, help in the preparation of ceramic membranes with required characteristics of structure and function. Thus, the goal of having a thin membrane with optimum pore size is achieved.

Ceramic membranes are also prepared in composite membrane configurations. Polymer coatings are given to the ceramic support to achieve the goal of a thin selective layer with optimized pore size. The polymer coating over the ceramic support fulfills the stated requirements. This will further increase their performance and efficiency in oily wastewater treatment. Similar advancements are required for developing better performing and efficient ceramic membranes for the treatment of oily wastewaters. The successful modifications of ceramic membranes with the stated advanced technologies resulted in >99.95% of oil purity in the filtrate.

6.3.3 Heavy Metal Removal

The better thermal and chemical properties of ceramic membranes make them suitable for the removal of heavy metals from different feeds, such as industrial wastewater, ground water, or municipal wastewater. In the following sections, removal of harmful heavy metals using ceramic membranes is discussed in detail.

6.3.3.1 Arsenic

Arsenic (As) is a ubiquitous and highly toxic element present in the environment. It is a crystalline "metalloid," with properties intermediate to metals and nonmetals. It is one of the most occurring trace element in the earth's crust (20th), seawater (14th), and human body (12th), respectively. The arsenic solubility highly depends upon the pH and ionic environment of the medium it is present in. It has four oxidation states: arsenate (As^V), arsenite (As^{III}), arsenic (As^0), and arsine (As^{-III}). Among the four oxidation states, As^V is the most stable. Arsenic can be present in any of the given chemical forms in the environment, including monomethylarsonic acid (MMA; $CH_3AsO(OH)_2$), arsenobetaine (AsB; $(CH_3)_3As + CH_2COOH$), trimethylarsine oxide (TMAO; $(CH_3)_3AsO$), dimethylarsinic acid (DMA; $(CH_3)_2AsOOH$), arsenocholine (AsC), arsenosugars (AsS), and arsenolipids. In general, inorganic arsenicals are more toxic than organic ones. Also, As^{III} is more toxic than its counterpart As^V. Similarly, dimethylarsinous acid (DMAAIII) and monomethylarsonous acid (MMAAIII) are more toxic compared to their parent compounds.

Arsenic rich minerals are the primary source of arsenic in the environment. These sources, both natural and anthropogenic, include dissolution of arsenic into water (especially groundwater) from pyrite ores; mining and smelting; semiconductor industries; coal combustion; and use of pesticides, phosphate fertilizers, and so on. The recommended limit of arsenic in drinking water is 0.01 mg/L. Arsenic concentration lies in the range of 0.5 to 5000 μg/L in groundwater, 0.15 to 0.45 μg/L in freshwater (rivers and lakes), 2 μg/L in seawater, and 24 mg/L in soil. These concentrations may increase or decrease depending upon the source, catchment geochemistry, and other pollution sources.

In humans, arsenic poisoning side effects include melanosis, leucomelanosis, keratosis, hyperkeratosis, dorsum, nonpetting edema, gangrene, and skin cancer. The presence of such a toxic material in the environment is not good for the ecosystem. Therefore, it is important to eradicate it from the environment. Various techniques are used for the removal of arsenic from the arsenic poisoned waters, including oxidation, phytoremediation, coagulation-flocculation, adsorption, electrocoagulation, ion exchange, electrokinetics, and membrane processes. Processes other than membrane processes are cumbersome and energy intensive. Therefore, the potential of membrane processes is explored for the removal of arsenic from arsenic-contaminated waters.

The pressure-driven membrane processes of microfiltration, ultrafiltration, nanofiltration, and reverse osmosis are explored for the removal of arsenic from arsenic-contaminated waters. Membrane processes like membrane distillation is also explored for the same purpose. These processes are able to bring down the arsenic concentration to 10 to 50 mg/L, but this arsenic concentration is not under acceptable limits. This is because membrane processes solely cannot remove arsenic from the waters due to the very small particle size of arsenic. Therefore, membrane processes give better results when integrated with other arsenic removal processes like flocculation and coagulation, ion exchange, or adsorption. Also, it is important to note that there is need to develop better membranes for arsenic removal with optimum pore size; functionalized for pH, temperature, or other external stimuli; and adsorption properties.

Researchers have combined membrane processes with electrocoagulation, ion exchange, and adsorption by either in situ or ex situ methods. In situ means that the desired property is added within the membrane, for example, charged membranes are prepared, which adds ion-exchange capability to the membrane process. Ex situ means the feed is pretreated with another process and then fed to a membrane process, such as electrocoagulation of the feed, which results in the increase of arsenic particle size and thus can be removed with the use of membrane processes. The synergistic effect of the processes results in the removal of arsenic up to acceptable ranges. Still, membrane processes are not on the exact level for proper arsenic removal and need a lot of development and advancements. Polysulfone, polyvinylidene fluoride, polyethylene, polytetrafluoroethylene, and polypropylene-based membranes are used either in nascent forms, like for membrane distillation, or modified for in situ arsenic removal.

6.3.3.2 Mercury

Mercury is extensively used for temperature measurement instruments. Also, gaseous mercury is used for the production of fluorescent lamps. Other uses of mercury include production of various chemicals, and in dental amalgams, antiseptics, and vaccine preservatives. It is also widely used in a mercury cell (Castner-Kellner process) and for cooling of nuclear reactors. Mercury is also a toxic heavy metal with very high tendencies to bind with proteins, affecting renal as well as nervous systems. Therefore, it is very important to reduce the presence of mercury to the acceptable and dischargeable limits in drinking and wastewaters, respectively.

The major source of mercury discharge into the environment is the chlor-alkali industry. Mercury is a threat to aquatic life and drinking water sources. Mercury is more dangerous in its organic forms, such as methyl mercury, ethyl mercury, and butyl mercury. Volcanoes are also major contributors of mercury discharge into the environment. Like the other heavy metals, there are many methods and processes available for the removal of mercury from water, including precipitation and adsorption, ion exchange, and flocculation and coagulation. The problem with these processes is that the acceptable limit of mercury (0.002 mg/L) is very low. This is unattainable using present-day technologies with low cost and energy utilizations. Therefore, membrane processes are the only alternatives that can be explored for effective and efficient removal of mercury from waters.

Pressure-driven membrane processes are used for the removal of mercury from waters along with the use of other processes or aides. The main issue in separating mercury is small particle size; therefore, some aides are used to increase the size of the mercury particle for its effective and efficient removal. Micellar-enhanced ultrafiltration, polymer-enhanced ultrafiltration, use of adsorbents either in situ or ex situ, or flocculation and coagulation are used.

These modified membrane processes have given almost 100% mercury retention. This is good but the next problem that arises is that of regeneration of the surfactant, polymer, or flocculent from the retentate or rejection of the membrane process. Therefore, it is important to have membranes that are able to separate mercury from the water without the aid of other processes to avoid secondary waste or regeneration of a compound. This demands further research and development in the field of membrane processes and membrane materials.

6.3.3.3 Chromium

Chromium is an essential component and holds the sixth position in terms of its availability in the earth's crust. Chromium is among the 14 most noxious heavy metals. Chromium is a very toxic heavy metal, commonly present in its two oxidation states: Cr^{III} and Cr^{VI}. Among the two species Cr^{VI} is highly toxic as compared to Cr^{III} because it is highly soluble and mobile in an aqueous system. It is highly carcinogenic and mutagenic for living organisms. On the other hand, Cr^{III} is toxic only to plant species and is less toxic to other living species. However, if it is present in high concentrations, then it also will be harmful to living species other than plants. On the Earth's surface mostly all water bodies are aerated; therefore chromium exists in its Cr^{VI} form in these water bodies. Since aeration is almost absent in groundwater, chromium exists in its Cr^{III} form. Though Cr^{III} species is less toxic to species other than plants, it also should be removed from contaminated waters for the reason that it can be oxidized to Cr^{VI} species. Therefore, it is important to reduce or remove both the chromium species from contaminated waters.

Chromium compounds are widely used in different industrial processes, including metal extraction and fabrication, surface finishing, leather tanning, textile, paints and pigments, and batteries.

The World Health Organization has set 0.05 mg/L as the acceptable limit of chromium in drinking water. The dischargeable limits of chromium are below 2 mg/L, according to the United States Environmental Protection Agency and the European Union. Several methods are developed for the removal of chromium from water, such as reduction, ion exchange, precipitation, adsorption, and membrane separations. Again, the problem associated with these processes is the low efficiency, high reagent costs, high process costs and energy consumptions, high amounts of toxic sludge generation, and generation of secondary waste products or by-products. Therefore, there is a need of alternative process that will be simple and cost effective without the generation of extra waste products or by-products.

This can be achieved by membrane processes. The membranes are either modified with metallic nanoparticles, polymers, or carbon nanotubes for the effective and efficient removal of chromium from contaminated water, or other processes like adsorption or coagulation are used. Carrier-mediated liquid membranes are also used for the effective and efficient removal of chromium from the contaminated waters. The modified membranes have shown great promise in the removal of chromium from contaminated waters with more than 90% rejection. The best part is that the reject is in its pure form, and thus can be reused and the same goes for permeate. The purified water is almost free of chromium and thus can be reused in the industry for various purposes. Therefore, membrane processes have the potential to be low on energy and cost, and high on chromium removal, efficiency, and water purification.

6.3.4 Protein Fractionation

Bovine milk is among the main sources of proteins for humans. It consists of proteins (3.2 w/w%), fat (3.4 w/w%), lactose (4.9 w/w%), and ash (0.7 w/w%). These milk constituent values vary with the type of diet, age, and breed of the animal, and the season of the year. Also, milk is used in the preparation and development of various other edible products. On the basis of physical and chemical properties, milk proteins are divided into two categories, namely, casein proteins and whey proteins. The casein proteins amount to 2.6 w/w% of the total 3.2 w/w% of the total milk proteins, and whey proteins amount to 0.6 w/w%. The distinguishing property of the two types of milk proteins is that casein protein contains phosphorus content and coagulate at its isoelectric point of pH 4.6; on the other hand, whey proteins are soluble at pH 4.6 since they do not contain phosphorus. Also, casein proteins are present in the form of casein micelles with spherical shape of diameter 0.05 to 0.5 μm, which are much larger in size than the whey proteins. Therefore, these physical and chemical property differences are used for the fractionation of milk proteins. Skim milk, milk after the removal of cream, is the best option to start the separation of milk proteins. The fractionation of casein and whey proteins further provides fractions rich in a specific protein or protein component, such as α-lactalbumin, β-casein, and immunoglobulin-g (IgG). These fractionated proteins or protein components can be utilized to add them into products targeted for infants, elders, or people with immunocompromised system. This will help in their apposite growth and development. Thus, protein fractionation helps in utilizing milk, and its fractionated products can be used in the preparation and development of various other edible products.

Membrane processes are used to separate the casein and whey proteins and their fractionation into further crucial proteins and protein components [5]. Skim milk is fed to a microfiltration membrane process; the casein proteins along with lactose and other serum proteins are retained by the membrane in the form of retentate. Later, the serum proteins are removed from the retentate by using additional membrane process stages. Ceramic membranes are preferred for the protein fractionation due to their good thermal as well as chemical properties. A typical ceramic membrane can withstand a temperature of more than 100°C and pH in the range of 0.5 to 13.5. The development of uniform transmembrane pressure (UTP) further enhances the protein fractionation efficiency of ceramic membranes. This development greatly reduces the membrane fouling and maintains a constant membrane flux. For protein fractionation, the ceramic membranes using UTP perform better than the ceramic-graded or spiral-wound polymeric membranes.

Ceramic membranes are used to separate milk proteins from skim milk at 50°C to 55°C. This helps in increasing the membrane flux and reduces potential microbial growth in the system. Temperatures higher than 55°C are not preferred since they result in the loss of membrane flux. There are two reasons for this membrane flux loss at this temperature: first the start of calcium phosphate precipitation and second the start of whey protein denaturation. Therefore, it is recommended to keep the operating temperature below 55°C. Membrane pore size is also an important parameter to be taken care of during protein fractionation. Ceramic membranes with a pore size of 0.05 μm are preferred and are able to retain all of the casein proteins present in the feed. Optimization of these parameters plays a vital role in the success of the overall process. Therefore, it should be done clearly based upon the feed type, membrane material, membrane properties, and other parameters, such as temperature.

6.3.5 Treatment of Dairy Wastewater

Dairy wastewater is similar to municipal wastewater that is high in organic content and biological and chemical oxygen demands (BOD and COD, respectively). Therefore, it is important to reduce the amount of total organic content present in dairy wastewater for its better treatment in the secondary as well as tertiary wastewater treatments. Continuous applications of stringent rules and regulations by environmental agencies regarding the discharge of wastewaters in the environment make it more necessary to treat dairy wastewater properly so as to address the compulsory limits.

In general, the dairy wastewater is treated in the same manner as municipal wastewater, that is, in three stages: primary treatment (physicochemical treatment), secondary treatment (biological treatment), and tertiary treatment (disinfection). Advancements in membrane processes result in another stage, that is, desalination for making the treated water reusable in the industry or for other applications, such as irrigation and washing. Membrane processes are used in the primary stage of dairy wastewater treatment to remove solids, oils, fats, and grease. The feed is pretreated by using processes such as coagulation and settling before feeding it to a membrane process. Mostly, microfiltration and ultrafiltration are used with ceramic membranes. Ceramic microfiltration membranes with a membrane pore size of 0.2 μm are used to remove most of the solids and fats from the dairy wastewater. This reduces the overall BOD and COD of the dairy wastewater to a great extent.

Ceramic membranes can be used in all three stages of dairy wastewater treatment for better results. The way they can be used is explained below.

- Primary treatment—Hollow fiber ceramic membranes in cross-flow configuration are used for the removal of organic content from the dairy wastewater to reduce the overall BOD and COD of the dairy wastewater. Membranes with optimized pore size and process parameters are used for effective and efficient primary treatment. The chemical and thermal properties of ceramic membranes help them withstand the harsh conditions of dairy wastewater feed. Therefore, use of ceramic membranes has been successful. Still, further improvements are required with membrane material and membrane pore size for enhanced process results in terms of the primary treatment of dairy wastewater.
- Secondary treatment—Secondary wastewater treatment is a biological treatment process, where aerobic or anaerobic microorganisms are used in the respective aerobic or anaerobic conditions. Ceramic membranes can be immobilized with the respective microorganisms and can be used for an effective secondary treatment of dairy wastewater. Membrane bioreactors are a good example of this and studies have been done on these reactors for the treatment of dairy wastewater. The outcomes of these studies are exciting. Therefore, this technique can be used for the utilization of ceramic membranes in the secondary treatment of dairy wastewater.
- Tertiary treatment—Tertiary treatment is the disinfection of the treated wastewaters so as to sterilize or remove the microorganisms used in the secondary treatment. Where ceramic membranes are used in the secondary treatment, in the form of a membrane bioreactor or by immobilizing microorganisms over the ceramic membranes, there will not be any need of tertiary treatment, and ceramic membranes can be used to further purify dairy wastewater for its reuse in various applications. When ceramic membranes are not used in the secondary treatment, then they will help in the removal of microorganisms from the feed used in the secondary treatment. Also, as said earlier, the ceramic membranes will further purify the treated dairy wastewater for its reuse.

The aforementioned information states that the use of ceramic membranes is quite possible for the complete treatment of dairy wastewater. Therefore, a single system with these three stages containing ceramic membranes with properties as per the demand of the treatment stage can be assembled and used to treat dairy wastewater in a single go. The treated water will be of such quality that it can be reused in the dairy industry itself or in some other applications. Thus, ceramic membranes are capable of complete dairy wastewater treatment.

STUDY QUESTIONS

1. If ceramic membranes carry so many advantages over polymeric membranes, then why are they not widely used on the industrial scale?
2. Polymeric membranes are widely used on the industrial scale and they are also used for juice clarification. Why are ceramic membranes also explored for juice clarification?
3. How is the enzymatic treatment of juice important and helpful for ceramic membrane processes?

4. Aslam processed mosambi juice by using ceramic membranes. To check the juice quality, what juice parameters should he analyze and why?
5. Is there any difference between the processing of various fruit juices, such as mosambi, orange, and pineapple, by using ceramic membranes?
6. How can ceramic membranes be modified for suitable oily wastewater treatment?
7. How would you clean a ceramic membrane fouled by oily wastewater?
8. How would you dispose of the heavy metals removed from a wastewater feed by using membranes?
9. Ceramic membranes are resistant to high temperatures, but most dairy proteins are prone to high temperatures. What is the use of these high-temperature-resistant ceramic membranes in food industry?
10. Write possible uses of diary wastewater treated with membrane processes.

REFERENCES

1. M. Mulder, *Basic principles of membrane technology*, Springer, 2007.
2. M. C. Porter, *Handbook of industrial membrane technology*, Westwood, NJ, Noyes Publications, 1989.
3. R. W. Baker, *Membrane technology and applications*, West Sussex, England, John Wiley & Sons, 2004.
4. Kang Li, *Ceramic membranes for separation and reactions*, West Sussex, England, John Wiley & Sons, 2007.
5. Vitaly Gitis and Gadi Rothenberg, *Ceramic membranes: New opportunities and practical applications*, Weinheim, Germany, Wiley-VCH, 2016.

7 Fouling Mechanisms and Remedies

7.1 INTRODUCTION

Membrane separation processes are widely used for the effective and efficient separation, purification, and concentration of different feed components [1]. The efficiency and efficacy of the membrane during the process decreases over time. This is due to the concentration polarization and fouling of the membranes. Concentration polarization and membrane fouling are the two most important drawbacks of membrane processes. Therefore, it is important to rectify these problems. Researchers all over the world are extensively working to overcome these problems of membranes. To subdue the effect of fouling measures, membrane enthusiasts are observing and maintaining effective cleaning, pore size, hydrophilicity, and antifouling nature of the membranes. Nowadays, more emphasis is given to the hydrophilicity and antifouling nature of the membranes [2]. Other factors such as temperature, pressure, or pH conditions can be controlled externally, but the innate properties of membranes, such as proneness to fouling due to hydrophobicity, have to be handled externally. Therefore, hydrophilic entities, such as hydrophilic polymers, charged polymers, surfactants, nanoparticles, and nanocomposites, are used to impart hydrophilicity to the membranes. These materials are either grafted or blended in the membranes. This will help in reducing the fouling of membranes by increasing their antifouling nature. The blending of these materials in the membrane is preferred as compared to grafting, since over time the grafted entities will run off from the surface of the membranes. The materials should remain with the membranes for longer periods so that the membranes will have longer life span and will able to give efficient performance for a longer period.

In this chapter, first, the factors affecting membrane performance are discussed along with mathematical models so as to better understand the phenomenon. Later, techniques are discussed to reduce or counter the fouling of membranes. Thus, this chapter will give great insight into the problem of membrane fouling and its counteractive measures.

7.2 FLUX DECLINE MECHANISMS

In general, the flux of a membrane is taken as the measure of its performance. Therefore, any decline in flux means a decline in the performance of a membrane. Thus, it is necessary to maintain an inclined flux over the period of a membrane process. In a membrane process, the first steps toward membrane fouling are pore plugging (R_p) and pore adsorption (R_a) resistances. These resistances arise either by plugging the feed components in the membrane pore, where the feed component size is equivalent to that of the membrane pore size, or by the adsorption of the feed components in the membrane pores, where the membrane pore size is bigger than the feed component size. Afterward comes the continuous accumulation of feed components near the membrane surface. This is generally known as concentration polarization and represented as R_{cp}, the resistance due to concentration polarization. This resistance reduces the mass transfer across the membrane and also reduces the membrane flux. Over time a gel layer is formed by the continuous accumulation of feed components over the membrane surface. This gel layer also offers a resistance known as gel layer resistance, R_g, which hampers the membrane performance and plays a vital role in the flux decline of a membrane. The reason for the resistances, such as concentration polarization, pore plugging, and pore adsorption, is the membrane resistance (R_m). This is the innate resistance of a particular membrane and totally depends upon the properties of the membrane and membrane material, such as

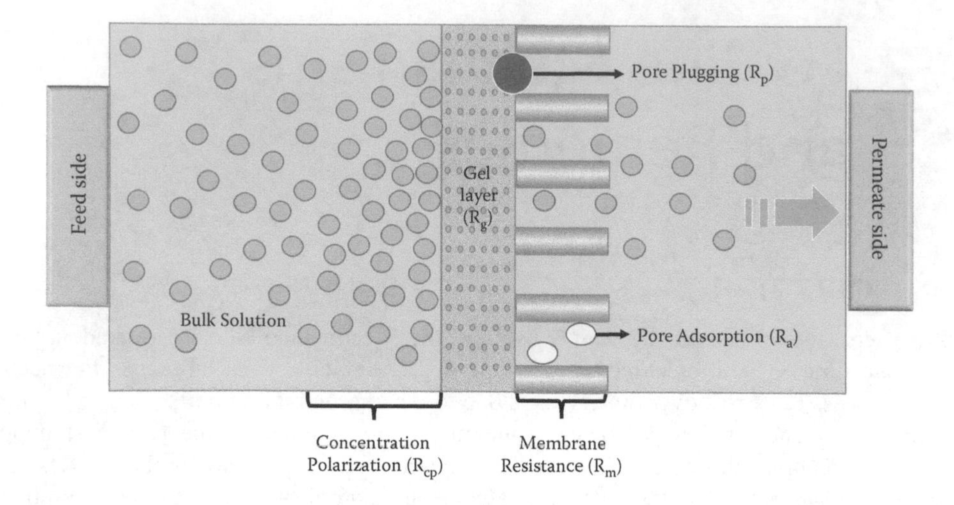

FIGURE 7.1 Schematic representation of different resistances of a membrane process.

hydrophobicity, hydrophilicity, charge, and response to external stimulus. All these resistances combined give the total membrane process resistance, R_t. These resistances can be seen in Figure 7.1.

The contribution of various resistances in the total membrane process resistance is different; some play a major role and some may not be present. These all depend upon the membrane properties and membrane process conditions. The resistances other than the membrane resistance can be controlled by using some physical means, such as mode of operation (cross-flow), use of stirrer, and pre-treatments of the feed, but the membrane resistance, which is the innate property of a membrane, cannot be controlled by these means. Therefore, in ideal conditions, the only membrane resistance involved in the decline of membrane performance is membrane resistance. Therefore, it is important to prepare a membrane with the least of this resistance. As explained in the "Introduction" of this chapter, this can be done by either using hydrophilic or hydrophobic materials, or making the membranes hydrophilic or hydrophobic as the situation demands. The membranes can be made hydrophilic or hydrophobic by coating, grafting, or blending hydrophilic or hydrophobic materials with the membranes. A lot of research is going on in this direction and has lots of scope for membrane enthusiasts to come up with better membranes and membrane materials.

7.2.1 Concentration Polarization Model

Membrane processes are famous for their selective separation of feed components from a feed. The separation may be based on size, charge, or affinity of the feed components. The selective separation means a particular feed component is given priority over other feed components for separation. The concentration polarization model can be explained by taking a simple example of the pressure-based membrane separation process, such as microfiltration, ultrafiltration, or reverse osmosis. In these membrane separation processes, considering a liquid feed consisting of a solvent and feed component, the feed component will be either rejected or permeated through the membrane depending upon the feed component and membrane properties. Assuming the feed component is partially rejected by the membrane, the feed component concentration in the bulk (C_b) will be high as compared to the permeate (C_p), taking into account that the solvent is freely permeated through the membrane. This is the basic concept of a membrane separation process as shown in Figure 7.2.

The rejected feed components with time accumulate, with a gradual increase in their concentration, over the membrane surface. This increasing concentration generates a back flow of the feed away from the membrane surface to the bulk, which attains steady state conditions over time. Also, the convective flow of the feed component to the membrane surface will be balanced by the combination

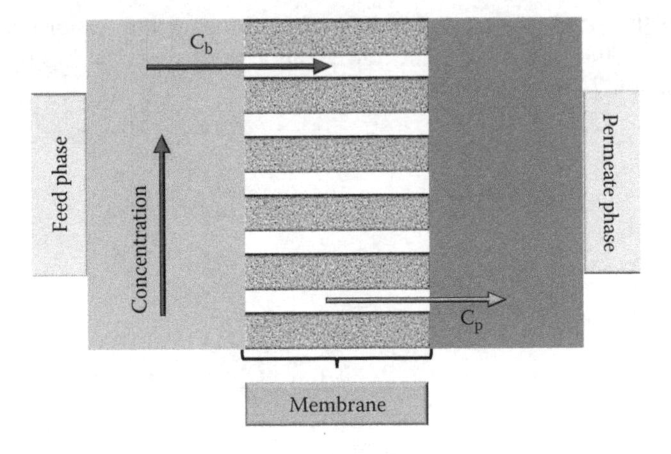

FIGURE 7.2 Basic concept of membrane-based separation.

of feed component flow across the membrane and diffusive back flow from the membrane surface to the bulk. Figure 7.3 shows the developed concentration profile in the boundary layer. It is also important to note that in this model membrane fouling is neglected and only concentration polarization is considered.

Now, assuming complete mixing at a distance δ from the membrane surface with given feed flow conditions (feed component concentration in bulk, c_b). Conversely, the feed concentration increases dramatically and reaches its highest value near the membrane surface (feed component concentration on membrane surface, c_m) due to the formation of a boundary layer. The feed component convective flow can be written as F_c, and considering the incomplete rejection of the solutes by the membrane, there will be a feed component flux equivalent to $F\ c_p$. The feed component convective flow balancing combination of feed component flow across the membrane and diffusive back flow to the bulk from the membrane surface, can be written as

$$F_c + D\frac{dc}{dx} = F\ c_p \tag{7.1}$$

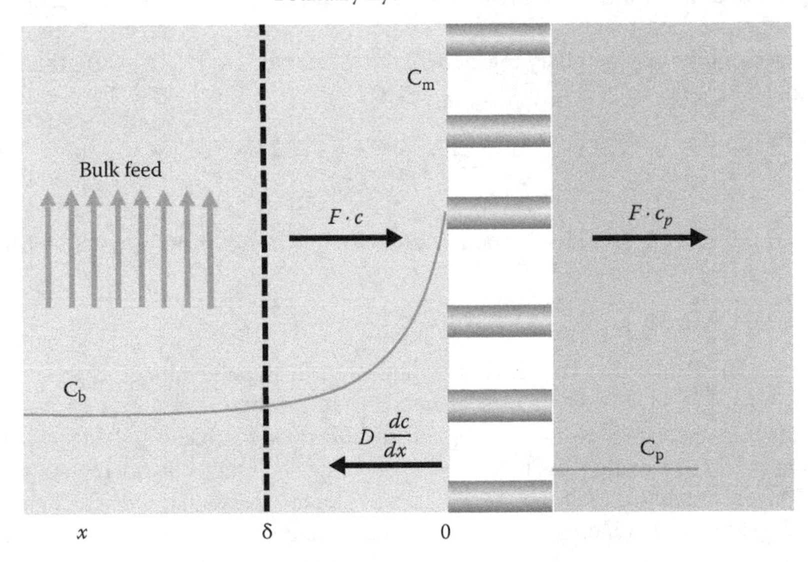

FIGURE 7.3 Membrane concentration polarization profile representation under steady-state conditions.

where F_c represents the membrane molar flux, D the feed component diffusion coefficient, F the total membrane flux, and c_p the feed component concentration in the permeate.

Integrating Equation 7.1 within the boundary conditions

$$x = 0 \Rightarrow c = c_m$$

$$x = \delta \Rightarrow c = c_b$$

gives

$$\ln \frac{c_m - c_p}{c_b - c_p} = F \frac{\delta}{D} \tag{7.2}$$

or

$$\frac{c_m - c_p}{c_b - c_p} = \exp\left(F \frac{\delta}{D}\right) \tag{7.3}$$

This ratio of the feed component diffusion coefficient, D, and the thickness of the boundary layer, δ, is known as the mass transfer coefficient, k_s.

Now, by considering intrinsic retention,

$$R_i = 1 - \frac{c_p}{c_m} \tag{7.4}$$

Equation 7.3 gives

$$\frac{c_m}{c_b} = \frac{exp\left(\frac{F}{k_s}\right)}{R_i + (1 - R_i)\ exp\left(\frac{F}{k_s}\right)} \tag{7.5}$$

This ratio of feed component concentration at membrane surface and bulk (c_m/c_b) is known as the concentration polarization modulus. The concentration polarization modulus increases with an increase in the membrane flux, F, and retention, R_i, and with a decrease in the mass transfer coefficient, k_s.

When the feed component is totally rejected by the membrane ($R_i = 0$; $C_p = 0$), then Equation 7.3 (and Equation 7.5) gives

$$\frac{c_m}{c_b} = exp\left(\frac{F}{k_s}\right) \tag{7.6}$$

Equation 7.6 modestly represents the concentration polarization in its simplest form.

7.2.2 Gel Layer Model

The gel layer model, also known as gel polarization or mass transfer model, is the first model that explained the polarization effects in membrane processes (micro- and ultrafiltration membrane processes). This film-theory-based model was first described by Michaels and later on developed by Blatt and Porter [3,4]. Basically, this model postulates that the membrane permeation rate is regulated by a gel layer, deposited over the membrane surface, at a particular value of applied pressure. This deposited gel layer also increases the overall membrane thickness and thus reduces the membrane hydraulic permeability. Another assumption of the model is that the osmotic pressure in the membrane process is negligible.

Figure 7.4 represents the schematics of concentration gradient during gel polarization during a membrane process. The three different kinds of membrane flux present in the membrane processes are convective transport of the feed component, which gives the first kind of membrane molar flux (F_c); back transport of the feed component, which yields the second membrane molar flux (F_a); and the membrane permeant molar flux represents the third type of membrane flux (F_p). These three types of membrane molar fluxes equalize one another at steady state and can be represented in the following form:

$$F_c = F_p + F_a \tag{7.7}$$

In terms of feed components, Equation 7.7 becomes

$$F_c = FC_p + D\left(\frac{dC}{dx}\right) \tag{7.8}$$

where F represents the membrane permeate flux, C the feed component concentration, C_p the feed component concentration in the permeate, D the diffusion coefficient of the feed component, and dC/dx the concentration gradient over a differential element in the boundary layer.

Now, integrating Equation 7.8 over the boundary layer of thickness δ gives

$$F = \frac{D}{\delta}\ln\left[\frac{(C_m - C_p)}{(C_b - C_p)}\right] \tag{7.9}$$

where C_m represents the feed component concentration over the membrane surface, C_b the feed component concentration in the bulk, and D/δ the mass transfer coefficient (k_s).

Assuming the feed components are totally rejected by the membrane ($C_p = 0$), then Equation 7.9 can be simplified as

$$F = k_s \ln\left(\frac{C_m}{C_b}\right) \tag{7.10}$$

It can be seen from Equation 7.10 that it is free from any pressure term. This shows that the presented model is only valid in the region where pressure plays no or a negligible role.

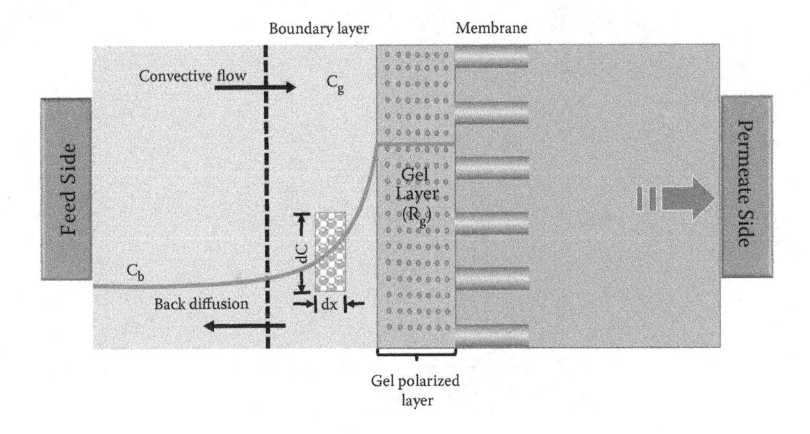

FIGURE 7.4 Schematics of concentration in a membrane process gradient during gel polarization.

7.2.3 Pore Blocking Model

The membrane process in ideal conditions follows the pore blocking model by assuming the porous membrane contains cylindrical pores. Therefore, the flux for a pure solvent or feed with negligible concentration can be given by the Hagen-Poiseuille law as [1,4–6]

$$F = \varepsilon_m \left(\frac{d_{pore}^2}{32\mu l_{pore}} \right) \Delta P \tag{7.11}$$

where F represents the flux (L/m^2h), ε_m the membrane porosity, d_{pore} the pore diameter, l_{pore} the pore length (membrane thickness), μ the dynamic viscosity, and ΔP the transmembrane pressure.

The hydraulic resistance (R_w) of a particular membrane is important to characterize its permeability. The R_w is a characteristic feature of a particular membrane defined for a pure solvent flux, which is further given by the following relation:

$$F = \frac{\Delta P}{R_w} \tag{7.12}$$

In the pore blocking model, it is presumed that the liquid is incompressible, thus a constant density Newtonian fluid with negligible end effects. Also, the flow is laminar through the membrane pores and time independent flux (steady state). This model also states that the membrane flux is directly proportional to the transmembrane pressure and inversely proportional to the feed viscosity. Generally, the viscosity of a fluid/liquid directly depends upon its composition, temperature, and velocity (in case of non-Newtonian liquids). Therefore, the flux should be increased by increasing the feed temperature and pressure, and by decreasing its concentration, but in reality this is limited to low feed pressures and velocities. If the membrane process differs significantly from any of the stated conditions, the flux becomes independent of pressure (Figure 7.5). Figure 7.6 shows that at low concentration polarization conditions (low pressure and feed concentrations, and high feed velocities), membrane flux will depend upon the transmembrane pressure. At higher pressures, there will be a deviance from the flux–pressure model irrespective of other membrane process conditions. This is due to the alliance of the gel-polarization layer. The membrane process will be independent of pressure when the pressure along with flow rate and temperature is low or with high feed concentrations.

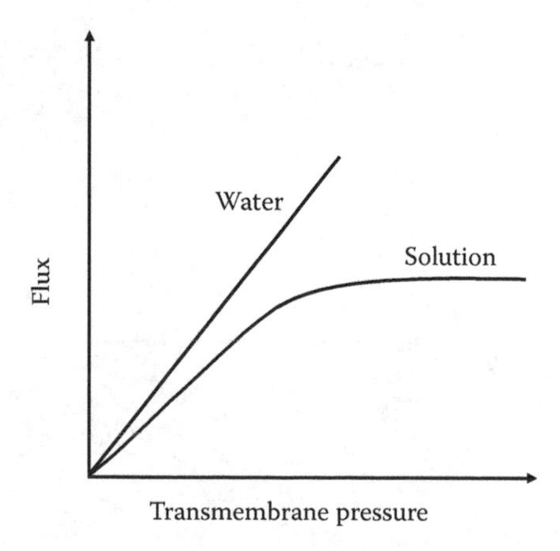

FIGURE 7.5 Pure water and a solution membrane flux represented as a function of transmembrane pressure.

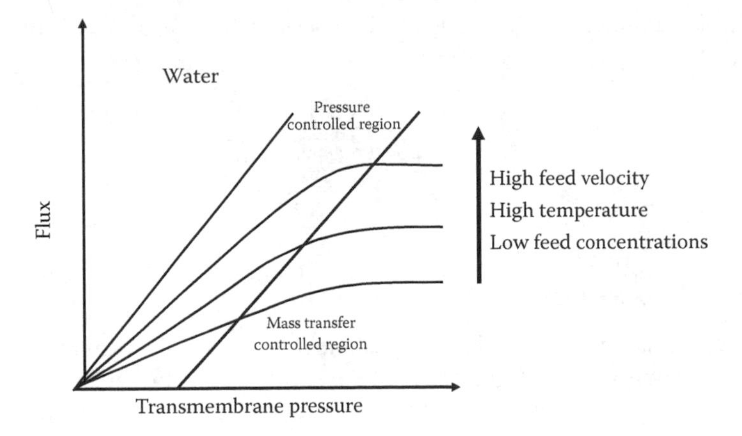

FIGURE 7.6 Schematic representation of the correlation between membrane flux and other membrane parameters.

7.2.4 Resistance in Series Model

The increase in concentration polarization in a membrane process due to the presence of feed components also increases the hydraulic resistance (R_w). Therefore, the Hagen-Poiseuille equation (Equation 7.11), which is only applicable to pure solvents, cannot be applied in this case. Thus, a number of hydraulic resistances in series regulates the membrane flux (F) in a resistance model. This can be given by the following relation [1]:

$$F = \frac{\Delta P}{(R_m + R_a + R_b)} \tag{7.13}$$

where R_m, R_a, and R_b represent the membrane, adsorption (including reversible and irreversible resistances), and boundary layer resistance. Membrane resistance is the innate property of a membrane that can be found out by pure water flux data of the membrane. The membrane resistance information play an important role in membrane modeling and tells about the effectiveness of a cleaning process altogether with the long-term stability of the membrane.

Hydraulic resistances, R_d and R_b, of the polarization layers are governed by the applied pressure and feed concentration in the layer. The feed resistances can be described by using specific resistances, which are related to the feed component properties by the Carman-Kozeny relationship [7] as

$$a = \frac{180\,(1-\varepsilon)}{r d_p^2 \varepsilon^3} \tag{7.14}$$

where ε represents the membrane porosity, r the density, and d_p the diameter of the polarized feed components.

The adsorption and boundary layer resitances can be calculated by using the following relations:

$$R_a = \alpha_a M_a \tag{7.15}$$

$$R_b = \alpha_b M_b \tag{7.16}$$

where M_a represents the deposited solute mass (per unit area) and M_b the solute mass in the boundary layer (per unit area). α_a and α_b are the individual specific resistances.

Assuming the adsorption/deposition as a kinetic process, where $M_a = M_a(t)$, then Equation 7.13 becomes

$$F = \frac{\Delta P}{(R_m + \alpha_a M_a(t) + \alpha_b M_b)} \tag{7.17}$$

7.2.5 Osmotic Pressure Control Model

Generally, in microfiltration and ultrafiltration membrane processes the osmotic effects are ignored due to negligible osmotic pressure shown by the feed concentrations. However, the process of concentration polarization results in the increase of the feed concentration over the membrane surface that definitely gives rise to the osmotic pressure effects to the membrane process. These osmotic pressure effects are effective until there is no gel layer formation or precipitation of the feed components and the boundary layer remains Newtonian.

Kedem and Katchalsky, in 1958, were the first to derive this osmotic pressure model [8]. This model relates the flux (F) of a membrane to the concentration polarization generated osmotic pressure difference ($\Delta\pi$) during a membrane process. It is given by the relation

$$F = \frac{|\Delta p| - |\Delta \pi|}{\mu R_m} \tag{7.18}$$

where Δp represents the pressure difference, $\Delta\pi$ the osmotic pressure difference, μ the solvent viscosity, and R_m the membrane resistance.

Here, the osmotic pressure difference is given by the following relation:

$$\Delta\pi = \pi(c_m) - \pi(c_p) \tag{7.19}$$

where c_m and c_p represent the feed concentrations on the feed and permeate side, respectively, and π the osmotic pressure, generally, given by the polynomial form

$$\pi = a_1 c + a_2 c^2 + a_3 c^3 \tag{7.20}$$

where a_1 represents the coefficient of van't Hoff's law for the infinitely dilute solutions, and a_2 and a_3 the nonideality of the solution.

Further, assuming total rejection, that is, $c_p = 0$, the osmotic pressure difference of Equation 7.18 will be

$$\Delta\pi = \pi(c_m) = a c_m^n \tag{7.21}$$

when $n > 1$, then c_m will be dependent on the flux as per the film model (Equation 7.21). This gives

$$F = \frac{|\Delta p| - a\, c_b^n \exp(nF/k_s)}{\mu R_m} \tag{7.22}$$

Therefore, the flux–pressure derivative will be

$$\frac{\partial F}{\partial|\Delta p|} = \left[\mu R_m + \frac{n}{k_s}(|\Delta p| - F\,\mu R_m)\right]^{-1} = \left[\mu R_m + \frac{n}{k_s}|\Delta\pi|\right]^{-1} \tag{7.23}$$

And this gives the asymptotes

$$\frac{\partial F}{\partial|\Delta p|} \rightarrow (\mu R_m)^{-1} \ for\ |\Delta p| \rightarrow 0,\ or\ |\Delta\pi| \rightarrow 0 \tag{7.24}$$

and

$$\frac{\partial F}{\partial|\Delta p|} \rightarrow 0\ for\ |\Delta p| \rightarrow \infty,\ or\ |\Delta\pi| \gg F\mu R_m \tag{7.25}$$

This shows that the flux–pressure profile begins at a low pressure difference. The slope of this pressure–flux profile is similar to pure water flux slope and it decreases with an increase in pressure difference. Similar, to the gel polarization model, this flux–pressure profile ceases to zero at high pressures.

The relationship of flux and concentration can be studied by rearranging and taking the logarithm and then differentiating Equation 7.22. It gives

$$\frac{\partial F}{\partial \ln c_b} = -\left(\left(\frac{1}{k_s}\right) + \frac{1}{n\left(\frac{|\Delta p|}{\mu R_m} - F\right)}\right)^{-1} = -k_s\left(1 + \frac{\mu R_m k_s}{n|\Delta p|}\right)^{-1} \tag{7.26}$$

Equation 7.18 shows that in the case of substantial polarization, that is,

$$|\Delta p| \gg F\mu R_m\ or\ \frac{n|\Delta\pi|}{\mu R_m k_s} \gg 1,\ then$$

$$\frac{\partial F}{\partial \ln c_b} \rightarrow -k_s$$

This is similar to the prediction given in the gel polarization model for the limiting slope in the case of F versus $\ln c_b$.

Here, the limiting concentration $c_{b,\lim}$ for $F \rightarrow 0$ can be obtained from Equation 7.22 when

$$\Delta p = a\ c_{b,lim}^{n} = \pi\left(c_{b,lim}\right) \tag{7.27}$$

In this condition osmotic pressure and applied pressure are equal. This condition also states that $c_{b,\lim} = f(\Delta p)$, which is an important variation from the gel polarization model according to which $c_{b,\lim} = c_g \neq f(\Delta p)$.

7.3 MEMBRANE FOULING REDUCTION TECHNIQUES

The membrane performance over time declines. The reason behind this fall in performance is the concentration polarization and membrane fouling. These two factors bring the difference in the pure water membrane flux and actual membrane flux; at any given time the membrane flux is always less than the original flux. Theoretically, when a membrane process attains steady state should show no further decline in the membrane flux, but in reality this is not the case and the membrane flux continues to decrease over time. Figure 7.7 shows this continuous decline in membrane flux with time.

The concentration polarization is a reversible phenomenon, and after the membrane process attains steady state it gets neutralized. Therefore, membrane fouling is the cause of the continuous

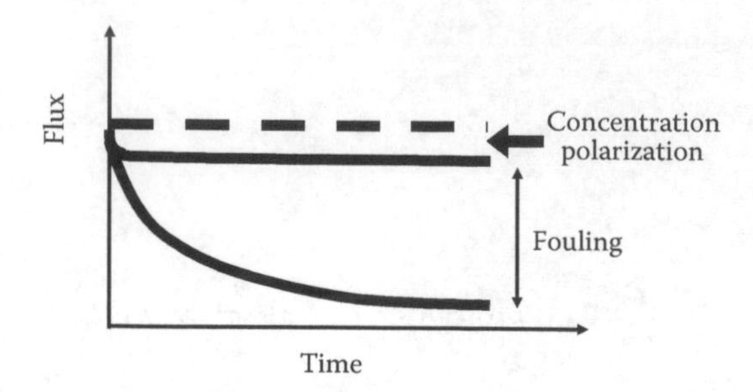

FIGURE 7.7 Schematic representation of the gradual decrease in membrane flux due to concentration polarization and fouling.

membrane flux decline with time. Membrane fouling can be defined as the reversible or irreversible deposition of rejected feed components over (surface or pore openings) or in the membrane (pores). There are different phenomena by which the feed components accumulate over or in the membranes, such as adsorption, pore blocking, precipitation, and cake formation.

It is difficult to describe the fouling phenomenon theoretically due to its complex nature. The fouling depends on the chemical and physical nature of the feed, such as concentration, pH, temperature, ionic strength, and types of interactions (e.g., hydrogen bonds, dipole–dipole interactions). However, it is necessary to know the dependable values if not accurate values of flux for process design. Therefore, to get a dependable flux value, resistances in series model can be used. In this model membrane resistance and cake layer resistance (R_c) are in series with each other. Thus, according to this model a membrane flux can be given as [1,2]

$$F = \frac{\Delta P}{(R_m + R_c)} \tag{7.28}$$

The cake layer deposited on the membrane surface over time gives the cake layer or filtration model. This cake layer model is utilized to obtain knowledge about the membrane fouling. In this model, the total cake layer resistance is equivalent to the product of the specific cake layer resistance (R_{sr}) and cake layer thickness (l_c). Also, it is assumed that the specific cake resistance is constant over the cake layer.

Therefore,

$$R_c = R_{sr} \cdot l_c \tag{7.29}$$

The specific cake layer resistance is usually given in terms of the Carmen-Kozeny relationship as

$$R_{sr} = 180\frac{(1-\varepsilon)^2}{(d_p^2\varepsilon^3)} \tag{7.30}$$

where ε represents the cake layer porosity and d_p the diameter of the feed component. Also, the cake layer thickness can be given as

$$l_c = \frac{m_c}{[r_p A(1-\varepsilon)]} \tag{7.31}$$

where m_s represents the cake mass, r_p the feed component density, and A the membrane area.

It is difficult to assess the mass of the cake. The cake is formed by the accumulation of several layers over the membrane surface, which adds up to make the cake layer thickness equivalent to several micrometers. It is too generous to say that the type, time to form, and thickness of the cake layer all depends upon the type of feed component and membrane process operating conditions. These individual cake layers developing over the time are responsible for the continuous membrane flux decline.

The cake layer resistance (R_c) can be estimated on the basis of mass balance. Assuming total feed component rejection (R_t = 100%), then

$$R_c = \frac{R_{sr}\ c_b\ V}{c_c\ A} \tag{7.32}$$

where V represents the permeate volume and c_c the feed component concentration in the cake layer.

Now, using Equation 7.32 the membrane flux can be written as

$$F = \frac{1}{A}\frac{dV}{dt} = \frac{\Delta P}{\eta\left[R_m + \dfrac{R_{sr}\ c_b\ V}{c_c\ A}\right]} \tag{7.33}$$

or

$$\frac{1}{F} = \frac{1}{F_w} + \left(\frac{\eta\ c_b R_{sr}}{\Delta P\ c_c}\right)\frac{V}{A} \tag{7.34}$$

F_w represents the pure water flux in the equation.

Membrane resistance, an important criterion of a membrane process, is neglected and Equation 7.34 is integrated within the limits $t = 0$ to $t = t$, which gives

$$t = \frac{\eta\ c_b\ R_{sr}}{2\ \Delta P\ c_c}\left(\frac{V}{A}\right)^2 \tag{7.35}$$

Equation 7.35 can be written in the form of membrane flux as

$$F = \left(\frac{\Delta P\ c_c}{\eta\ c_b\ R_{sr}}\right)^{0.5} t^{-0.5} \tag{7.36}$$

Equation 7.36 shows that the decline in membrane flux can be completely calculated or known by analyzing the cake formed over the membrane surface while neglecting the membrane resistance.

A number of theories were developed, but the complex nature of the membrane fouling mechanism makes it difficult to explain it with a single equation or a film theory. However, a very simple and realistic equation is available that presents various valued contributions through the variable exponential factor, thus is very useful and presented next:

$$F = F_o\ t^n \tag{7.37}$$

where $n < 0$. F and F_o represent the actual and initial membrane flux and n the exponent function of the velocity.

Membrane fouling is very crucial for any membrane process. Therefore, methods have been developed and researchers are working day and night to find ways to decrease or eliminate membrane fouling. Some of the widely used methods are discussed in the succeeding sections. Due to the complexity of the membrane fouling phenomenon, the discussion is kept general so that a clear picture of the methods can be presented. These methods have been quite successful in achieving the desired results of decreasing membrane fouling to a great extent.

7.3.1 Cross Flow

Membrane processes can be operated in either a dead-end or cross-flow configuration (Figure 7.8). In the dead-end configuration the feed flows perpendicular to the membrane and thus the feed components have direct impact on the membrane. Therefore, the membrane fouling and concentration polarization is highest in this case. On the other hand, feed flows parallel to the membrane in a cross-flow configuration. This results in decreased membrane fouling and concentration polarization, since

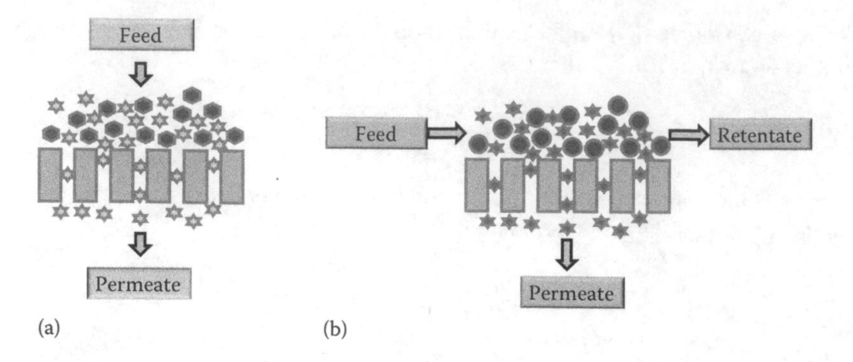

FIGURE 7.8 Schematic representation of (a) dead-end and (b) cross-flow membrane process configurations.

the feed component does not impact the membrane surface directly. Cross-flow configuration, therefore, is preferred for industrial-scale membrane processes. This will reduce the membrane fouling and concentration polarization to a great extent and give better results in terms of effectiveness and efficiency. In a dead-end configuration, the cake layer grows with time and results in the continuous decline of the membrane flux. The cross-flow configuration, on the other hand, does not face any such challenge and the membrane flux decline is less. Also, it can be very easily controlled by using suitable module and flow velocities in the cross-flow configuration. Therefore, the cross-flow configuration is used as much as possible due to its advantage of reduced membrane fouling and concentration polarization, and high membrane performance on an industrial scale. The cross-flow operation is also discussed in Chapter 2 in detail.

7.3.2 Turbulence Over Membrane Surface

As discussed in the previous section, the dead-end operation results in a continuous decline of membrane flux. If it is not possible to use any other configuration for the membrane process, then an option is available to reduce the continuous decline of membrane flux in the dead-end configuration. It is the creation of turbulence over the membrane surface during the membrane process. In the case of the cross-flow membrane operation, this is one of the reasons for reduced membrane fouling and concentration polarization. This will not allow the feed components to deposit on the membrane surface, and thus the membrane fouling and concentration polarization problems can be tackled to a great extent. This can be achieved by using a stirrer over the membrane surface, as shown in Figure 7.9. The feed will be stirred at an optimum stirring speed so as to reduce the membrane flux decline. To make this process more prominent, baffles or spacers, commonly known as turbulence promoters, can be used inside the module. It will increase the effectiveness of the stirring process and further help in reducing the membrane fouling and concentration polarization. The position, size, and shape of the stirrer, turbulence promoters, and impellers are important and should be optimized for a particular membrane process to achieve the best results.

7.3.3 Polymer-Enhanced Filtration

In general, polymer-enhanced filtration is a technique to separate very small feed components. Recently, it has garnered much attention as a separation aide for the separation of small-sized feed components (metal ions or organic compounds) using membrane processes, especially ultrafiltration. Polymer-enhanced filtration works on the principle (Figure 7.10) that by introducing a water-soluble polymer having binding sites for the target feed component to the feed will bind the feed components so as to form macromolecular complexes, which can be easily filtered or separated by using membrane processes, such as ultrafiltration [9]. This will assist in the otherwise difficult to separate

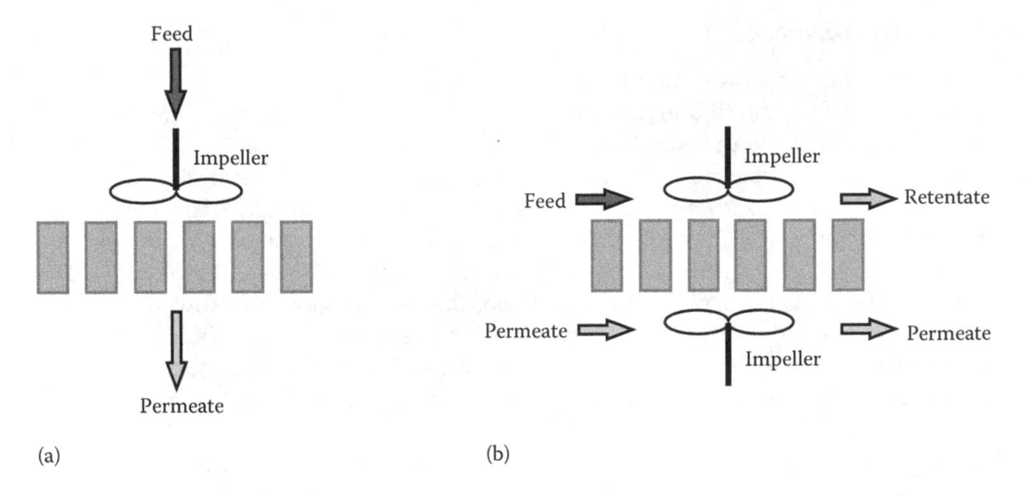

FIGURE 7.9 Schematic representation of a stirrer in (a) dead-end filtration and (b) cross-flow membrane processes to confront concentration polarization.

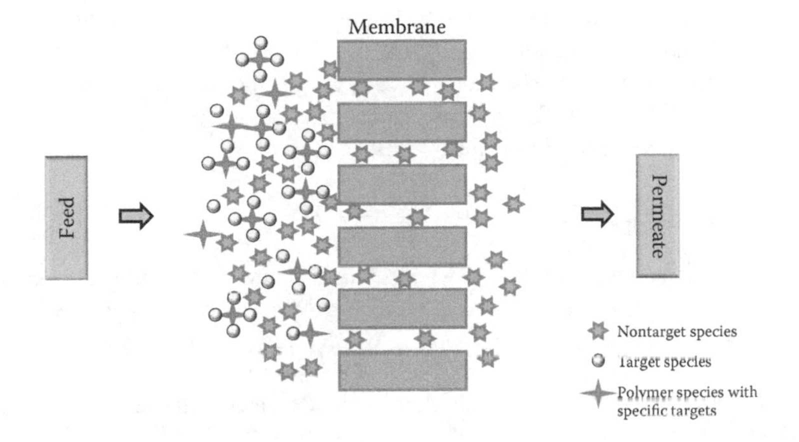

FIGURE 7.10 Schematic representation of polymeric-enhanced filtration.

and membrane fouling of small size feed components. The obtained polymer-feed component complex can be further treated for the retrieval of the polymer and pure feed component. Instead of being a new aide in the separation technology, polymer-enhanced filtration is gaining ground for its advantages and superiority. It has great potential in fields, such as wastewater treatment, heavy metal removal, and fractionation of proteins or ionic mixtures.

Polymer-enhanced filtration is helpful in the reduction of membrane fouling, since it binds the feed components with the help of polymers having the binding targets for the feed components and not allowing them to foul the membrane. The increased size of the polymer-feed component complex does not allow it to plug the membrane pores and the polymer restricts the adsorption of the feed components over the membrane surface. Thus, the cake layer formation is ceased and this results in high membrane flux along with increased separation efficiency. Therefore, polymer-enhanced filtration is a great aide to be used for the retrieval of small feed components and reduced membrane fouling.

Generally, polymers with hydrophilic groups, such as amine or carboxyl are used in this process. These polymers are hydrophilic in nature and thus reduce the membrane fouling. Still, there is a need for the development of better polymers for this process.

7.3.4 Micellar-Enhanced Filtration

Micellar-enhanced filtration is similar in concept to polymer-enhanced filtration with the only difference being the use of micelles instead of polymers for the successful separation of feed components and reduction in the membrane fouling. The working principle (Figure 7.11) of this method is that when a surfactant is added to an aqueous phase at a concentration higher than its critical micelle concentration (CMC), then it forms large amphiphilic aggregated micelles. The small feed components having affinity for the micelle binds to them and thus are separated and not allowed to foul the membrane. This happens when the feed components are attracted toward the micelle surface and later the micelle engulfs them to its interior and traps the feed components. Now, the feed components are not free to foul the membrane and can be easily separated. Later, the micelle feed component is broken to retrieve the feed components and the surfactant for reuse. The membranes after simple cleaning procedures, such as clean water flushing or backwashing treatment, attain their original flux values.

The micellar-enhanced filtration is widely used for the removal of heavy metal ions (such as Co^{2+}, Cu^{2+}, Zn^{2+}, and Ni^{2+}), dyes, and organic compounds. To further enhance the separation or binding efficiency of the micelles, a ligand can also be used, which has a higher affinity for the feed component. This will further add to the separation efficiency of the micelles and reduce the membrane fouling tendency of the feed components.

The polymer- and micelle-enhanced filtration techniques are always thought as separation aide techniques for the membrane processes, and for the first time in this book are considered as membrane fouling reduction techniques. Therefore, there is a lot to do in this field for developing these techniques as better antifouling techniques for membrane processes. There is a need of further development of better polymers and micelles.

7.3.5 Application of Ultrasound

Ultrasound gives rise to acoustic cavitation, when irradiated upon a medium. This cavitation is the powerhouse as it is the formation, growth, and collapse of the bubbles. This collapse of bubbles produces enormous amounts of energy in the form of high temperature and pressure. This process generates temperatures around 5000 K and roughly 1000 atm pressure. The heating and cooling rates used to be above 10^{10} K/s. Thus, this cavitation process creates extreme physical and chemical conditions, which are nowadays used for different processes [10].

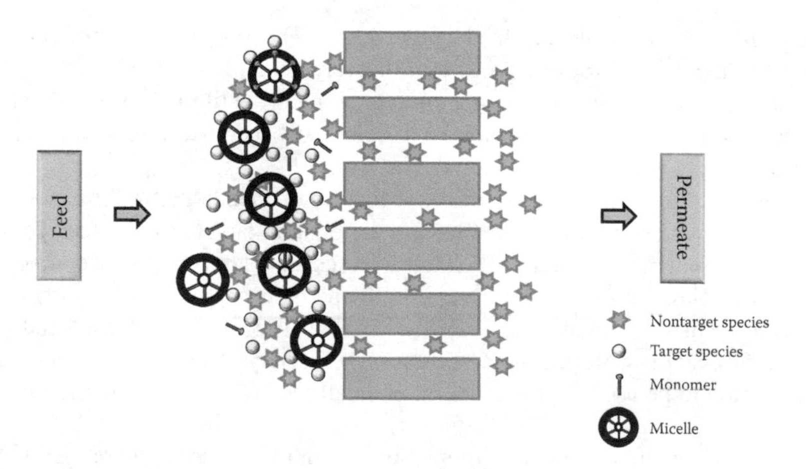

FIGURE 7.11 Schematic representation of micellar-enhanced filtration.

The current common methods of membrane cleaning are backflushing and chemical cleaning. These methods are not adequate. For example, in case of backflushing after a few cycles the membrane is not able to retrieve its original flux, and with chemical cleaning chances are there will be damage to the membranes. In addition to this, the membrane process has to be regularly stopped to carry out these cleaning procedures, which result in loss of cost, time, and product. Therefore, a technique is required that can overcome these limitations.

Ultrasound is one such technique that can provide the required results of membrane cleaning without hampering the membrane or membrane process. In ultrasound, upon collapse the cavitation process generates acoustic streaming, microstreaming, microstreamers, microjets, and shock waves [10]. This high amount of energy and different processes can be used for the cleaning of fouled membranes. Also, ultrasound can be applied to a membrane process under running conditions, which will avoid the deposition of feed components over the membrane surface. Thus, there will be no need to stop the membrane process for a cleaning overhaul of membranes. Other advantages of ultrasound over other conventional membrane cleaning methods are no additional chemicals are required or used, takes less time (a few seconds of ultrasound application is sufficient for cleaning of membranes), cost effective, can be applied effectively to both polymeric as well as ceramic membranes, and is an efficient process.

Ultrasound seems to be an efficient process, but there are also some challenges, including in the case of polymeric membranes the membranes deteriorate if the ultrasound is not appropriately applied. Similarly, upscaling of the process is a great problem. Therefore, these two sections need very much attention and have great scope to work upon.

7.3.6 Hydrophilicity of Membrane Material

The resistance offered by a membrane, generally known as membrane resistance (R_m), is an intrinsic property of a membrane. This resistance plays a vital role in the fouling of membranes and accounts for approximately 50% to 60% of membrane fouling in a membrane process. This resistance cannot be overcome by using any external phenomenon or reducing or eradicating it. Changes have to be made in the membrane itself. This resistance is due to the nature of the material of which membranes are made of. Therefore, it is important to either change the material or modify it.

Polymeric membranes are mostly made of hydrophobic polymers, such as polysulfone or polyvinylidene fluoride. Therefore, the hydrophobic feed components deposit on the membrane very easily. To eliminate this, techniques are explored to make the membranes more hydrophilic. For this purpose, hydrophilic entities such as hydrophilic polymers, nanoparticles, carbon nanotubes, and organic compounds are either coated/grafted or blended in the membranes, and a lot of research has been carried out in this regard. This will reduce the membrane fouling, since now the hydrophobic feed components are not able to deposit on the membrane surface.

To take this process a step further, responsive groups are also coated/grafted or blended in the membranes, which in response to the external stimulus bring changes to the membrane and thus helps in making the membrane antifouling in nature and reduces membrane fouling to a great extent. Figure 7.12 shows hydrophilic polymers grafted, coated, and blended in the membranes and Figure 7.13 shows a self-cleaning (pH or temperature) responsive membrane. The working principle of these self-cleaning responsive pH and temperature membranes is that the coated/grafted pH or temperature responsive groups remain in a condensed form over the membrane surface and in pore walls at a particular pH and temperature value and expand upon change in the pH and temperature value. This expansion of the responsive groups will detach the deposited feed components from the membrane surface and thus clean it. Therefore, the membrane will be cleaned on its own by just changing the pH or temperature conditions, depending upon the type of responsive groups used. This is a fascinating membrane field used in addition to cleaning for processes like drug delivery, product recovery, and racemic separations. These mechanisms are explained in detail in Section 5.4 of Chapter 5.

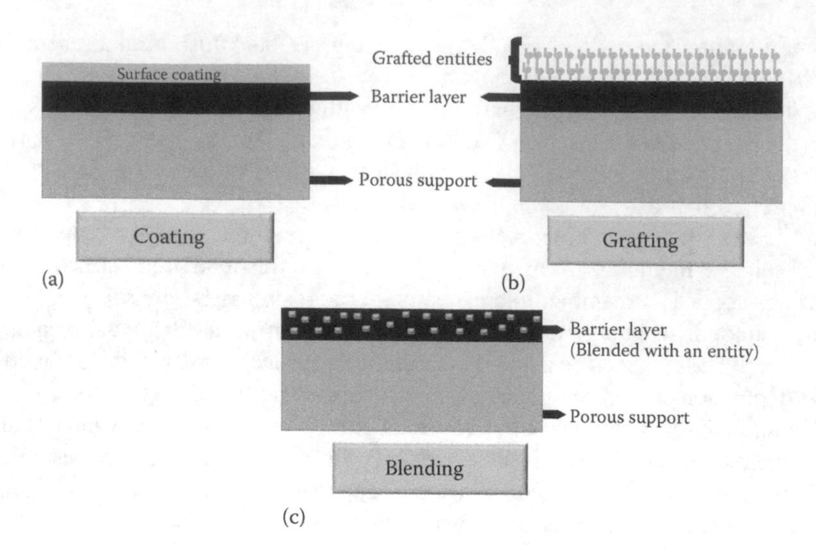

FIGURE 7.12 Schematic representation of (a) coating, (b) grafting, and (c) blending of (hydrophilic or responsive) materials with membranes to reduce overall membrane fouling.

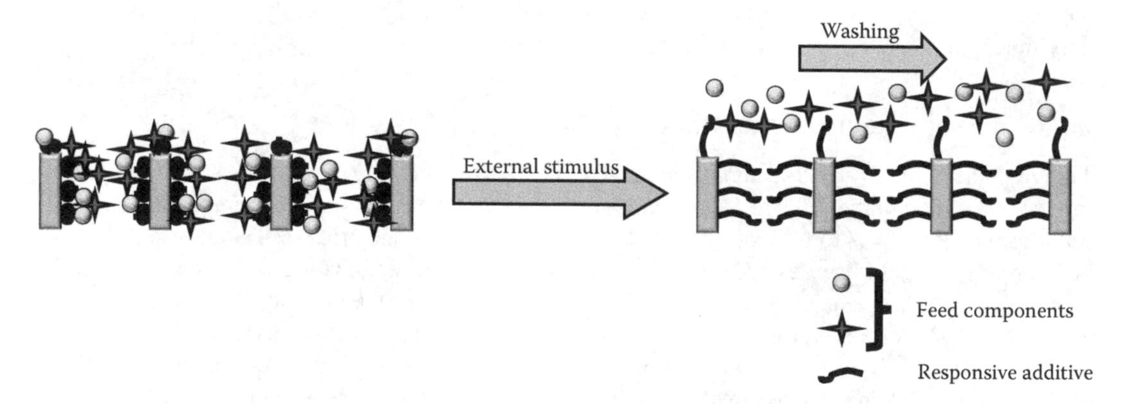

FIGURE 7.13 Schematic representation of self-cleaning responsive membranes.

7.3.7 Membrane Surface Charge

The membrane fouling can also be reduced by imparting a specific charge to the membrane in response to the feed [1]. The feed components, if charged, then deposit on the likely charged membrane surface. Therefore, it can be stopped by grafting or coating a charged species on the membrane surface. For instance, if the feed components are positively charged, then a negative charge should be imparted to the membrane for better flux output and reduced membrane fouling, as shown in Figure 7.14. Additionally, the membrane surface charge will also help in selective separation of a feed component. The unwanted feed components will not be allowed to pass through the membrane due to charge restrictions and the required feed component will pass through the membrane easily. Thus, this mechanism will reduce membrane fouling as well increase the selectivity of the membrane.

The problem with grafting and coating of different species over the membrane surface is that over time they come out and thus leave the membrane unsuitable for further use. Another problem with it is pore blocking, which further reduces the membrane flux. Blending is one option that can be used instead of coating or grafting, but in blending there are chances that the desired groups will not show up on the membrane surface. Therefore, there are uncertainties in every mechanism and there is a need for a strong antifouling mechanism.

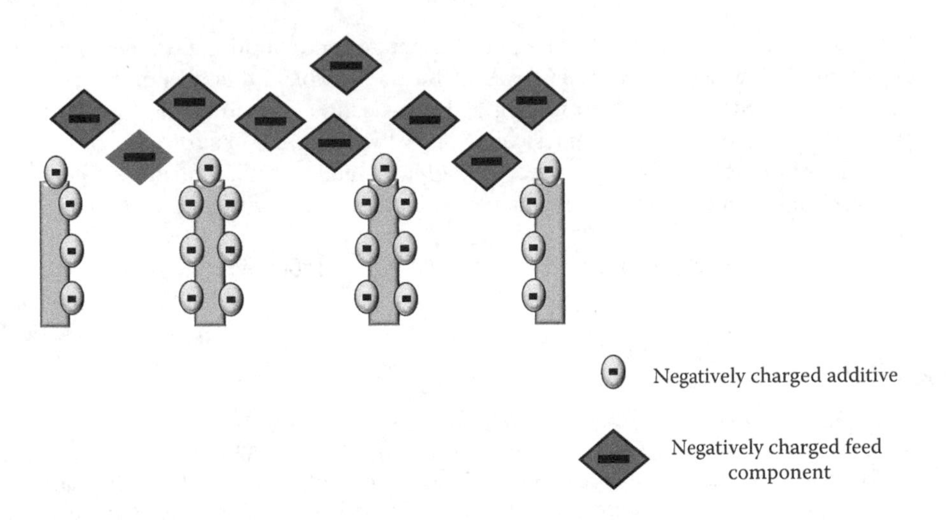

FIGURE 7.14 Schematic representation of a charged antifouling membrane.

Other than these methods, the commonly used method is the pretreatment of the feed. This includes premicrofiltration or ultrafiltration, pH adjustment, heat or chemical treatment, adsorption, chlorination, and addition of complexing agents. It is an important step in the reduction of membrane fouling. Generally, it is overlooked in many cases by seeing it as an unnecessary or negative approach in the way that it will further complex the process. Therefore, it is important to see and analyze each and every step in a process that can be beneficial for the same. Sometimes, this results in tremendous success and relieves the system from the burden of time, cost, and stress. For example, in case of protein separations or pervaporation, simple adjustment of pH or pretreatment of the feed reduces the membrane fouling to a great extent and gives superior results.

STUDY QUESTIONS

1. What is fouling and how does it affect the membrane performance?
2. Alex was facing flux decline with pure water permeation. What might be the reasons for it?
3. Explain the phenomenon of concentration polarization.
4. Ali has prepared antifouling membranes by grafting hydrophilic entities over the membrane surface, but over time the membranes show gradual flux decline. What will be the reason behind this and how could this be overcome?
5. Pammi wants to separate a protein by using a polysulfone membrane. So, how do the properties of the membranes and the proteins affect the overall separation process? What measures should be taken to overcome these hurdles for an effective and efficient separation of proteins to take place?
6. After using the ultrafiltration membrane process for the separation of a protein, Narendren was not satisfied with the overall membrane flux and thought of increasing the applied pressure. But after a slight increase, the membrane flux suddenly drops to a negligible value. What might be the reason for this membrane process behavior and what is this effect known as?
7. Sonam was analyzing the antifouling nature of an ultrafiltration membrane by using bovine serum albumin. She successfully noted the pure water flux of the membrane and start permeating the bovine serum albumin solution through the membrane. After the successful run she cleaned the membrane and again calculated the pure water flux, but she noticed a huge difference in the initial pure water flux and the latest flux values. Why did this difference in the pure water flux of the membrane occur? What measures should she take?

8. Is it possible to see osmotic pressure effects in micro- and ultrafiltration membranes? If yes, then explain how and also explain the model used to explain this phenomenon.
9. Reena, frustrated by membrane fouling and flux decline, got rid of the said problems to a great extent without using any physical device or chemicals. What did she do or use?
10. Write the positive and negative aspects of polymer- and micellar-enhanced filtrations.
11. Is ultrasound a better option of membrane cleaning? Write your views listing the advantages and disadvantages.
12. How useful are the charged membranes in reducing membrane fouling?

REFERENCES

1. M. Mulder, *Basic principles of membrane technology*, Springer, 2007.
2. Richard W. Baker, *Membrane technology and applications*, 2nd ed., John Wiley & Sons, 2004.
3. E. F. Blatt, A. Dravid, A. S. Michaels and L. Nelsen, Solute polarization and cake formation in membrane ultrafiltration: Causes, consequences and control techniques, in *Membrane science and technology*, edited by J. E. Flinn, Plenum, 47–97, New York, 1970.
4. M. C. Porter, Ultrafiltration of colloidal suspensions, *AIChE Symposium Series*, 68, 21–30, 1972.
5. S. P. Sutera and R. Skalak, The history of Poiseuille's law, *Annual Review of Fluid Mechanics*, 25, 1–19, 1993.
6. B. J. Kirby, *Micro- and nanoscale fluid mechanics: Transport in microfluidic devices*, Cambridge University Press, 2010.
7. P. C. Carman, *Fluid flow through granular beds*, Transactions of the Institution of Chemical Engineers London, 15, 150, 1937.
8. O. Kedem and A. Katchalsky, Thermodynamic analysis of the permeability of biological membranes to non-electrolytes, *Biochimica et biophysica Acta*, 1958.
9. Shri Ramaswamy, Hus-Jiang Huang, and Bandaru V. Ramarao (Eds.), *Separation and purification technologies in biorefineries*, John Wiley & Sons, 2013.
10. F. Priego Capote and M. D. Luque de Castro, *Analytical applications of ultrasound*, Elsevier, 2007.

8 Current Trends in Membrane Science

8.1 INTRODUCTION

It is important to be up to date in this time of rapid growth in fundamental and applied research. In this chapter, the recent advances and development in membrane science are elaborated upon in the fields of health care, environment, textiles, and military.

8.2 DEVELOPMENT OF ARTIFICIAL ORGANS

Artificial organs are man-made devices that are used to replace a defective natural organ. The artificial organ performs the functions of the natural organ and give new life to the patient. There are situations in which a transplant is required, but due to the unavailability of an organ nothing can be done. In general, thousands of patients are on a waiting list for an organ, but donors are not ready to contribute. On the other hand, even if a donor is available, it is possible that the donor's organ and the patient are incompatible. Therefore, there comes a need to develop artificial organs that replicate the functions of natural organs and can be used universally without the concern of compatibility. Membrane science has played a vital role in this area by helping in the development of artificial organs, such as the artificial kidney, lung, liver, and pancreas. Membranes of various properties are used for the successful development of these organs.

8.2.1 ARTIFICIAL KIDNEY

In human beings, there is a pair of kidneys with each kidney generally measuring 11 cm and weighing 160 g. The function of a normal kidney in the human body is to regulate the acid–base balance of blood, blood pressure, amount of calcium in blood, and help in the accumulation of urine and its disposal through urinary tract.

Kidney failure halts all processes carried out by a working kidney and results in the accumulation of harmful waste and fluids in the body. Kidney failure could be due to infection, hypertension, diabetes, or medications. The solution for kidney failure is the transplant of a healthy kidney from a healthy donor, but, generally, it is not possible due to the nonavailability of a suitable donor. Therefore, often the only option available is the use of an artificial kidney.

The artificial kidney process is known as dialysis. In this, blood is passed through a special membrane so as to remove the waste and excessive fluids. The clean blood is then returned back to the body. The working principle of this membrane process is discussed in detail in Chapter 2.

Dialysis was reported to be used for the first time in the early 20th century by Abel et al. [1] for the purification of blood. They used cellulose-based handmade collodium tubes for the purpose. It was the year 1925 when dialysis was first used on a patient. It is done by Haas [2]. Later, in 1943, the artificial kidney was developed by Kolff et al. [3]. They developed an artificial kidney comprised of a rotating drum dialyzer, which was further well-found with a cellophane tubing membrane. The blood from a patient's body comes to the wooden drum wrapped with cellophane tubes, and the wooden drum rotates in or around a dialysate solution. Later, the blood is collected in a glass cylinder from where it is returned back to the patient's body. Nowadays, dialysis is a well-developed, worldwide-used blood purification treatment procedure.

8.2.1.1 Developments in Dialysis Membrane Material

The dialysis membrane should encompass the properties of biocompatibility, thin selective top layer for high solute fluxes, high porosity for high permeability, narrow pore size distribution for sharp molecular weight cut-off, low surface roughness for low interaction with blood components, mechanically strong to withstand required pressure, and optimal chemical and thermal strength to withstand the process parameters (such as sterilization). Since the membrane (especially a synthetic membrane) contains both hydrophobic and charged domains, they tend to adsorb bacteria and other endotoxins. This is not good for membrane performance and thus has to be corrected either by introducing more hydrophilic domains or modified materials.

In general, dialysis membrane materials are classified into cellulosic and synthetic materials. Up to the late 1960s, cellulose materials were widely used for the preparation of dialysis membranes. It was discovered that the free hydroxyl group (-OH) available for cellulose membranes had low or poor blood compatibility. Thus, it was either substituted with benzyl groups or acetylated, or the cellulose membrane was coated with poly(ethylene glycol) or vitamin E. Therefore, cellulose acetate replaced regenerated cellulose as a membrane material. On the other hand, synthetic membranes are prepared from hydrophilic or hydrophilized copolymers, such as polymethyl methacrylate, polyethylene vinyl alcohol, and polyacrylonitrile. Hydrophilic blends are also used, in which hydrophobic polymers (for example, polysulfone or polyarylether sulfone) are mixed with hydrophilic polymers (for example, polyvinyl pyrrolidone, poly(ethylene glycol), and aliphatic/aromatic polyamides).

Dialysis membranes are usually prepared from the phase inversion method in tubular or hollow fiber configurations. The important factors required in a dialysis membrane material are low thrombogenicity and coagulation potential, low or no activation of the immune system, without allergic or hypersensitive reaction, lacking interaction with administered drugs or other blood components, and missing hemodynamic effects. Membranes prepared from such materials are necessary in the dialysis process. The artificial kidney was developed because of developments in membrane science and there is still room for improvement.

8.2.2 Artificial Lung

The natural lung is the organ responsible for the oxygen and carbon dioxide exchange between the blood and its environment. The lung has a huge branched capillary network, which helps this process to take place efficiently. The total lung membrane area for this exchange is equal to 80 m^2 and with a thickness of 1 μm. This total lung capacity is larger than required for a normal person; therefore, a person with nominal lung impairment can live a normal life. In case of lung impairments where it is difficult for the patient to carry out the process of gaseous exchange efficiently, a blood oxygenator or artificial lung is required. Also, they are required during surgery when a patient's lungs cannot work properly.

In the past, several models were developed for the supply of oxygen to the blood with more or less success, leading to the development of clinically employable blood oxygenators. The first concept of a blood oxygenator or artificial lung dates back to 1885. In the 1930s, Brukhonenko of the USSR and Juhn Heysham Gibbon of the United States independently demonstrated the working of a direct contact blood oxygenator [4–6]. These preliminary studies faced the problem of bubbles and froth upon contact of blood with oxygen or air. Therefore, a solution is required to cease the problem of bubbles and froth formation. This problem was tackled by using a membrane as a barrier between the blood and air. In 1955, Kolff (already the inventor of the artificial kidney) built the first membrane-based blood oxygenator [7]. Kolff used polyethylene membranes, which were rolled in the form of a reel, on a laboratory-scale blood oxygenator. In 1956, disposable blood oxygenators were developed, which relieved users from the time-consuming and laborious cleaning process required for the reuse of the blood oxygenator. Later, better performing membrane materials were developed for the preparation of membranes, including cellophane, Teflon, and silicone rubber. The development of silicone rubber hollow fiber membranes in the 1960s–1970s made membrane blood oxygenators

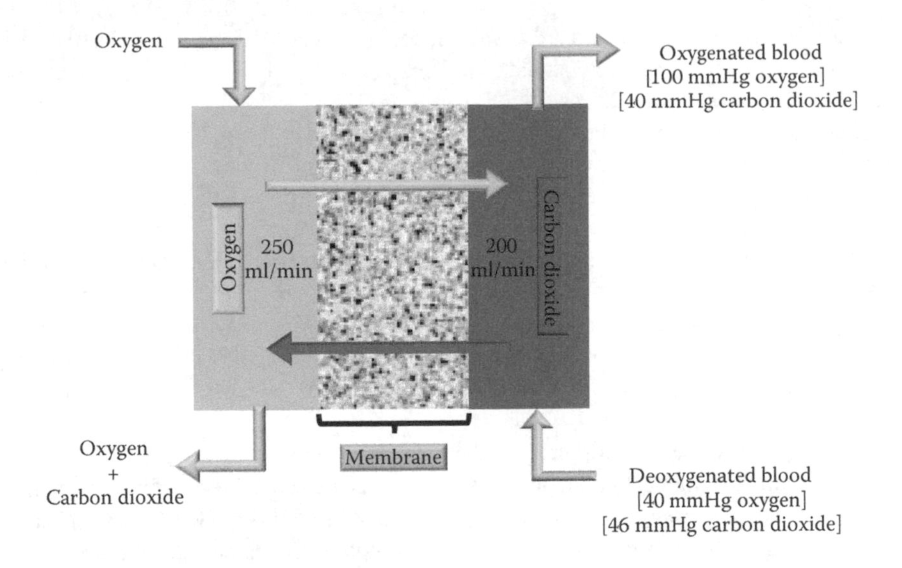

FIGURE 8.1 Principle of a blood oxygenator.

commercially successful. The development of low-resistance microporous hollow fiber membranes further increased the performance of blood oxygenators. Nowadays, membrane materials like polyolefin polymers are used for artificial lung membranes. Especially, poly(4 methyl pentene-1) is often used as a membrane material for the artificial lung due to its good properties. Asymmetric hollow fiber membranes of this material are prepared by using thermal-induced phase separation.

Nowadays, blood oxygenators are well developed and frequently used during surgeries. The main requirements of a blood oxygenator are listed in the following:

- It should be able to oxygenate venous blood at a rate of 5 L/min for a period of 30 minutes to several hours.
- In addition to the oxygenation of blood, it should also be able to simultaneously remove carbon dioxide from blood to a certain level (~40 mmHg).
- It should have reasonable blood priming volume (1–4 L).
- It should not initiate hemolysis or protein denaturation.
- It should be simple; easy; and safe to use, clean, and sterilize.

Membrane blood oxygenators are a boon for the process of blood oxygenation. The risk of air embolism is minimized due to the barrier created by the membranes between the blood and air. The large membrane surface area provides ample space for the successful oxygenation and carbon dioxide removal from blood without the need of any gas removal system. Figure 8.1 schematically represents the principle of a blood oxygenator.

Present-day blood oxygenators are impressive and progressing technically in terms of pumps, oxygenators, and biocompatibility. However, clinicians would still like to see the use of blood oxygenators simplified and the cost lower. Altogether, a search for new membrane materials and optimized membrane modules is ongoing and is the need of the hour for further developments of blood oxygenators.

8.2.3 Artificial Liver

The liver is a vital organ of the human body. It is responsible for many physiological functions, including metabolism and detoxification of several pharmacological and endogenous substances. The mature cells of a liver are known as hepatocytes and are responsible for its functions.

Hepatocytes are parenchymal cells and are always in contact with nonparenchymal cells, such as bile duct, sinusoidal endothelial, kupffer, stellate, and pit cells. The liver keeps a human body safe, healthy, and free from toxic substances.

Autoimmune dysfunction, viral hepatitis, hepatocellular cancer, long duration exposure to toxins (such as alcohol and drugs), or trauma may lead to acute or chronic liver failure. This leads to various life-threatening consequences like hepatic encephalopathy, multiorgan failure, cerebral edema, and severe hypotension. In general, the most common way to restore the functions of an acutely or chronically failed liver is a liver transplant. The problem with this is the same as in other transplant cases, that is, the availability of a suitable donor or organ. Due to this acute shortage of organs, many patients die while waiting for a suitable organ. Therefore, it is important to develop techniques and devices for an artificial liver, which could save the life of a patient until the regeneration of his liver or give him time to wait for the delivery of suitable organ and its successful transplant.

In earlier days, before the development of the artificial liver, various processes, such as hemodialysis, hemofiltration, hemoperfusion, and plasmapheresis were used. It has been seen that there is no or very low impact of these processes on a patient's health and it is very difficult for the patient to survive solely based on these processes. Therefore, attempts were made with membranes prepared from cellophane or polyacrylonitrile, having low or less permeability for toxic substances of small molecular weights. Recently, many membrane-based processes or devices have been developed based on the adsorption and filtration principles, such as the molecular adsorbent recirculation system (MARS), single-pass albumin dialysis (SPAD), and the Prometheus system [8]. In these processes, various new and modified materials, such as charcoal and polymers, are used for detoxification of the blood. Nowadays, albumin is the choice of material for the selective removal of toxic materials from the blood. Mostly, high flux hollow fiber membranes are used as filters in the aforementioned systems. For example, high flux polysulfone hollow fiber membranes are used for albumin dialysis. The MARS and SPAD systems also use the same membrane with a 50 kDa molecular weight cut-off, which will not allow the passage of albumin A sections. Contrary to these systems, the Prometheus system uses a specific albumin-permeable polysulfone membrane with a molecular weight-cut off of 250 kDa. These inorganic devices have shown promising results in acute liver failure cases for a short period of time. However, the studies conducted on these systems were short, and they provided limited clinical data, which was not sufficient to accurately assess the rate of success. The other factor that limits the success of these systems is that these systems are not able to replace the other important functions carried out by a healthy liver. Therefore, further research and development operations are being carried out in terms of developing a successful bioartificial liver support (BALS) system using cellular hepatic components, which can fully exploit and support a patient's failing liver.

In the BALS system, hepatoma cells (primary hepatocytes) or a cell line from humans or other animals are taken into a bioreactor either in suspended form as aggregates or encapsulated in biomaterials or in attached form over a flat or hollow fiber membrane surface. It is important to maintain the differentiated hepatocyte functions for a longer period so as to get a functional BALS system. In the case of a bioreactor consisting of a flat membrane, a high-density hepatocyte culture is obtained under well-regulated oxygen concentrations, which closely imitate the in vivo microenvironment. Here, a flat membrane is used to culture porcine hepatocytes in sandwiched form with nonparenchymal cells between collagen layers. This results in maintained liver-specific functions, as the cells remain polarized in in vitro conditions. The permeable membrane helps in maintaining the crucial oxygen and carbon dioxide exchange in the culture. The gas exchange is important for the culture. Membrane materials, such as polytetrafluoroethylene, polyethersulfone, and modified polyether ether ketone, are widely used for the successful culture of human primary hepatocytes. However, the flat sheet membrane bioreactor used as a extracorporeal liver support system also has disadvantages, such as exposure of the cultured cells to the shear stress, large dead volumes, and a complex scaling up process of flat plates. However, any improvements in these drawbacks will make the flat sheet membrane bioreactor a clinical success. In place of flat sheet polymeric membranes, various polymeric hollow fiber membranes are used for the successful culture of human or animal hepatocytes and are an integral part of a BALS system. The membrane surface works as the attachment and growth area for the cells and the semipermeable nature of the membranes helps in the exchange of gases and nutrients. The membranes are made hydrophilic for successful mass transport, and hydrophobic membrane capillaries are used for successful gaseous exchange. The hepatocytes in a bioreactor work as a functional liver. There is still need of improvement and development in the successful formation of an artificial liver. Membranes with better selective capacities of mass holding and permeability are required. Similarly, improvements are required in the overall BALS system for the development of a better artificial liver.

8.2.4 Artificial Pancreas

The pancreas helps in the regulation of blood glucose levels by the production and release of insulin hormone. The insulin is produced by the β-cells of the pancreas when there are irregularities in the blood glucose levels and released by the pancreas in the blood stream. These cells are located in areas of the pancreas known as pancreatic islets or the islets of Langerhans.

The pancreas of a diabetic patient functions abnormally and does not secrete or produce insulin properly. Therefore, diabetic patients require periodic insulin shots for the proper regulation of glucose. For patients with type I diabetes, an insulin shot is a must, but is also important in some cases of type II diabetes. This injection therapy of insulin is not permanent and thus must be repeated periodically and is not capable of replacing the pancreas's function of sensing and regulation of blood glucose. Also, if an injection with high dose or longer duration insulin effect is given to the patient, then it results in diabetic complications, such as kidney failure. Therefore, a good option in this case is a pancreas transplant, but again with organ transplant the availability of a suitable organ or organ donor is a problem. Thus, an artificial pancreas is a perfect solution if a transplant organ is not available.

The development of an artificial pancreas can be brought to fruition by grouping or integrating pancreatic islets (contains insulin producing β-cells) into a membrane. The idea behind this concept is that the membrane, whether flat or hollow fiber, separates the pancreatic islets from the blood stream with the membrane having permeabilities for glucose and insulin. In addition, the membrane will be completely impermeable to the immunoglobulins and lymphocytes. This concept was first tested in the 1970s–1980s, when a polyacrylonitirle-vinyl chloride copolymer hollow fiber membrane with molecular weight cut off of 80 kDa was used to protect the transplanted pancreatic islets from the blood stream in a dog [9]. Similarly, an acrylic coiled tubular membrane was used to house pancreatic islets having a 50 kDa molecular weight cut-off and 60 cm^2 surface area. This membrane was tested both in vitro and in vivo in diabetic dogs. Since then, the artificial pancreas has seen lots of development and many devices have been proposed using flat sheet and hollow fiber membranes. For hollow fiber membranes, the pancreatic islets were either loaded inside the lumen of the membrane or on the outer surface and the blood stream is allowed to pass through the outer or inside the lumen, respectively. The high surface area makes the hollow fiber membrane a more attractive option as compared to flat sheet membranes. Depending upon the place of the pancreatic islets' integration into the membrane, the artificial pancreas can be of three types:

- Intravascular—In this case, the pancreatic islets are integrated into the membrane and the blood stream of a host is used for the implantation of the membrane.
- Extravascular—In this case, the pancreatic islets are integrated into the membrane and the membrane is implanted in an extravascular site.
- Microencapsulated—In this case, the pancreatic islets are encapsulated into a polymeric membrane restricting their contact with the host immune system and thus allows implantation of the device (artificial pancreas) without the use of an immunosuppressive therapy to suppress the host immune system.

Recent advances in the field have focused on the development of devices that are more secure and safe. Membranes with better permeabilities for glucose and insulin and attachment for the pancreatic islets are required. The area is advancing but still require much to achieve success.

8.3 CONDUCTIVE MEMBRANES IN TEXTILES

The textile industry is one of the oldest industries, and is presently a multibillion dollar business. Over time, the textile industry has became one of the most well-developed industries due to the increasing demand from various sources, such as clothing (for various people and commodities ranging from common people to advanced war wearables) and packing. Regular improvements and developments were seen in the material, fabrics, and the performance of the fabrics, as well as the ability of the textile industry to improve the processes to use and develop new materials and fabrics.

Before going on to discuss conductive membranes and their role in textiles, let's have a look at the present-day improvements and developments in the textile industry, mainly its products. Presently, the textile industry emphasizes the use of smart materials for the development of a smart product, which adds new properties and features to the product [10]. This smartness is the feature of the product to perform advanced functions, such as textiles with conductivity, shape memory, flame protection, and breathable properties. This added smartness increases the comfort, safety, and life of the textile. There are advanced technologies available so as to make the textile improved and advanced in its overall features, such as microencapsulation, coating, blending, and lamination of the textile with improved materials and processes. Some of these technologies are discussed in the following sections.

8.3.1 Microencapsulation Technology in Textiles

In general, microencapsulation is the encapsulation of particles with sizes in the range of nanometers and millimeters. The basic reason to encapsulate a particle or substance is to keep it safe by restricting its direct interaction with the environment and to add additional properties and features to the overall product. These additional features or properties are

- Protection of the encapsulated material, which gives the material longer life
- Controlled or targeted release of the encapsulated material to enhance the performance and to decrease the cost, due to less material consumption
- Smart release of the material under the virtue of external factors, such as pH, temperature, and pressure
- Fast production of the ultimate product, since the encapsulated material is safe and protected
- Protection to the user as well as the environment in case the material is toxic

These are some of the positive points of microencapsulation or encapsulation in general. Therefore, it is widely used and accepted on the industrial scale. This complete process totally depends upon the material used for the encapsulation. Thus, it is better for the growth of this technique to strive for new and better materials.

Microencapsulation is prominent and widely used in the textile industry because of the following advantages: the encapsulated fabric can survive multiple washing and drying cycles; temperature-phase change materials, such as thermal protection against extreme temperature fluctuations to astronauts; controlled delivery of substances, such as fragrances, insect and animal repellants, antimicrobials, and drugs; flame-retardant fabric is created by encapsulating it with a flame retardant material.

8.3.2 Intelligent Breathable Fabrics

Breathability is important to fabrics. This is due to the fact that it is necessary to allow perspiration to be passed through fabrics from the wearer's body, especially in case of all-weather fabrics or garments. These fabrics are permeable to water vapors, resistant to rain, and protect the wearer against wind. Due to their properties, these fabrics are widely used in the fabrication of sportswear, protective gear, military clothing, mountaineering suits, diving suits, and survival suits. In addition to wearables, these fabrics can be used in the fabrication of tents, sleeping bags, and packing materials.

A lot of research is ongoing in the development of advanced materials for the production of improved and developed fabrics. Materials like polymers, nanoparticles, and composites are being explored for the development of better breathable fabrics. Progress has also been made in the formation of smart breathable fabric, such as temperature-responsive breathable fabrics. These fabrics adapt to changing environmental conditions like in case of hiking or diving. For temperature-responsive materials, poly(N-isopropylacrylamide) is being used and tested. The new materials can be developed by either modifying available materials or by developing new materials by copolymerization or other chemical modifications. Therefore, this field provides a lot of scope for further development and improvement for the betterment of fabrics with more comfort and safety for users.

8.3.3 Shape-Memory Fabrics

Fabrics fabricated using shape-memory polymers can memorize a macroscopic shape. Under specific stimuli and conditions of temperature, pressure, magnetic, or light they attain a temporary shape, which can be reverted by changing the stimulus or condition. The shape-memory polymers have great ability to recover the original shape as compared to memory alloys. Also, they are light in weight, low in cost, and carry better molding abilities. These properties make them highly sought after materials for various applications, such as temperature-responsive membranes, self-deployable sun sails, smart medical devices, and drug delivery systems. Similarly, these polymers are used to develop smart fabrics with shape-memory properties. The change in shape is due to the smart properties available with the polymer, which is elastic entropy of polymer chains. A shape-memory polymer consists of domains and segments, where the domains are responsible for the permanent shape of the polymer and segments are responsible for the elasticity of the polymer. Figure 8.2 represents schematics of the domains and segments present in a shape-memory polymer. Shape-memory polymers are often

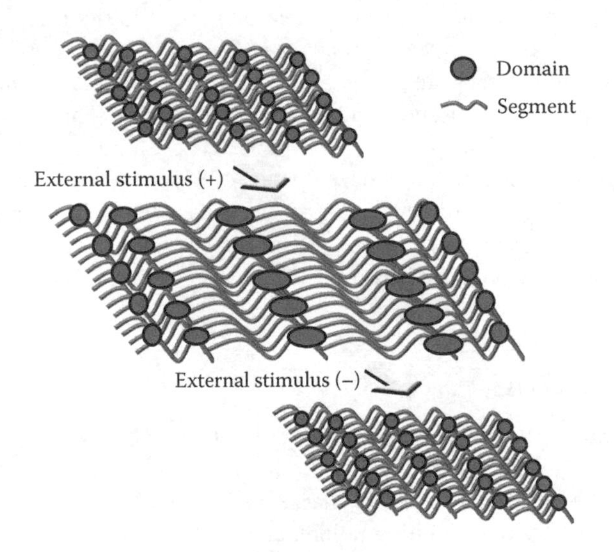

FIGURE 8.2 Schematic representation of domains and segments.

temperature responsive but can also be triggered by other stimuli including pH, magnetic, electrical, or electromagnetic. Presently, shape-memory fabrics are widely used for the fabrication of sportswear and in various medical applications, clothing, bedding, and hygiene products (sanitary napkins and diapers). Research in the field of shape-memory fabrics can be divided into two categories: supporting technologies (materials, processes, and techniques) and future applications. Currently, two-way shape memory polymers are being explored and techniques developed to produce them. Therefore, shape-memory polymers provide a great material to use for various advanced and improved applications. This will result in the production of better products for comfort and safety of users.

8.3.4 Phase Change Materials

Phase change materials are the ones that release or absorb a high sum of latent heat while going through a change in their physical state, such as from solid to liquid, or vice versa. The moment a phase change material reaches its specific phase change temperature, it changes its respective phase from one form to another. By controlling the release or absorption of latent heat of these phase change materials, these materials can be used for the improvement and development of various thermoresponsive products. Some examples of phase change materials are paraffin and salt hydrates, such as calcium chloride hexahydrate and sodium hydrogen phosphate dodecahydrate.

In the 1980s, the U.S. space agency, the National Aeronautics and Space Administration (NASA), developed the technology to use phase change materials to textiles so as to produce space suits with improved thermal performance. Nowadays, phase change materials are widely used in the production and development of clothing and other textile products, such as heat protective gloves, chemical protective clothing, smart thermal skiwear, and in architectural membranes for improving the energy efficiency of buildings. Improvements, especially in terms of novel and exclusive phase change materials, are the need of the hour for further developments in this field.

8.3.5 Smart Flame-Retardant Fabrics

In the aerospace industry and military applications, fabrics are required that can tolerate ultrahigh temperatures. There are lots of flame-retardant materials available, but they are not useful for the aforementioned two organizations, since the limit of temperature is very high in these cases, for example, explosions, space, or space flights. On the other hand, smart flame-retardant fabrics are the fabrics that provide multilayer protective and fire-resistant levels to the user. These are prepared by blending conventionally available materials such as a blend of cotton and polyester.

Typical flame-retardant materials are phosphate, chlorine, halogens, and ammonium based. These chemicals are not environmentally friendly and thus restricted in use. Therefore, there is need for the use and development of smart flame-retardant materials. This can be done by either using conventional materials in a smart or modified way, development of a material composition that will burn in a controlled manner, or adding a flame-retardant property to a product in a smart way. Therefore, to make this technique widely acceptable, materials and processes have to be developed with smart properties. Presently, work is ongoing in the said three ways for the development of smart flame-retardant materials.

8.3.6 Conductive Textiles

In the case of conventional textiles, static charge is built up due to contact, and later separation between two different materials, for example, by rubbing and separating two garments. If the garments are insulated, then the amount of static charge build up will be very high and the tendency of dust pick up and the possibility of igniting a flammable or explosive substance may damage electronic appliances, or transfer charge to a nonearthed appliance or object. Therefore, there is a need to have conductive textiles for various antistatic applications. Conductive textiles will also be useful in sensing, monitoring, corrosion protection, and electromagnetic shielding applications.

Nowadays, conductive textiles are much more eminent because of the development in the present-day electronic industry and the zeal to improve the comfort level of life. Embedding electronic gadgets, circuits, and other devices into the clothing will make textiles more than a piece of cloth, they will be smart electronic textiles. They can be utilized for various purposes, such as physiological monitoring, communication, energy harvesting, thermomanagement, protection, and entertainment.

Research is ongoing to make comfortable, lightweight, flexible, durable, and economical conductive textiles. For this there is a need to explore and develop new materials, especially electromagnetic materials. Also, various ways of imparting conductivity to the textile is also being explored, for example, coating, lamination, and blending. Conductive (charged) polymers, metallic nanoparticles (such as silver, gold, and platinum), carbon nanotubes, and conductive composites are used. Present-day research has shown that coating is the best method for the fabrication of conductive textiles, due to the continuous availability of the conductive material over the fabric or textile surface. Still, there is a need to explore the combination and development of other novel and improved methods for the fabrication of conductive textiles. Therefore, this area of application is also demanding and provides ample opportunities to develop new and novel products for the better and comfortable life of the users.

8.4 ROLE OF MEMBRANES IN MILITARY WARFARE

The military is the pride and protective shield of any nation. Soldiers are always there for their countrymen, whatever the situation. They can be found in any corner of a country irrespective of the terrain, climate, vegetation, and living conditions. Their strategic positions in far-flung remote areas make it difficult for them to have a comfortable life, for basic day-to-day life materials, such as potable water, food, and housing, may be hard to come by, and not only during times of war. Thus, it is important to provide service members with the best of basic dry and wet rations as well as housing, which should be safe and comfortable.

Membranes are used for various applications for military use. The basic example is a filter for the purification of water for drinking. Similarly, membranes are used in energy generation (fuel cells), body armor (smart, safe, and protective textiles), and air purification (in case of chemical and biological warfare). Membrane processes are continuously in development for the betterment of a soldier's life in peacetime as well as war. Following are some of the recent and significant developments in the membrane science for the military.

8.4.1 Water Supply

At military stations that are stationed in far-flung areas or on sea carriers, it is difficult to have a continuous supply of pure water. Also, it is noted that at such remote places or during war, soldiers are often admitted to hospitals due to water-related diseases. Therefore, researchers have developed onsite water treatment plants by using membranes, which are small, portable, energy efficient, and effective in giving proper water supply to the users. Advanced technologies are used so as to have low fouling and highly permeable membrane systems for water purification.

Presently, reverse osmosis is widely used by militaries of different countries to meet their daily water needs. Current developments in the field of membrane science, especially reverse osmosis, made it easy and affordable for soldiers to have an uninterrupted water supply. The U.S. Army has used reverse osmosis technology to meet its water requirement for many years. It uses a reverse osmosis plant mounted over a trailer with a generator and continuously utilize it to meet daily water needs. These reverse osmosis systems are available in various capacities; some are huge and some are very small.

Present membrane science developments for water purification are still not completely free from drawbacks, and thus require a lot of research in the improvement and development of better membrane processes, systems, and technologies to achieve better results.

8.4.2 Clothing

For the military, it is very important to have protective and all-weather clothing for various working conditions. Soldiers are stationed from deserts to snowy mountains and grassy plains to rainy forests; these places provide totally different climates and atmospheric conditions. Similarly, in present-day war situations, the use of chemical and biological weapons is a possibility. Therefore, for a safe and comfortable life at these places and situations, smart protective clothing is required.

Developments are going on in this regard and researchers have developed smart fabrics for smart clothing by using advanced and functional materials. For adverse climate conditions, breathable textiles have been developed, as explained in Section 8.3, and for chemical warfare, fabrics are developed with materials capable of decomposing chemical warfare agents. Materials such as enzymes, metal oxide nanoparticles (e.g., MgO, CaO, ZnO, Fe_2O_3, and TiO_2), ortho-iodobenzoic acid, alkalis, and nanocomposites are used as decomposing agents of chemical warfare agents in the fabrication of smart clothing. Still, there is much to be done in terms of development in the technology, material, and processes of smart clothing fabrication.

8.4.3 Membrane Reactors

In this era, wars are not fought based on conventional terms and strategies. Present-day technology has evolved so much that now almost every army on this planet has target-guided supersonic missiles, where you do not even have to go to the warfront to fight a war. These missiles can be launched from anywhere inside the country and will hit its target anywhere along the enemy lines. Now, the next level of technology is to use chemical and biological agents in war. These agents cannot be seen and will mix with the surrounding atmosphere and will cause millions of deaths. Also, it is difficult to stop them from spreading in the environment. Therefore, techniques are being developed to stop, retain, or decompose these agents before they do any harm. Membrane reactors are proposed as a vital option for the decomposition of these agents. The membrane present in the reactor will consist of material that can adsorb and decompose the agents, such as activated charcoal impregnated with various metals or metal oxides. Catalytic decomposition of the toxic agents will take place in these reactors.

These reactors can be installed in rooms, cabins, and buildings. This will keep residents safe from toxic agents. Similar to other techniques for this process, novel and new materials are required for efficient and effective results. Researchers are working on a combination of various materials to develop effective materials to be used in such membrane reactors.

8.4.4 Monitoring and Sensing

As discussed in the preceding heading, chemical and biological agents are nowadays a big threat in case of a war. These agents cannot be detected by conventional methods, and advanced smart techniques are required to detect them.

Membranes can be used as an advanced sensor for the detection of chemical and biological warfare agents. Immobilized microorganisms, enzymes, or cells on a membrane can be used for the detection of a chemical or biological warfare agent, for example, use of *Rhodococcus erythropolis* and *Issatchenkia orientalis* to detect the presence of a biological warfare agent by monitoring the oxygen consumption. These devices can be implanted in the clothing or buildings for the detection of any possible chemical or biological attack.

Presently, research is focused to find more such combinations of materials that can be used for successful detection of chemical or biological warfare agents. Potential chemical and biological warfare agents are also being explored and studied to have a proper knowledge against them, so as to develop techniques for their successful detection and decomposition.

8.4.5 Energy

Energy is the fourth most important aspect of a soldier's life after air, water, and food. It is not possible for any army to win a war without energy; it is as important for the devices, technologies, and machines to run as are air, water, and food for a soldier. In general, it is difficult to provide electricity, by conventional means, to the remote locations where the soldiers are stationed. Therefore, novel, portable, and cost- and energy-effective techniques are required to provide the desired electricity to the soldiers at their remote places. Fuel cells are the best option for this purpose and are also time tested (in case of space missions). Therefore, membrane fuel cells can be used for the development of devices so as to provide power to soldiers. Fuel cells are discussed in detail in Section 5.3.6 of Chapter 5.

8.5 ROLE OF MEMBRANES IN THE FIELD OF HEALTH

Membranes are widely used and turned out to be very useful in the field of health care. Membranes have potential for various biological and medical applications, such as sensing, sorting, releasing, and separating biological molecules. Membranes with properties similar to natural filtration systems are actively being developed for their use in the successful development of smart membranes for sterilization, particle removal, drug delivery, artificial organs, and other novel medical devices. Recent advancements in the field of membrane science have made it possible to precisely regulate and control the morphology and other physical and chemical properties. This makes the membrane eligible to be successfully used for in vivo and in vitro membrane applications. Some of these major membrane applications are discussed in the following. They include sterilization and particle removal, enzyme concentration, drug delivery, tissue engineering, hemodialysis, blood oxygenation, antibody production, and health care industry waste management.

8.5.1 Ultrapure Water

Water is the most essential component for all organisms present on this planet. Similarly, it is important and forms a crucial part of any industrial process or application. For the health care industry too it is a basic requirement of many processes and applications. The requirements of the health care industry are somewhat stringent if the quality of water is considered. It is important to have water in its purest form, free from microorganisms, solids, and pyrogens. A microorganism's presence in water may cause infections or add to the deterioration of the already deteriorated health of a patient. The smallest blood vessels have a diameter of around 3 μm; therefore, if any solid larger than this size is present, it may block the blood vessel, which may result in blood clotting, granulomas, and might be life threatening. Therefore, pyrogens present in the water may result in the unregulated increase of body temperature and may cause heat shock or sudden death. Due to these reasons the water used in the health care industry for various purposes should be free from these impurities and should be of purest form.

There are conventional processes, such as boiling, that will achieve water of ultrapure form, but they are costly and time consuming on an industrial scale. Therefore, membranes are the best options to achieve the desired results efficiently as well as economically. Nowadays, membranes are used widely to provide ultrapure water to the health care industry. Microfiltration and ultrafiltration membranes are usually used for this purpose. Microfiltration membranes are capable of effective removal of solids and microorganisms. However, microfiltration membranes are not able to remove pyrogens; for the complete removal of pyrogens, a membrane with molecular weight cut-off of 10,000 kDa is totally effective. Thus, membranes provide a complete solution for the delivery of ultrapure water.

8.5.2 Sterilization and Particle Removal

Pharmaceutical reagents, media, or products, such as vaccines, tissue culture media, human plasma fractions, parenteral drugs (solutions), and injectables, are mostly heat and pressure sensitive. It is also important that these compounds are sterilized before their administration to humans, but their heat and pressure sensitivities make conventional techniques of sterilization unusable. Therefore, they have to be sterilized with techniques that do not use heat and high pressures for the sterilization process. Membranes are the best option for this due to their nature of operation and selectivity.

Filtration sterilization should be perfect, if only a single microorganism passed the filter. However, a single microorganism can multiply overnight or grow into thousands. This may result in serious or fatal infections. It is also evident from clinical data that the presence of particulate matter in IV solutions indicates a serious health risk, for example, clot formation because of the adherence of red blood cells to particles, direct blockage of blood vessels, or partial occlusion of arteries. Therefore, the U.S. Food and Drug Administration (FDA) has specified that a large volume parenteral drug should always be filtered through a filter with final mean porosity equivalent to 0.45 μm.

Membranes are also used to sterilize the air and gases by removing any particulate matter or microorganism present in them before their use in the health care industry for various applications, for example, in the case of filtration of air or nitrogen (which is used for solution fillings or transfers, in fermenters and reactors), for the sterilization of the compressed air used in sterilizers, and as vent filters. It is also approved and recommended by the FDA to use air vents or vent filters at all places where control of microorganism is required, for example, in all stills and tanks containing media, water, or solutions required in the processing and manufacturing of various products in the health care industry. This clearly states why vent filters should be used as retentive filters for fermenters and bioreactors.

In the case of gases, the sterilization filters should be of hydrophobic nature. Membranes prepared from polysulfone, polytetrafluoroethylene, or polypropylene are naturally hydrophobic and non-wetting. One important thing other than the nature of the filters is the sterilization of the filters themselves, that is, the filters should be sterilized and installed aseptically. The filters should be tested prior to their use. Therefore, it is required that the filters be sterilizable for better use and reuse. Autoclavable and steam sterilizable filters are therefore in demand for sterilization and particle removal applications.

8.5.3 Drug Delivery

Drug delivery is an important aspect in the treatment of a particular biological problem in a human body. In conventional methods of drug delivery, a drug (in the form of a tablet or injection solution) is administered orally or intravenously and provides delivery at a certain time. This results in high doses for a particular time period and the patient has to take the drug repeatedly over a period of time or until the problem is cured. Therefore, an efficient drug delivery system that delivers or releases constant drug levels in the blood and avoids the need of multiple or repeat of drug doses is required. In general, an efficient drug delivery system is the one that delivers a drug at a specific time and site in a particular release pattern.

There are two types of membrane-based drug delivery systems: osmotic membrane drug delivery systems and diffusion-controlled membrane drug delivery systems. Diffusion-controlled membrane systems are widely used and have found commercial application in the field of drug delivery.

8.5.3.1 Osmotic Membrane Drug Delivery Systems

In an osmotic membrane drug delivery system, a membrane that is permeable to water but not to the drug is used as a drug reservoir. The concentrated drug is released from the reservoir as it is filled with water. By using this system, various kinds of drugs can be delivered with high flux values. In some cases, this drug delivery system is used as a one-time source of a drug, where due to the absence of an orifice in the reservoir, the osmotic pressure rises to a level where the reservoir bursts and the drug is released at once, instead of the continuous release of a drug. Many osmotic membrane drug delivery systems were developed over the time, for example, the Rose-Nelson system in 1955, the Higuchi-Leeper and Higuchi-Theeuwes system in 1970, and the Theeuwes system of 1987 [11]. Recent systems are usually regulated by the membrane porosity and the orifices are absent. Membranes are made up of hydrophilic materials (e.g., cellulose acetate) and are in the form of capsules for the successful delivery of a drug by following the principle of an osmotic membrane system.

8.5.3.2 Diffusion-Controlled Membrane Drug Delivery Systems

Diffusion-controlled membrane drug delivery systems work depending upon the property of a drug to diffuse through the membrane instead of the membrane semipermeable property. In addition to the drug diffusion, the membrane thickness plays a role in efficient and successful drug delivery. The drug delivery follows Fick's law of diffusion. The membrane may be porous or nonporous, biodegradable or not. This system is used for various applications by using tablets, patches, or implants.

Research in this system is mainly for the best drug and polymer combination that successfully satisfies the desired criteria or requirements of the system. Various predictive methods are used to calculate the drug diffusion through a particular membrane prepared from a specific polymer. Some of the predictive systems in use are Hildebrand's theory to analyze the solubility of a drug in a solvent and the theory of Flory-Huggins to predict the solubility of a drug in a polymer. These theories, for example, help in formulating a relation between the melting temperature of drugs and their permeability through polymers. Thus, this system is based on diffusion properties when it efficiently and with control delivers a drug through a membrane to a patient.

In the case of tablets, a drug is pressed to give it a form of a tablet and then coated with a hydrophilic membrane. Once inside a patient's body, it gets hydrated and a viscous gel barrier is formed through which the drug diffuses in a controlled manner. The diffusion of the drug depends upon the type of drug and polymer used for the membrane preparation.

Patches are one of the widely used drug delivery systems. They are mostly used in case of transdermal and ocular drug deliveries. In this case, a drug is sandwiched between two polymeric membranes and released in a controlled form when applied to the patient. Diffusion is the basic force of drug delivery in this system, for which the driving force may be a concentration difference or any other force like electrical current. Thus, a drug is delivered successfully in a controlled form to the affected area.

Implants are the sources of a drug, made up of membrane reservoirs containing the drug in a liquid or powder form. A constant thickness membrane stores the drug, then releases it in a controlled form via diffusion based upon the properties of the drug and membrane polymer. In the case of biodegradable membranes, the drug should be released completely before the membrane degrades and in the case of nonbiodegradable membranes the implant is removed surgically after a particular time period. The major drawback of implant use is accidental rupture of the membrane. In that case a sudden release of the drug will take place and the dose will also be very high, which might in many cases not be good for the patient.

In these two discussed systems, the membranes are used in various forms, for example, as reservoirs or a matrix delivering a drug by passive or active diffusion. In the passive diffusion transport mode, concentration difference is usually the sole driving force and in active diffusion, an electric current or any other driving force is used for the delivery of a drug. In the case of an electric current, the process is known as iontophoresis. Nowadays, stimuli-responsive membranes are developed and used for efficient and successful drug deliveries. In these cases, a membrane is made stimuli responsive, which behaves differently upon getting a particular stimulus. The stimulus may be temperature, pH, pressure, magnetic, or electrical based. Upon the application of the particular stimulus, the membrane pores may open or close, and thus regulate the delivery of a drug. These systems are very efficient and successful but much work is still to be done in their development. Development is required in the materials and membranes to have affinity or mutual properties with a particular drug. The interaction between the drug and membrane material in these drug delivery systems is very important. Therefore, research in the development of better materials is important so that there should be enough options available for use in efficient and successful drug delivery.

8.5.4 Tissue Engineering

Organ transplantation is a growing area in the field of health care. There is always a high demand for organs for transplantation purposes, but the lack of donors and other medical complexities make it difficult. Therefore, it is important to find a way of replacing malfunctioning organs with functional organs without any delay and medical complexities. The field of artificial organs shows great potential to solve the problems of donor and organ shortages. There are proposals for in vivo engineering of the damaged organ or tissue with an in vitro developed artificial organ, tissue, or device. This will help in addressing the difficulties, problems, and complexities involved in direct organ transplantation, rejection, and transmission of pathogens. Therefore, there is great possibility to replace direct organ transplantation techniques.

The widely used theme for this purpose is the fabrication of a scaffold. A scaffold is a temporary 3D support for the attachment and growth of the respective tissue cells prior to their transplantation to a host. The functionality of the artificial tissue depends to an extent upon the design of the scaffold. The scaffold should be porous with well-connected pores for efficient nutrient transport to the cells and should have good cell adherence properties with appropriate mechanical properties. Also, the scaffold should be prepared from biocompatible and biodegradable materials in tandem to the tissue regeneration and extracellular matrix remodeling. Membranes contain all of the required properties and therefore are the best option for the tissue scaffold. Thus, membranes can also be used for the regeneration of functional tissues or a complete organ itself.

Polymeric membranes are often used for applications related to soft tissues, such as skeletal muscle. On the other hand, for hard tissues such as bones, ceramics or metals are used. Both natural and synthetic polymers and copolymers are used for efficient and successful regeneration of tissues. Some examples of the natural polymers used in soft tissue regeneration are collagen, fibrin, gelatin, polysaccharides, and poly(hydroxybutyrate), and examples of synthetic polymers are poly(lactic acid), poly(glycolic acid), poly(ε-caprolactone), poly(anhydrides), poly(esters), and poly(orthoesters). On the other hand, some of the commonly used materials for hard tissue regeneration are hydroxyapatite, tricalcium phosphate, silica, titanium, steel, aluminum, and composites of various polymers, metals, or nanoparticles.

8.5.5 Hemodialysis

Advancements in the field of hemodialysis have made it a safe and secure technique to be used for renal failures. Improvements in dialysate systems, the development of more reliable monitoring equipment, and the automation of safety devices have made hemodialysis easy and more reliable to use. Further, other technical improvements and developments in membranes, water quality, computer controls, and the dialysate made it an advanced and mostly automatic technique. Due to these advancements hemodialysis can be fully automated, and the feedback controls employed maintain the desired conditions at optimal values, for example, the ultrafiltration rate and dialysate temperature. Also, the automation helps in real-time monitoring of the process. These developments and advancements further improved the survival rate.

Developments in membranes and membrane materials further made the technique more effective and economical. Improvements not only enhanced biocompatibility or permeability but also in reproducing the physiological processes of the glomerular (ultra)filtration. In early days, the main focus was on increasing the permeability of the membranes prepared mainly from hydrophobic base materials, such as polyamide, polysulfone, polyarylethersulfone, or polyacrylnitirle, with large pore sizes. Later, importance was given to the biocompatibility and development of an implantable device for carrying out in vivo hemodialysis. Three-dimensional scaffolds using membranes were developed for implantable hemodialysis devices. Nowadays, researchers are concentrating on living membranes to imitate the role of a kidney, in terms of not only filtration but also for other physiological functions carried out by a kidney. Working on this theme, researchers have developed a renal tubule cell assist device, commonly known as RAD. It consists of human renal cells adhered to a membrane surface mimicking the roles and functions of a kidney. This device not only performs the basic filtration function of a kidney but also carries out functions like metabolic activities and endocrine processes effectively. RAD, along with a hemofilter, forms an artificial kidney that performs filtration, metabolic and endocrine activities, and reabsorption. This device has been successfully tested on animals and tests are going on for humans. Further, developments are also required in this field for the development of highly effective devices.

8.5.6 Blood Oxygenation

Present-day advancements in the field of membrane science have helped in the development of artificial blood oxygenation techniques, for example, a blood oxygenator. Membranes with excellent biocompatibilities and gas permeabilities are now available. However, there are some areas where much improvements and developments are required, such as the handling, manufacturing, and assembling of these membranes into the artificial blood oxygenator device. Research is ongoing to address these problems, and membranes and membrane modules are being developed to cater to the needs of an artificial blood oxygenator.

Nowadays, off-pump coronary artery bypass (OPCAB) grafting is becoming a widely used technique. It is important to develop an emergency cardiopulmonary bypass procedure for the safe and successful execution of an OPCAB procedure. Therefore, development of a small-scale membrane blood oxygenator will make the OPCAB technique easily and effectively executable. Though an artificial membrane oxygenator is widely used for severe respiratory and heart failure cases, the original outcome, in terms of lives saved, is not satisfactory. Therefore, there is a need to develop systems that can be used for longer periods of time. This needs improvement in terms of membranes and the ancillary systems of an artificial blood oxygenator. Membranes with low fouling, better flux, biocompatibility, and longer life spans are required. Similarly, automation and regulation of the ancillary systems are required. Recent advancements in these fields show promising growth that ensures that in the near future these objectives will be achieved.

8.5.7 Plasmapheresis

Plasmapheresis is a technique used to separate plasma from blood using membrane filtration. Membrane filtration has own advantages over conventional ways of plasmapheresis, including as efficiency, ease, safety, and economy. Centrifugation was the technique used in earlier days for plasmapheresis. Blood was collected from donors in plastic bags and then centrifuged so as to separate the plasma from the blood. The main purpose for separating plasma from blood is to get purified components, such as albumin or factor VIII (antihemophilic factor). Another important advantage of membrane-based plasmapheresis is that it can be done continuously, unlike with other conventional techniques. In addition to this, it is rare that the blood and separated plasma will remix with each other, and it is less time consuming as compared to other conventional techniques.

The early attempts of plasmapheresis with membranes were not very successful due to the problem of fouling. Measures were taken to avoid fouling in the process by using stirrers, but stirrers faced the problems of the hemolysis of the red blood cells and low flux. Later, attempts were made to make the membranes of antifouling nature by using novel membrane materials. This antifouling membrane strategy has seen some success, but there were other areas requiring improvements, such as the optimization of transmembrane pressure and determination of the best pore geometry for the process. Hemolysis of red blood cells was still a major problem associated with the membrane plasmapheresis. Red blood cells at a pressure higher than the critical pressure hemolyze as well as through distortion in the pores. Therefore, these areas still require a lot of improvements and developments. Nowadays, plasmapheresis is not only used for the separation of plasma from blood but also therapeutically, for example, in plasma exchange therapy. In plasma exchange therapy, a patient's plasma is purified by removing the toxic metabolites or compounds from it and then returning the purified plasma to the patient's body. In general, this therapy is required in the cases of autoimmune diseases, such as Goodpasture syndrome, rheumatoid arthritis, autoimmune hemolytic anemia, and systemic lupus erythematosus, where an antibody or a component of self-immune system attacks one's own body. For example, in Goodpasture syndrome an antibody attacks the basement membrane of lungs and kidneys, which results in bleeding and ultimately in respective organ failure. It is also successfully used for many other metabolic and immunological diseases/disorders. The technique of plasmapheresis in many cases is also used to detoxify the body of a patient in case of drug overdose or poison. This technique is also used by many pharmaceutical companies for animal blood pooling on a very large scale for yielding desired components/products from the collected blood, such as antibodies. Therefore, this technique also has a great future and will further develop with the developments in the field of membrane science.

8.5.8 Antibodies and Enzymes

In the health care industry, antibodies and enzymes are widely used for various applications, such as detection or sensing of toxic chemicals or metabolites in the human body or products consumed by humans, like food or medicinal products, air, or water; in the diagnosis or prognosis of a medical condition; and are either a constituent of a medicinal product or help in production of many medicinal products. This also adds to the efficiency and effectiveness of various processes by making them swift and cost effective. Therefore, these two constitute important entities of the healthcare industry. Previously, these were produced on an industrial scale by using animals, which are costly, time consuming, less productive, and unethical. Thus, new techniques with effective and efficient production of these two were needed. Therefore, membranes were explored for this purpose and as always they did not disappoint. Production of antibodies and enzymes by using membranes in the form of membrane bioreactors made the process of antibodies and enzymes production a great success.

In a membrane system, antibodies or enzymes are produced by immobilizing cells or microorganisms on the membrane surface. This membrane system is known as a membrane bioreactor and ceases the use of animals as reactors and also other processes previously used for the production of these products. The cells and microorganisms have a tendency to grow on a surface and thus a membrane surface gave them optimum space to grow and proliferate. Therefore, a good yield of antibodies or enzymes can be achieved by using a membrane bioreactor.

Presently, there is the need for extensive research for the improvement and development of advanced membranes with better characteristics in terms of cell or microorganisms adhesion, growth, and product separation, for the development of systems with better regulation and automation of the process, development of improved and specific culture medias, and further developments in the process itself for further increasing its efficiency, ease of use, and reducing the overall cost of the process.

8.6 MEMBRANES FOR A BETTER ENVIRONMENT

The fast increase in the population, urbanization, and industrialization has also increased the demand of resources, deteriorating and exhausting them. This rapid growth has had a harsh impact on the environment and is worsening the climate. Problems like climate change, poor availability of much needed resources like air and water, poor crop yield, and low quality of life are some of the deleterious effects of these problems. To counter these world problems, various measures have been taken by employing advanced technologies for various applications and processes to reduce the effect of the problems [12]. Membrane processes are one of the processes that are used for sustainable use of energy, resources, and to reduce the impact of pollution on the environment. In the last century, membranes have seen tremendous growth and development. Due to their properties, membranes are used in almost every industry. The key developments that made membranes the user's choice are development in terms of better membrane material, development and better understanding of membrane preparation techniques, availability of better membrane modules, and development of different effective and efficient membrane processes. However, membranes still need development and improvements in membrane performance, membrane systems, antifouling, and other processes intensive to specific membranes.

Presently, membranes are widely used for processes like desalination, wastewater treatment, purification and disinfection of drinking water, manufacturing and separation of food, pharmaceutical, and biotechnology products, artificial organs development, fuel cells, and gas separations. Some of the recent developments in membrane science with respect to the environment are discussed next.

8.6.1 Ozone-Resistant Membranes

In wastewater treatment, the ozone plays a vital role due to its strong oxidizing powers. It is very effective for the disinfection, decoloration, and odor removal from wastewater. It is widely used in advanced wastewater treatment processes to treat wastewaters. On the other hand, membranes are good for filtration processes to remove turbidity, solids, and microorganisms from wastewater. But fouling and flux decline are major problems with membranes. Therefore, it is required to pretreat the feed before feeding it to a membrane process; this makes the overall process costlier. Also, the results of membrane processes are less competitive as compared to other conventional processes. Thus, it will be good if the process of ozonization was combined with membranes. The overall process quality and efficiency would be increased and the cost reduced. The oxidation powers of ozone will help to reduce membrane fouling by decomposing the organic materials present in wastewater and prevent them from stacking at the membrane surface. By the virtue of it, a constant high membrane flux can be maintained. Therefore, different ozone-resistant membranes and membrane modules specific for ozone use are developed by using ozone-resistant materials. The combination of ozone and membranes consistently provides high flux, low fouling, and better treatment of wastewater. Similar technologies are required for further development and wider use of membrane processes on the industrial or commercial scale.

8.6.2 Membrane Bioreactors

Membrane bioreactors are wastewater treatment systems used to separate solids (activated sludge) and liquids (water). The characteristic features of a membrane bioreactor make it suitable for use in efficient and effective wastewater treatment. It helps in saving overall space requirements since final settling tanks are not required, reduces treatment time, is easy to maintain and manage, and produces high-quality treated water. This makes membrane bioreactors the choice for a better and cost-effective wastewater treatment.

Presently, developments are being made in the types and shapes of membranes and modules for membrane bioreactors. Membranes with tough physical and chemical properties, high permeability, easily cleanable, and compact designs are in development for membrane bioreactors. Membrane bioreactors are widely used for the treatment of wastewaters from various industries, including municipal, food, pharmaceutical, biotechnology, and electronics. Long-running operations are possible with the use of membrane bioreactors due to ease of maintenance and management. Also, the various developments in membrane preparation techniques, membrane materials, and membrane modules make membrane bioreactors a widely used wastewater treatment technique.

8.6.3 Blue Energy

Increase in the demands of power in recent times has put a lot of burden on the energy sector. Presently, energy sources, especially fossil fuels, are consumed at an alarming rate. The excessive use of fossil fuels is resulting in the deterioration of climate and showing deleterious effects on the health and growth of species on this planet. These negatives associated with the use of fossil fuels and their limited availability generate a resurgence for the use and development of alternative energy sources, especially renewable energy sources, such as solar, wind, geothermal, and tidal. The complexities associated with these renewable energy sources, such as underdeveloped technology and the limitation of being site specific, make them unpopular for wider use.

Recently, a new membrane-based technology has attained a bit of commercial success. This technology is pressure-retarded osmosis. The energy available or used in this technique is widely known as blue energy. The basic principle of this technique in simple words is if there are two water solutions with different salt concentrations separated by a semipermeable membrane, then as per the principle of osmosis, water will move from the less concentrated salt solution to the one with higher salt concentrations. This flow or movement of water across the membrane generates a pressure on one side of the membrane; this generated pressure can be used to move turbines and generate electricity. This process generates maximum energy at higher flux as compared to that of lower flux. This low flux can be achieved by taking out the salted water so that the pressure restricts the flow of freshwater from the other side of the membrane. Thus, this process is named pressure-retarded osmosis.

Presently, this process is used to harness energy at places where a river meets sea. This location generates the required conditions for harnessing blue energy effectively and efficiently. In the year 2009, Statkraft, an energy sector company, opened an osmotic power plant that harnessed blue energy by using the pressure-retarded osmosis process with a total energy generation capacity of 4 kW, which is miniscule as compared to other energy generation processes and demand. Therefore, the plant was shut down in 2013. Thus, there is still need of development and improvements to make this process viable for larger use. Mostly, this process is related to rivers and oceans, but it is possible the process can be used with wastewaters coming from different industries, as the concentration difference in the feeds can be used for the generation of pressure as is the case with river and seawater. The generated power will definitely be low but can be used to carry out small works of industry.

Reverse electrodialysis is similar to the pressure-retarded osmosis with the difference that instead of water, salt ions flow across the membrane to generate power. Charged or ionic membranes are used for the process to allow either positive or negative ions to flow through them. This flow of ions produces an electrical voltage that can be directly used to create an electrical current flow, that is, also without the need of pressure-driven turbines. Theoretically, this process seems to be effective and efficient to generate energy, but still many improvements and developments are required to avail it on a commercial scale.

The third process to harness blue energy is known as capacitive mixing, where seawater and freshwater are fed alternately to a chamber containing two electrodes, and this electrode-containing chamber works as a charge storing device. This results in the generation of voltage.

Nevertheless, these processes have their limitations, such as low energy generation, site specific, and less cost effective. Though, as said earlier, the site specificity problem can be solved by using wastewater from different industries, such as sugar or food, textile, leather, pharmaceutical, and electronics. This statement is backed by a recent development of capacitive mixing in which the efficiency of capacitive mixing can be increased (almost doubled) by using warm water (~50°C) as compared to normal water. To keep this process cost effective and energy intensive, it is suggested that warmed wastewater from industries can be used, for example, cooling water from power plants instead of heating it by using extra energy. Similarly, water with dissolved carbon dioxide can be used instead of seawater in capacitive mixing process. Since carbon dioxide has a high dissolving tendency in water, it can be captured from fossil fuel power plants or flue gases by using membrane processes and can be dissolved in water for use in the capacitive mixing process for energy generation. This not only helps in generating energy but also reduces the overall concentration of carbon dioxide in the environment, which further reduces the global warming effect. Thus, there is a need for novel ideas, devices, and processes for further development of various processes and to save the environment.

8.6.4 Gas Separations

The rapid increase in the global industrialization and economy resulted in a pollution increase. Airborne pollution is the severest of them all since it is not bounded to a specific place; molecules can be generated miles away. Therefore, if a toxic gas is generated by an industry, then its effects can be seen miles away. Major gaseous pollutants are carbon monoxide and dioxide, sulfur oxides, nitrogen oxides, and chlorofluorocarbons. This gases cause global warming, acid rain, difficulty in breathing, low visibility (due to the presence of particulate matter), and diseases (such as respiratory, skin, and allergies). Thus, it is necessary to control this daily increase in air pollution. Nowadays, membrane-based systems are widely employed for the control of air pollution. Membrane systems are employed at the point of emissions of these gases, so as to capture them before they enter the atmosphere. In addition to this, membrane systems and processes are also used in the separation of gases present in the various products in the form of impurities, such as the recovery of hydrogen from refinery of gases, carbon dioxide removal from hydrogen produced in steam reforming and coal gasification plants, separation of olefin–paraffin mixtures, and for enhancing the quality of gas used in oil production and of biogas generated from landfills and animal excreta.

Nowadays, most research is concentrated in the capture of carbon dioxide from the atmosphere or its point of emission. This is due to its deleterious effects on the environment, such as global warming. Other important areas, in reference to environment safety, where membranes are employed are in the separation of acid gas from syngas and atmosphere. Developments and improvements are made on a regular basis in terms of process design, membrane modules, membrane material, membrane structure, and membrane preparation techniques. Membranes with better selectivity and permeability are needed by the gas separation industry because membranes with low selectivity and permeability are neither yield effective nor cost effective (since initial costs are high and with these kinds of membranes the main objective cannot be achieved). The main classes of polymers presently used for gas separations are polyimides, polyethers, thermally rearranged polymers, polymers with intrinsic microporosity, and substituted oxyacetylene. Researchers add various novelties to these classes of polymers so as to obtain the desired gas separation results. Recently, metal organic frameworks were included in the gas separation membranes and have shown promising results. Similarly, researchers are working on the development and optimization of mixed matrix and zeolite membranes by using novel materials of polymers, ceramics, nanoparticles, carbon nanotubes, and nanocomposites. Still, a lot of advancement and development is required to make membrane gas separation a widely successful and acceptable technology.

STUDY QUESTIONS

1. Name the best past and present achievement of membrane science.
2. What are the membrane properties that made it possible to use membranes for the development of artificial organs?
3. If Ramesh wants to develop an artificial kidney, then what are the areas he should work upon to achieve his goal?
4. Is it possible to use a gas separation membrane for blood oxygenation? Based on your answer explain why.
5. Neha needs to develop an artificial liver. What is the major area she should careful about and why?
6. Recent advancements in membrane science are not enough to prolong the use of an artificial liver in the case of an acute liver failure patient. What advancements and developments should be done to improve the longevity of an artificial liver?
7. Nimish wants to develop an artificial liver by culturing hepatic cells but has been unsuccessful. What are the possible reasons for his failure?
8. There are various integration methods to integrate pancreatic islets into membranes. Which one do you think is best and why?
9. The Indian military needs smart warfare gear. The company awarded the contract of developing a membrane-based textile that can be used not only by military personnel but also by other people with active lifestyles, such as sportspersons, adventurers, and scientists, so as to generate maximum profit. What and how should the company design and prepare something like this?
10. How can membranes be used to develop smart firefighting wearables?
11. What are the advantages of smart conductive textiles?
12. In how many ways can membranes be used by the military?
13. Paul is suffering from diabetes and thus has to take regular insulin shots. How can he use membranes to relieve himself from the pain of insulin shots?
14. Peter wants to develop a membrane drug delivery system for older people in which they won't have to remember the timings of their pills. What method of drug delivery should he use? How can he develop a system using membranes and what are the requirements to develop such a system?
15. What is the technique used for autoimmune diseases? What is its working mechanism?
16. Sham has a concentrated solution and pure water along with a reverse osmosis membrane. Can he generate power by using these materials? What else does he require and how can he do it?
17. Is it possible to separate hydrogen and carbon dioxide gas from each other by using membranes? Based on your answer explain why.

REFERENCES

1. J. J. Abel, L. G. Rowntree, and B. B. Turner, Plasma removal with return of corpuscles (Plasmapheresis), *Journal of Pharmacology and Experimental Therapeutics*, 5(6), 625–641, 1914.
2. G. Haas, Versuche der Blutauswaschung am Lebenden mit Hilfe der Dialyse, *Klin Wochenschr*, 4(1), 13–14, 1925.
3. W. J. Kolff, H. TH. J. Berk, M. Welle, A. J. W. van der Ley, E. C. van Dijk, and J. van Noordwijk, The artificial kidney: A dialyser with a great area, *Journal of International Medicine*, 117(2), 121–134, 1944.
4. S. Brukhonenko and S. Tchetchuline, Experiences avec la teteisolee du chien, *Journal of de Physiologie et de Pathologie Generale*, 27, 31–79, 1929.
5. S. Brukhonenko, Circulation artificielle du sang dans l'organisme entire d'un chin avec Coeur exclu., *Journal of de Physiologie et de Pathologie Generale*, 27, 251–272, 1929.

6. J. H. Gibbon Jr., Artificial maintenance of circulation during experimental occlusion of the pulmonary artery, *Archives of Surgery*, 34, 1105–1131, 1937.
7. W. J. Kolff, D. B. Effler, L. K. Groves, G. Peereboom, and P. P. Moraca, Disposable membrane oxygenator (heart-lung machine) and its use in experimental surgery, *Cleveland Clinic Quarterly*, 23(2), 69–97, 1956.
8. M. C. Annesini, L. Marrelli, V. Piemonte, and L. Turchetti, *Artificial organ engineering*, Springer-Verlag, London, 2017.
9. W. L. Chick, A. A. Like, and V. Lauris, Beta cell culture on synthetic capillaries: An artificial endocrine pancreas, *Science*, 187(4179), 847, 1975.
10. W. C. Smith (Ed.), *Smart textile coatings and laminates*, Woodhead Publishing/CRC Press, 2010.
11. Richard W. Baker, *Membrane technology and applications*, 2nd ed., John Wiley & Sons, 2004.
12. M. Mulder, *Basic principles of membrane technology*, Springer, 2007.

9 Future Perspectives of Membrane Science

9.1 INTRODUCTION

In recent years, membranes turned out to be potential processes for many industrial as well as commercial applications. Their properties, such as selectivity, permeability, efficiency, effectiveness, simplicity, flexibility, stability, compatibility with other processes and surroundings, easy control, low energy requirements, low cost, and scope for easy scale up make them a perfect process to be used and explored for present and future applications. In the previous chapters, these properties of membranes and their use in various applications were discussed in detail along with their pros and cons. In this chapter, the various advances and future challenges are discussed in detail for some of the potential areas of membrane applications. These potential applications are on the verge of important developments and advancements in terms of technology and processes. Many new materials, processes, and technologies are introduced in the field of membrane science for its further betterment and enhancement in efficiency and effectiveness.

In spite of various advancements and developments in the field of membrane science, it still seems that membrane science is in its infancy. However, developments in some of the fields, such as health care, wastewater treatment, food, textiles, pharma, and material science has shown some scope and hope for the bright future of membrane technology. Membrane use in the field of health care, especially in the development of artificial organs, is breathtaking and still growing. Importance is also given for the development of many other organs (such as artificial brain and retina) by the use of membrane science and technology as it is used for the development of artificial kidney, liver, and pancreas. In addition to this, research is ongoing for the growth and development of conventional membrane processes and areas, such as membrane encapsulation and packaging. Studies based on the transport phenomenon and mechanisms will definitely result in a positive outcome for the development of these areas. Membrane processes are also applied to the advanced areas of space and space exploration programs. Also, research based on the development of novel nanomaterials with specific properties and structures holds great promise for the development of membrane science.

The current progress with future perspectives is discussed in the subsequent sections of this chapter related to some of the potential areas of membrane science. These areas are important for the sustainable development and advancement of membrane technology and will help in the commercialization as well as growth of the human race.

9.2 ADVANCED MATERIALS FOR MEMBRANES

Nowadays, membrane technology is a well-accepted industrial technology due to regular improvements and developments. Membranes are used for a myriad of applications as an effective and efficient technology. Still, many of the remarkable applications need further improvements and developments in the field of membrane science, as they cannot be done effectively and efficiently by using present-day membranes. Therefore, membranes with better properties and especially membrane materials need to be developed for their utilization for the separation of challenging mixtures or modern-day applications related to health care, pharmaceutical, food, and other industries. Membranes with effective and efficient properties of separating biomolecules, enantiomers, or isomers are required. Similarly, there is a need for biomimetic and smart membranes to carry out functions analogous to the biological membranes. This can be achieved by developing advanced membrane materials, such as responsive polymers, hydrogels, nanocomposites, and carbon nanotubes. The advancements made in these materials are

- Responsive polymers—Responsive polymers are swiftly replacing conventionally used polymers for the modification or betterment of membranes and add a niche to the prepared membranes. There are types of responsive polymers based on the stimuli they respond to, such as temperature, pH, chemical, light, magnetic, and electrical responsive [1]. They are used to prepare smart membranes and add functions to the prepared membranes analogous to biological membranes. Presently, there are not many responsive polymers available; thus, there is a need to explore new responsive polymers. Researchers all over the world are into this and continuously thriving for new responsive polymers. Also, copolymers and nanocomposites are prepared with existing responsive polymers to bring new characteristics to them. Thus, this is a field for material scientists to explore and work.
- Hydrogels—A hydrogel is a 3D network of physically or chemically cross-linked polymers with swelling properties (on addition of water), potential material with life-forming capabilities. Hydrogels are highly biocompatible with high storage capacities and low interfacial tensions [2]. Therefore, hydrogels are excellent platforms for the development of smart membranes for the health care and biotechnology industries. They have great scope for their use in the development of biomimetic and stimuli-responsive membranes. Hydrogels were developed with various properties for the betterment of membrane science, including mechanical, chemical, pH, and light-responsive hydrogels. Still, there are a lot of hurdles to be addressed for the full development of hydrogels and their applications. It is still difficult to develop a hydrogel system with enough quickness to respond to an external stimulus of minimal magnitude. Further, there is scope to develop a single hydrogel responsive to many stimuli and its use on the industrial scale for various applications. Hydrogels present a diaspora of applications where their characteristics can be used efficiently if the hydrogels can be developed appropriately, such as in the development of membrane scaffolds for the development of artificial organs, advanced biodegradable materials, drug deliveries, and textiles.
- Aptamers—For common stimuli-responsive materials, such as polymers or hydrogels, the stimuli that are needed for a response to be generated from a material out of the whole system, for example, the swelling or shrinking of thermoresponsive polymer brushes in the presence of a temperature change. Here, the stimuli not only activate the responsive polymers but also affect the overall system, having an effect on the membrane or hydrogel and the feed solution. Therefore, where the response is required in a local location, it is given to the bulk too. Thus, it will be good if localized responses can be generated, as it will make the process more energy efficient and cease the effects of the stimuli over the bulk. The answer for this is aptamers, which are oligonucleic acids (single-stranded DNA or RNA olegonucleotides) having properties of molecular recognition. In nature, they are part of riboswitches that carry out gene expression. These are widely used because of their exceptional properties of specificity for bioanalytical as well as drug delivery [1]. They are better than other affinity or selective biomolecules, such as antigens, antibodies, or proteins, because of selectivity, stability, easy modification, easy to use, and scope for automation for their selection for a specific target molecule. Therefore, aptamers provide a new energy and scope to the use of membranes for various advanced applications in the health care and biomedical fields. The aptamers as discussed can be used by easily incorporating them with membranes, and membranes with advanced specific selectivities can be developed. The use of aptamers has crossed its initial stage of use in membrane science and needs further developments for their inception as a truly membrane-modifying material. Though it seems to be an exciting field for further development of membrane technology, aptamers are not free from hurdles. It is very difficult to work with aptamers due to their inconsistent and difficult-to-handle functional properties. It is not easy to work out the functional properties of DNA or RNA hybrids. Specialization with proper

knowledge of their interactions and forces involved are required. The characterization techniques for the characterization of functional properties of aptamers are also a challenge to face because there are negligible techniques to characterize the functional properties of aptamers. Progress is needed in the field of DNA- or RNA-based nanotechnology for the development of membrane technology. Researchers are working on the theory of these biomolecules so as to understand them better and make maximum use of them in the development of membrane technology. Similarly, there are many potential materials available in nature that need to be identified. Molecules like this will definitely make a better tomorrow for the membrane science.

- Nanomaterials—The field has grown by leaps and bounds in recent times. Advancements in nanotechnology have helped to advance other fields of science as well as the development of future technologies. Nanotechnology has also played a vital role in membrane science [2,3]. Recent developments in the field of nanotechnology have been applied with success in the modification of membranes for better operation and performance. The most remarkable advancements are the development of carbon nanotubes, graphene, and the technology to mass-produce them. Nanomaterials, such as metallic, polymeric, ceramic, and carbon nanotubes are used on a very large scale for the improvement and development of membrane technology. They are grafted, coated, and blended in the membranes for developing membranes with improved properties and functions. Mixed matrix membranes are the result of using nanomaterials in a polymer matrix. Mixed matrix membranes have improved properties, such as mechanical strength, better flux, and antifouling nature. Nanomaterials are still used and will be used in large numbers for bringing various advancements in membrane technology. Nanomaterials have helped to grow the strength and functionality of the membranes. The problems associated with the use of nanomaterials in the preparation of membranes are well understood and solved. The future is bright for nanomaterials in the development of membranes, since they are now a strong part of the membrane field. In the future, advanced materials will be developed by incorporating nanomaterials with different materials, and the nanomaterials will add positive energies and functions to the developed material for its use in the field of membrane science. Therefore, a membrane future without nanomaterials is dull. Presently, it is required to work upon the use and development of novel nanomaterials for their use not only in membranes but in other ancillary applications. The novel materials may help in reducing the overall use of material, energy, sources, and labors. There will also be an increase in the overall surface area of the membranes. It will also be feasible to knit and embroider the membranes or membrane pores with nanomaterials for exceptional results.
- Nanocomposites—The combination of polymers and nanomaterials gaves rise to very efficient, strong, and functional materials called nanocomposites. These materials, if used, add versatility, strength, and life to the final material or overall process. Novelty is brought to nanocomposites by using functional or polymers with distinguished properties [4]. The synergy of the nanomaterial and polymeric properties brought out a beautiful material with superb qualities suitable to be used in membranes, such as reinforcement; chemical, mechanical, and temperature resistance; electrical, optical, or magnetic properties; barrier properties; and antibacterial or bactericidal properties. This synergistic effect makes membranes effective and efficient in their performance. Therefore, nanocomposites are playing a dual role by making membranes resistive to external forces and adding functional features to them. These two features enhance the overall membrane flux, antifouling nature, and life span. Recent advances in nanotechnology have made it possible to produce thin-layer nanocomposite membranes, which can be used in different processes, such as reverse and forward osmosis, since these nanocomposites help in reducing the resistance and fouling to a great extent and increase the overall membrane flux manyfold. Nanocomposites are a boon especially for polymeric membranes. In the future, knowledge as well as technology has to be developed for the advancement of nanocomposites. The push to search for new materials has to be increased and future is bright for the development

of materials or nanocomposites from waste materials, such as fly ash, slag, and sludge from different industries. These materials carry useful and important properties that have potential to be used in the improvement or development of better nanocomposites. Similarly, knowledge of interdisciplinary subjects and interdisciplinary collaborations is required for fast and effective growth of the nanocomposite field in the right direction.

9.3 SUPERHYDROPHILIC, -HYDROPHOBIC, -OLEOPHILIC, -OLEOPHOBIC MATERIALS AND MEMBRANES

Industrial organic pollutants as well as oil spills have intensified water pollution to a great level. Consumption of such water is totally prohibited; this affects all organisms and vegetation near or around the polluted water body. In the case of oceans, oil spills are disasters that add toxic materials to the ocean. Therefore, to address water pollution by oil and organic pollutants, materials with superwettability are developed. These materials are efficient and effective in the separation of oil and water mixtures. Superhydrophobic and superoleophobic surfaces, such as lotus leaf, ignited the idea for the development of such materials [5]. Further, the development and advancements in the field of colloid and interface science as well as bionics in the last decade made the idea a reality. The overall properties of these super materials are due to the super surface wettability, which is due to the chemical composition on the material's surface and topography. There are different types of super materials based on the type of materials used for their preparation, such as metal, fabric, sponge and foam, and carbon-based materials.

Superhydrophobic and superoleophilic materials separate oil from water, as they easily absorb the oil and let it enter the pores, while repelling the water phase. On the other hand, superhydrophilic and superoleophobic (underwater) materials function inversely as compared to the superhydrophobic and superoleophilic materials by allowing water to permeate and totally repelling oil. These materials are inspired from fish scales, which protect fish from oil pollution in the sea. These materials are better than the superhydrophobic and superoleophilic materials in terms of less fouling, since in the case of superhydrophobic and superoleophilic materials, the oil fouls the membrane surface and is hard to clean. Superhydrophilic and superoleophobic materials contain the properties of both a superhydrophilic (hydrophilic) surface and a superoleophobic (hydrophobic) surface. Based on these materials, smart materials are also developed with switchable wettability, where the wettability of a material can be controlled by external stimuli. Work is also done on the separation of water and oil emulsions by using all of these aforementioned special wettability materials.

Special wettability materials possess great potential for the development of membrane processes for the effective and efficient separation of water and oil mixtures. However, at present, these materials are in their initial stages and thus facing problems on the industrial scale for the separation of water from oil. First, there is a need for special wettability materials with good design, stability, and durability. Second, the developed superwettability materials cannot be used directly in the field for water and oil mixture separation, thus this area needs development and advancement so that the superwettability materials can be used directly at the site of oil spill or oil pollution. Third, the technology to prepare these materials on a commercial scale is underdeveloped and therefore for the success of these materials on a larger scale, mass production technology has to be developed for these materials. Fourth, the swiftness and high output are lacking in the process of water and oil separation by using these materials due to the contradiction between membrane pores and the oil droplet size. Consequently, it is necessary to develop techniques to realize the separation of a wide range of oil and water emulsions. Finally, the current and past research in this field carried out by using water and oil mixtures of very low viscosity is far from reality, and thus research has to be done with high-viscosity water and oil mixtures. It is difficult to handle high-viscosity water and oil mixtures due to increased membrane fouling and difficulty in cleaning. Therefore, advancements and developments are required in this field for suitable use of superwettability materials.

9.4 SMART GATING MEMBRANES: SMART SEPARATIONS

Smart gating membranes are inspired from the ion channels present across cell membranes. Smart membranes are prepared by using responsive materials that form responsive gates in the membrane pores [6]. Therefore, under the influence of specific physicochemical stimuli, such as temperature, pH, electrical, or magnetic, these smart gating membranes regulate the transmembrane transport. Nowadays, smart gating membranes are playing a critical role for sustainable development due to their beneficial properties, such as no chemicals and additives requirements, energy efficiency, quality of the product, and low costs. Smart gating membranes are suitable for numerous applications, such as water treatment, smart separations, sensors, controlled release, health care, and food industry.

The practical success of these smart gating membranes depends upon their easy preparation, high membrane flux, effective response, and chemical and mechanical stability. These properties confirm the overall low cost, easy mass production, and effective performance of smart gating membranes and make them practically suitable for various applications. Smart polymers, such as poly(N-isopropylacrylamide), poly(N-vinylcaprolactam), poly(acrylic acid), poly(benzyl-methacrylate), and supramolecular di- or triblock copolymers, are often used to prepare smart gating membranes. The responsive domains can be introduced into the membranes in two ways, either by grafting/coating or by blending. The former method introduces responsive domains to the membrane by grafting, coating, or pore filling. There will be chemical and physical forces between the responsive domains and the membrane surface, which keep them together. There are advantages and disadvantages of these methods in terms of membrane responsiveness and flux. If there is a high number of responsive domains over the membrane surface, the membrane responsiveness will be high and low flux. Similarly, there is a low number of responsive domains over the membrane surface, the membrane will be have low responsiveness but high flux. Furthermore, making a smart gating membrane with this procedure is a two-step process, which makes it more complex. First the membranes are prepared and later the responsive domains are grafted over the membrane surface. This has disadvantages including the requirement of a high amount of responsive domains, low probability of proper grafting of responsive domains over the membrane surface, and it is difficult to mass-produce and scale up this process of membrane modification with effectiveness. Therefore, the one-step blending option is being explored by researchers for better and effective results. In this method, the responsive domains are introduced to the membrane casting solution before the membrane casting. Usually a nonsolvent-induced phase separation process is used to cast/prepare the membranes. The advantage of this one-step method is that it can be used to mass-produce membranes with currently available technologies. However, nonsolvent-induced phase separation has its own limitations. Due to its promptness it is difficult to control the distribution of the responsive domains in the prepared membranes. Therefore, it is quite difficult to obtain desired smart membrane gates because of which either the membrane responsiveness or flux is limited. In addition, in the nonsolvent-induced phase separation method the responsive domains are also responsible for pore formations in the membrane, thus the membrane responsiveness and mechanical property is questionable. Therefore, these limitations of the blending process restrict the application and popularization of this method as well as the prepared smart gating membranes. This shows that there is great demand for a fabrication technique that is easy and can provide smart gating membranes with both responsiveness as well as high flux. Researchers have explored other phase separations processes for the fabrication of smart gating membranes, such as vapor-induced phase separation that is slower than the nonsolvent-induced phase separation process, therefore the membrane morphology can be controlled confidently. Also, self-assembling responsive domains are used to prepare smart gating membranes by this method so that the membrane responsiveness and mechanical strength are not compromised. The responsive domains assemble themselves in the growing membrane pores and matrix due to the energy changes taking place during the phase inversion process.

Further, the gating in a membrane can be positive or negative. In the case of positive gating, the membrane performance, in terms of permeability or flux, increases with an increase in the specific stimulus. On the other hand, the membrane performance decreases with an increase in a specific stimulus in the case of negative gating. Positive and negative membrane gating is schematically shown in Figure 9.1. Temperature, pH, chemical, ionic, light, magnetic, and electronic are the most commonly used stimuli in smart gating membranes. These stimuli made the membranes work as or mimic biological, chemical, and sensory functions. Smart gating membranes are useful in case of ionic transports, since they help in the selective passage of specific ions across them. The ionic transport is important in living organisms, since ions like potassium (K^+) are important biologically and on the other hand, ions like arsenic (As^{3+}) are harmful. Therefore, the gating membranes can be used to control the transport of these ions in a positive way by allowing them to pass through (in the case of K^+) or in a negative way by blocking them (in the case of As^{3+}). Also, gating membranes can be used as sensors for the detection of chemicals like glucose. Since glucose is an important indicator and plays a role in diseases like hypoglycemia and diabetes, it will play a great role in the timely detection and cure of these diseases. Thus, gating membranes are also an important candidate for drug delivery application as explained in Chapter 5. Similarly, gating membranes with magnetic, electronic, and light responsiveness are used for different applications, such as remote sensing and control.

Presently, smart gating membranes are used for various applications in different modes, such as self-adjusting membrane hydraulic permeability (ethanol production) and diffusional permeability (controlled glucose release); stimuli-responsive separations, such as sieving effect (size)-based separations and affinity-based separations; and self-cleaning membranes. These areas have been explored to a great extent, but some limitations are still present and there is always scope for using current technology for advanced processes and applications by either modifying the present technologies or developing new technologies by considering the present technology as example. Smart gating membranes are not only useful for traditional membrane applications but also for the development of new and advanced membrane applications in various fields. There is use for smart gating membranes in health care, food, pharmaceutical, and biotechnology for controlled drug release, production of new products, and real-time monitoring effectively and efficiently. Still, there are a lot of challenges due to be addressed, such as mechanisms of mass transfer in a smart gating membrane, and smart gating membranes are not yet tested for long-term use for industrial applications. There is a lack of advanced knowledge and technology for the application of smart gating membranes on an industrial scale. There is a prerequisite of better biocompatible materials for the application of smart gating membranes in the biomedical industry. Therefore, future research should focus on the development of novel materials, novel gating functions, in-depth knowledge of mass transfer and

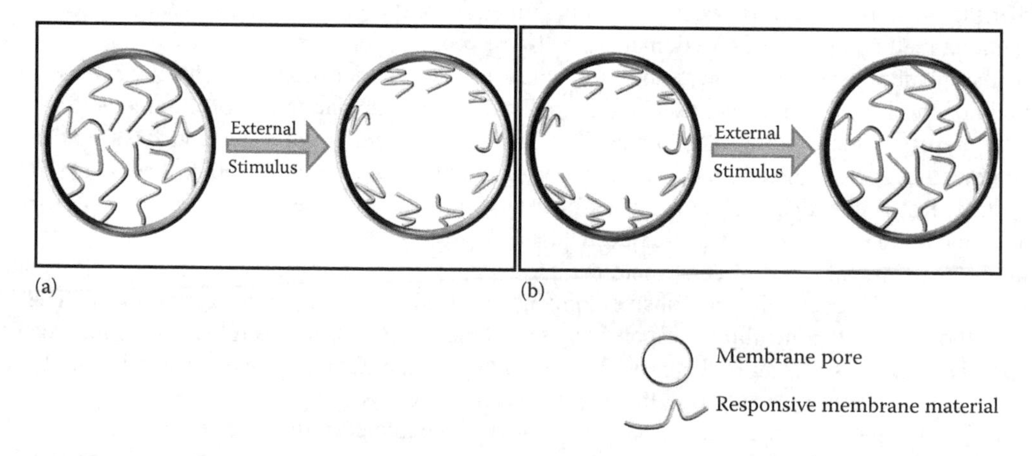

FIGURE 9.1 Schematic representation of (a) positive and (b) negative membrane gating.

separation mechanisms, methods of fabrication for industrial-scale production, and membranes with increased responsiveness.

9.5 HEALTH CARE: SMART WOUND HEALING AND ADVANCED DRUG DELIVERY SYSTEMS

Membrane applications play a vital role in the health care industry. This is one of the main fields for research, exploration, and revenue generation for membrane science. There is vast scope for the use and development of membrane science in health care. Membranes usually prepared from biocompatible and, in some cases, biodegradable materials are used for applications, such as hemodialysis, drug delivery, artificial organs, and tissue engineering.

Wounds are a common occurrence, but the process of wound healing is very complex. With a grievous injury, the healing process does not result in proper healing and thus results in scarred abnormal tissue with reduced mechanical strength. Biomaterials, such as chitin and chitosan, are known to promote wound healing. Therefore, artificial skin grafts are made by using such materials. Skin cells, namely keratinocytes or fibroblasts, are used to grow artificial skin grafts by using membrane scaffolds. These artificial skins will help in the fast and effective wound healing of patients [7], especially patients with burns. Materials with antibacterial or bactericidal properties are also added and the wettability, porosity, and mechanical strength are maintained to completely imitate the original skin. This will help the wounded patient heal fast with a lower chance of infection, scar formation, and need of autografts or skin donors. An example of such artificial skin is Apligraf. Similarly, an example of a biocompatible and biodegradable artificial skin is Suprathel. Presently, there is need for the development of technology to mass-produce and increase the stability or life span of the artificial skins. New materials are also needed for the advancements in this field. The rate at which research is going on in this field will definitely result in bright outcomes in the near future. Therefore, the future holds potential growth and development in this field.

Drug delivery is a well-explored and interesting area of membrane science with lots of scope for future developments. Membranes are a great option for drug delivery due to their properties such as selectivity, permeability, and low cost. The use of membranes for drug delivery is further enhanced by the developments made in the field of novel polymers, especially functional or responsive polymers. These responsive polymers are capable of responding to an external stimulus and therefore it is advantageous to use such polymers in the preparation of stimuli-responsive membranes for drug delivery. Initial success made these membranes an instant hit and funding for these projects poured in from every direction. The extensive research came with the successful development of different materials that can be used for the preparation of membranes for drug delivery. Materials, such as hydrogels, responsive polymers, nanomaterials, and nanocomposites, are being explored with great success. Along with these materials membranes have also been developed by leaps and bounds. The specificity, selectivity, and permeability are well regulated due to the better understanding of the underlying principles, phenomenon, and mechanisms of membrane preparation, transport, and fouling. Nowadays, emphasis is given on the development of biocompatible and biodegradable membranes for drug delivery. In addition to these advanced technologies, lab-on-a-chip is also being explored for making the drug delivery system easy, effective, and efficient.

The current membrane delivery system faces challenges such as high cost, limited materials, low feasibility of up-scaling, and low accuracy. Therefore, membrane drug delivery systems need advancements for the development of technology to overcome these challenges. More studies are required to better understand the underlying mechanisms and transport phenomenon in a membrane drug delivery system for a better outcome and future developments. Better understanding of the drug stability and drug interaction effects with the polymers (materials) are also required. Advanced monitoring devices and administration procedures along with better targeting features will also help the membrane drug delivery systems with optimum delivery of the drug to a target. Improved drugs for better stability and life span if developed will definitely take this industry to a further level.

9.6 MEMBRANE SENSORS

Sensors are important for many applications to monitor and regulate various parameters. The high demands for quality sensors resulted in the development of advanced ultrasensitive sensors. The in-depth knowledge about the micro and nanostructures of materials has made it possible to develop such swift and responsive sensors. Sensors are developed by various chemical methods using the domains of chemistry and electronics. Membranes are also used as sensors due to the availability of high surface area, better protection to the analytes, selectivity, and low costs.

Recently, the focus is on sensors having high surface-to-volume ratios with reduced cross sections. Therefore, one-dimensional sensors are very popular nowadays. The electrospun nanofiber mats provide an excellent option for this purpose due to high surface-to-volume ratios with reduced cross section. The high surface area provides enough reaction or adsorption sites for the analytes. Also, the reduced cross section and short length make it easy and fast to transfer a charge or electron across the fiber length. In addition to these, the high porosity and interconnectivity further adds to the distinguished and suitable properties for the development of an excellent sensor. Therefore, electrospinning is widely used for the preparation of ultrasensitive membrane-based sensors. As explained, the high surface area and high porosity make electrospun membranes productive for this purpose. The electrospun membranes are prepared or immobilized with metal oxides, enzymes, conductive polymers, and responsive polymers for the detection of gases, liquids, biological agents, light, temperature, pressure, and electromagnetic entities. Similarly, membrane sensors are used in the development of smart textiles for the development of smart wearables (discussed in Section 8.3 of Chapter 8 and Section 9.8 of this chapter).

There is potential as well as scope for future developments in the field of membrane sensors. For example, immobilized entities, such as enzymes or metal oxides, are not stable and therefore work is needed to be done for increasing their stability so as to increase the shelf-life of membrane sensors. Also, the response of developed membrane sensors is not satisfactory and therefore demands further exploration and improvements. There is a lot of scope for development of membrane sensors for the detection of various toxic materials present in different forms and states. Similarly, there are areas that need further improvements for making membrane sensors widely acceptable. Thus, the future is bright for the membrane sensors and a lot of work is going on in this direction.

9.7 BIOINSPIRED AND BIOMIMETIC MEMBRANES

The metaphor "standing on the shoulders of giants" aptly fits for bioinspired and biomimetic membranes. Nature is the best teacher and teaches us something new and novel now and then. Nowadays, it is very popular to follow and learn from nature. It is well established that to imitate something already proven increases the chances being successful. Nature gives vivid examples for the development of novel and exceptional technologies to be used in various applications. Similarly, membrane enthusiasts have followed nature and imitate some functions of nature and tried to incorporate them into membranes for solving different problems associated with membranes, such as fouling. The initial success has made bioinspiration and biomimetics ever-growing research fields in the forefront of not only membrane science but also other fields, such as material science, chemistry, and chemical engineering. The problem with bioinspiration and biomimetics is that they are easy to explore but hard to implant due to technological limitations. Basically, they are the connecting bridge amid fields of basic sciences and applied engineering. Bioinspiration membranes are not only limited to imitating natural concepts, but can also use the latest developments in engineering and technology [5]. On the other hand, biomimetic membranes are mostly limited to imitating natural concepts. This directly shows that the scope of bioinspired membranes is better than the biomimetic membranes. Also, due to their nature and development procedure, bioinspired membranes are lacking in progress as compared to biomimetic membranes.

Bioinspired and biomimetic membranes are prepared by mostly imitating the cell membranes, such as surface layer proteins, lipid bilayers, ionophore-based membranes, and biologically antifouling surfaces; and properties associated with some organisms, such as gecko and mussels. However, there are still many challenges facing bioinspired and biomimetic membranes as next-generation membranes. Hence, there is a need of extensive research and development to fully explore and utilize this field. Technological advancements are required to be pursued, so that the membrane researchers have enough options to imitate the natural concepts in membranes. This needs the in-depth knowledge of the biological systems regarding their working principles and control approaches. The biomimetic groundwork will work as the basis of bioinspired technology development. The technological developments should be of a level so that the imitations seem perfect and the actual benefit of the natural concept could be extracted in terms of membrane selectivity, permeability, or antifouling nature. There is also the need to develop processes and methods to prepare bioinspired and biomimetic membranes with limited elements or materials and the overall cost should be reduced. The present processes for the preparation of bioinspired and biomimetic membranes are lengthy, material intensive, and costly. And lastly, the collaboration of interdisciplinary researchers is the most important aspect for the advancement and development of bioinspired and biomimetic membranes. Thus, material scientists, biologists, chemists, physicists, and chemical engineers from around the globe should work together on a common goal of sustainable development of bioinspired and biomimetic membranes.

9.8 MEMBRANES FOR SMART TEXTILES

The last decade has seen great improvements and advancements in smart materials for the development of smart textiles. Smart textiles are used in a myriad of applications, such as space programs, military warfare gear, sportswear, and health care [8]. Some of the major achievements are development of smart clothing lines integrated with sensors, which are capable of measuring, collecting, and transmitting health data of the wearer remotely to a hospital; textiles integrated with sensors used for the health monitoring of concrete structures of (national) importance; smart textiles with high performance characteristics for special missions, for example, space, sea, or military warfare; and smart textiles with sensing properties capable of detecting temperature change, gases, biological or chemical warfare agents, and stress/strain. Similarly, smart textiles are used for the development of artificial organs, such as nose, ears, liver, and pancreas by growing the organ-specific cells on the biodegradable smart textiles. Therefore, smart textiles are very important for the development of novel techniques and products for safe, comfortable, and efficient clothing for users of wide range.

Embroidery is the latest edition to the smart textile industry where different embroidered patterns are used to prepare textiles with improved properties of strength and responsiveness. The embroidery makes it possible to use two smart textile fibers with their specific properties together to give a novel textile with properties of both fibers. The embroidered smart textile fibers could help or induce fast wound healing. Similarly, by embroidering different smart materials and patterns, textiles can be prepared with ultrahigh strength and durability. Similarly, new techniques have to be developed from the available techniques, which have their own uniqueness and may help in achieving the desired goals. The most important thing to develop any smart material is interdisciplinary collaboration and research, so that a mutual understanding and knowledge can be used for the development of novel and new materials, and products for a better, comfortable, and safe life of the users. The developed techniques should not be localized but should be global and multifaceted. For example, a smart textile only capable of regulating temperature but not able to withstand harsh weather conditions will only be useful as sportswear and not military warfare gear. But if the smart textile is also durable and has the strength to bear great stress and strain, then it can also be used as military warfare gear. Another example is of using wearable information processing systems, which monitor the health of the wearer. It is available in two types: one is a vital signs monitoring system and the other mobile wearable information processing system. Therefore, it will be good to converge both systems, as that can help us monitor and regulate the symptoms of the wearer. The developed system should be user friendly so that a person without expertise in computers can process and use the available information. Therefore, much work is needed to develop new smart materials with improved desired properties, techniques to utilize the smartness of the developed material, and technology to make it possible to use the developed smart textile on an industrial scale.

STUDY QUESTIONS

1. Why it is taking so long to develop effective and efficient membranes for industrial use?
2. What is that single entity that needs the most attention for the development of membrane science?
3. Amiya has developed a novel material that is capable of repelling oil and permeating water. Which category does this material belong to? How can she use it to develop a membrane and what potential role could the prepared membrane play?
4. Namrata got a mixture of two solutes that have different structures at different pH. Can she use membranes to separate the two solutes? Based on your answer explain how.
5. Harry developed a thermoresponsive membrane by blending a thermoresponsive material with the membrane casting solution. But when he analyzed the thermoresponsiveness of the membrane, satisfactory results were not found. What might be the possible reason for this and how he can tackle this?
6. Jimmy always gets hurt by his reckless behavior. He never cares for his wounds and does not go for regular dressings. How can membranes be used to fulfill his desire to not go for regular dressings and also help the wound heal?
7. Ram is studying carbon dioxide evolution. He has developed a system where he generates carbon dioxide from various materials. He wants to study the mechanism and the material state at the time of carbon dioxide generation. He wants to develop some sort of sensor to detect the generation of carbon dioxide the moment it starts. Can membranes help him? If yes, how?
8. Suggest some membrane-based techniques for the enantiomeric separations.

REFERENCES

1. D. Bhattacharyya and T. Schäfer (eds.), S. R. Wickramasinghe and S. Daunert (co-eds.), *Responsive membranes and materials*, John Wiley & Sons, 2013.
2. Richard W. Baker, *Membrane technology and applications*, 2nd ed., John Wiley & Sons, 2004.
3. M. Mulder, *Basic principles of membrane technology*, Springer, 2007.
4. M. F. Montemor (ed.), *Smart composite coatings and membranes: Transport, structural, environmental and energy applications*, Elsevier, 2016.
5. Y. Zheng, *Bio-inspired wettability surfaces: Developments in micro- and nanostructures*, CRC Press/Taylor & Francis Group, 2016.
6. Liang-Yin Chu, *Smart membrane materials and systems: From flat membranes to microcapsule membranes*, Springer, 2011.
7. Q. Wei (ed.), *Functional nanofibers and their applications*, Woodhead Publishing Limited, 2012.
8. W. C. Smith (ed.), *Smart textile coatings and laminates*, Woodhead Publishing/CRC Press, 2010.

Index

D

E

F

G

H

I

J

K

L

M

N

O

P

U

V

W

X

Z

CPSIA information can be obtained
at www.ICGtesting.com
Printed in the USA
JSHW041239010622
26549JS00005B/161

9 780367 572105